self-study workbook

Version 7.1.1
for UNIX® and Windows NT®

ARC
Macro Language

Developing ARC/INFO® Menus and Macros with AML™

GeoInformation International
a division of Pearson Professional Limited
307 Cambridge Science Park
Milton Road
Cambridge
CB4 4ZD
and associated companies throughout the world

Distributed in the Americas by
John Wiley & Sons, Inc., 605 Third Avenue, New York, New York 10158-0012.

Address inquiries relating to ARC Macro Language (AML™) software to Environmental Systems Research Institute, Inc., 380 New York Street, Redlands, California 92373-8100, or to your local ESRI software distributor.

First edition 1993. Printed with changes 1994. Second edition 1997.

British Library Cataloging-in-Publication Data
A CIP record for this book is available from the British Library.

ISBN 1 86242 044 0

Library of Congress Cataloging-in-Publication Data
A CIP record for this book is available from the Library of Congress.

ISBN 0 470 24404 6 (Americas only)

Printed in the United States of America.

Contents

How to use this workbook

Chapter 1   AML building blocks

Chapter 2   Menu-driven ARC/INFO

Chapter 4   Branching out

Chapter 5   Actions worth repeating

Chapter 13   Threads: Sewing your application together

Chapter 14   Enhanced pulldown menus

Chapter 15 Building generic tools: ESRI's ArcTools

Chapter 16 Communicating with other software applications

How to use this workbook

Who should use this workbook

This workbook is for every ARC/INFO® software user. It's for programmers and people who have never programmed. It's for people who already use ARC Macro Language (AML™) in their work and for people who are new to AML. This workbook is for everyone who wants to build ARC/INFO applications.

This workbook is designed to help you get work done. AML provides a wide variety of operations to complement ARC/INFO. AML programs organize ARC/INFO commands into sophisticated geoprocessing operations and AML menus provide the basis for combining these operations into a structured, easy-to-use graphical user interface.

AML is a programming language, but this isn't a programming manual. Instead, this workbook teaches you the programming concepts and techniques necessary to accomplish ARC/INFO tasks and solve application problems.

How the workbook is organized

The workbook has sixteen chapters, each divided into three lessons. Each lesson includes numerous examples as well as questions and exercises. The answers to the questions follow the lesson's summary, and the solutions to the exercises are in appendix D.

Most examples in the workbook are programs that do something useful in ARC/INFO. Exercises reinforce the concepts presented in the lessons and many exercises give you the opportunity to create or complete an AML program that accomplishes an ARC/INFO task.

Chapter overview

Chapters 1 through 9 cover the basics of programming with AML. These chapters will be particularly useful to those who either have never programmed or don't know AML. Chapters 10 through 16 cover advanced topics that even those who already program with AML will find useful. Here's a brief summary of each chapter:

Creating menus and macros with AML

Chapter 1 introduces AML and describes how you can use AML to create custom applications and increase productivity. This chapter introduces the building blocks of AML—directives, functions, and variables—and teaches you how to use them to write programs. You'll learn that programming can be as easy as typing ARC/INFO commands at the keyboard.

Chapter 2 examines the eight types of menus that AML supports, how they're created, and how they appear on the screen. This chapter also presents the AML functions that create menus automatically.

Chapter 3 shows you how to access programs and menus stored in various directories on your system. It also teaches you how to encode AML programs so that others can execute but not read them.

Writing programs with AML

Chapter 4 teaches you the programming technique of branching. Branching is used to test conditions and choose from alternative courses of action to prevent errors and provide a friendly interface in ARC/INFO.

Chapter 5 presents another programming technique, called looping, that's used for executing repetitive ARC/INFO tasks.

Chapter 6 shows you how to use AML to store coordinates and pass them to ARC/INFO commands as needed.

Chapter 7 introduces AML's character manipulation functions and shows you how to use them for performing such tasks as extracting elements from and formatting character strings for ARC/INFO to use.

Chapter 8 teaches modular programming and the techniques you need to pass data between modular programs. It also introduces programming with routines.

Chapter 9 presents time-saving techniques for debugging programs. It shows you how to anticipate errors and design programs that react to errors intelligently.

Chapter 10 teaches you how to read and write ASCII files with AML and shows how AML can create menus from AML programs as needed. This chapter introduces the file-handling tools that AML provides.

Chapter 11 introduces the database-related functions in AML, and teaches when and why they're useful. This chapter also explores AML tools for accessing other software and how ARC/INFO Cursors gives AML direct access to INFO™ and external database management systems (DBMSs).

Building menu interfaces

Chapter 12 builds on the information presented in chapter 2 to explore form menus in detail. It includes an online tutorial using the sophisticated FormEdit™ interface, used for designing and implementing form menus.

Chapter 13 shows you how to use an AML tool called threads to display and manage multiple menus in an application interface.

Chapter 14 teaches you how to create enhanced pulldown menus using the MenuEdit interface. You'll learn how to get enhanced pulldown menus to dynamically interact with form menus.

Chapter 15 is devoted to ArcTools. It teaches you from start to finish how to build an ArcTool and how to incorporate it in the ArcTools™ menu system. This chapter also shows you how to modify the functionality of an existing ArcTool.

Chapter 16 covers Inter-Application Communication (IAC). IAC enables ARC/INFO to act as a client or a server with other applications. You'll see how ARC/INFO establishes a dialog with ArcView® GIS and other software programs.

Workbook icons

Familiarizing yourself with the following icons used in this workbook will help you navigate the lessons:

 FYI

For your information (FYI).

These boxes include additional information, asides, tips, and tricks. Found in every chapter.

 STYLE

Style.

These boxes include style types for writing programs that are easy to read and modify. Found primarily in chapter 4.

Beware!

Read this information. It reveals the errors that will cause your programs to fail. Found in every chapter.

Questions.

The questions in boxes like this relate to the information you're reading. Found in most chapters and numbered sequentially in each.

Answers.

These are the answers to the questions found in the lessons. Answers follow each lesson summary.

Hardware and software

It isn't necessary to have a computer to learn the material presented in this workbook. A computer is necessary, however, if you want to complete the online tutorial presented in chapter 12, or if you want to execute any of the AML programs and menus presented in the book. If you do, the following hardware and software are necessary:

Hardware

• Workstation configured for ARC/INFO software
• X Windows™ graphics terminal (preferably color) that supports ARC/INFO graphics and menus

Software

• ARC/INFO Version 7.1.1 or higher

A note about workstations

ARC/INFO takes full advantage of the user interface tools provided by your system's resident window manager. The look of display windows and menus in ARC/INFO conforms to either the Common Desktop Environment (CDE) user interface or the Microsoft® Windows NT® user interface specifications. The menus you see in this workbook were created in CDE.

What you should know before starting

This workbook assumes that you have some knowledge of your computer's operating system, specifically,

- Directory structure and file management

    Changing directories

    Listing directories and files

    Creating, copying, renaming, and deleting directories and files

- Text editor

    Creating a new file

    Editing an existing file

    Displaying the contents of a file

- Windowing system

    Creating, moving, and resizing windows

- Mouse

    Using the mouse buttons

CD-ROM

The CD–ROM packaged with the workbook is for Microsoft Windows NT and UNIX® platforms supported by ARC/INFO Version 7.1.1. It contains all the AML programs used as examples in the lessons and the exercise solutions. It also includes sample data needed to run the example programs.

The AML programs shown in the book and provided on the CD are indexed in the subject index that follows the alphabetical index in the back of the book. Keep in mind that there are many solutions to every programming problem. The solutions provided in the workbook and on the CD represent only one solution, and not necessarily the best solution for your situation.

Training data installation

The data used in this workbook is included on the accompanying CD–ROM disc. The training data requires approximately 30MB of hard disk space to install. Both PC and supported UNIX versions are included on the CD. The two platforms have their own installation instructions.

Each chapter contains a similar directory structure. Under each chapter directory is a subdirectory for each lesson. Under the lesson subdirectories are example scripts and an exercise subdirectory that holds the solutions for all the exercises in that lesson. This structure is reflected in the contents of chapter 6, lesson 1, shown below:

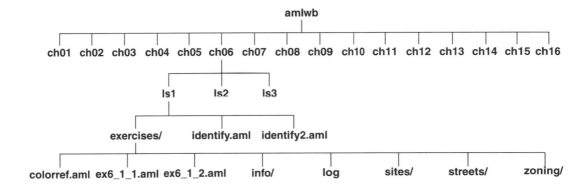

PC installation

The setup program will copy the training data directory (*amlwb*) to a specified location on your hard disk.

Load the CD-ROM

Run Windows® as you normally would. Put the CD disc into your CD–ROM drive.

Start the setup program

For Microsoft Windows NT: From the Program Manager File menu, choose Run. In the Run dialog box, type the following command (use the drive letter that corresponds to your CD-ROM device):

```
<drive>:\Nt\setup
```

The setup program will create an installation wizard for your computer. The wizard will present several menus.

Choose a destination menu

You may use this menu to specify an installation directory. By default, the installation wizard will copy the training directory to *D:\amlwb*. After setting a new directory or accepting the default, choose the Next button to start the file transfer process.

UNIX installation

The install program will copy the training data directory (*amlwb*) to a specified location on your hard disk. The install program is a series of textual prompts. Your inputs are shown in bold typeface.

Mount the CD–ROM device

Put the CD disc into your CD–ROM drive. *Note: Solaris 2.3 and greater includes Volume Management™ software that automatically mounts the CD–ROM drive and provides immediate access to it. You only need to insert the CD disc into the drive, so skip to "Run the installation script."*

Become the root user:

```
UNIX> su -
```

If a */cdrom* directory doesn't exist, create one as follows:

```
UNIX> mkdir /cdrom
```

Mount the CD–ROM drive with the command appropriate for your UNIX workstation (where *dev* is the device name for your CD–ROM drive):

```
             DG®> mount -0 noversion,ro -t cdrom /dev /cdrom
   Digital UNIX> mount -t cdfs -r -o nodefperm,noversion /dev /cdrom
             HP®> mount -rt cdfs /dev /cdrom(or use sam)
            IBM®> mount -v 'cdrfs' -p'' -r'' /dev /cdrom(or use smit)
Silicon Graphics®> mount -o setx -t iso9660 /dev /cdrom
   Sun Solaris 1.x> mount -rt hsfs /dev /cdrom
```

Exit from being root user:

```
UNIX> exit
```

Run the installation script

The installation script is a series of textual prompts. Default selections are noted in brackets ([ ]). Type a question mark (?) at any prompt for a list of options or help. Quit the installation script at any time by typing quit or q. Return to a previous prompt by typing a caret (^).

The installation script will create a directory called *amlwb* in the specified directory. You may move or copy the *advai* directory, but don't change its name. You may have to change privileges for the new *amlwb* directory to read–write.

The dialog is presented below. Your responses are shown in bold typeface. Run the installation script with the command appropriate for your UNIX workstation.

```
             DG > /cdrom/unix/install.dg -load
   Digital UNIX> /cdrom/unix/install.digunix -load
             HP> /cdrom/unix/install.hp -load
            IBM> /cdrom/unix/install.ibm -load
Silicon Graphics> /cdrom/unix/install.sgi -load
   Sun Solaris 1.x> /cdrom/unix/install.solaris1 -load
   Sun Solaris 2.3> /cdrom/cdrom0/unix/install.solaris2 -load
```

```
Press <Return> to take the default, '?' for help, '^' to return
to the previous question, or 'q' to quit.

Enter CD-ROM mount point: [/cdrom/cdrom0] /cdrom/cdrom0/unix

Enter pathname to install directory: [your current directory]
<your install directory>
```

```
Developing ARC/INFO menus and macros with AML
-----------------------------------------------------------

AML workbook
Package numbers to load: [all] <press Return>

Package selection complete
-------------------------
You have chosen the following packages to be loaded
AML workbook
        Training Database

Is this correct? [yes] <press Return>

Total size of chosen packages in megabytes: 29.3
Available space in megabytes: (free space on your disk)

List file names while loading? [no] <press Return>

Loading package(s), Please wait...
Finished loading package(s)
Exiting...
```

Getting help

Be sure to visit the discussion forum for the *ARC Macro Language* workbook on the World Wide Web at http://www.esri.com/esribooks. You'll find user discussions and ESRI support information related to the book and CD.

1 ▽ *AML building blocks*

ARC Macro Language (AML™) gives you the power to tailor ARC/INFO®
software's suite of geoprocessing operations to fit your particular applications.
Anyone who uses command-level ARC/INFO can write AML programs. Even if
you've never thought of yourself as a programmer, after completing this chapter
you'll be able to write AML programs that expedite your work with ARC/INFO.
You'll see how easy it is to apply AML's functionality—you can even create
AML programs automatically as you type commands at the keyboard.

This chapter introduces AML and some reasons for using it. In addition, you'll
learn the building blocks of AML—directives, variables, and functions—and
begin using them to write your own programs. The remainder of this workbook
presents a detailed discussion of these elements and teaches you how to structure
them to build macros and menus for accomplishing specific tasks in ARC/INFO.

This chapter covers the following topics:

Lesson 1.1—Introducing AML

- A brief history of AML
- How you can use AML

Lesson 1.2—Getting started with AML

- Creating AML programs without programming
- AML directives
- Getting help

Lesson 1.3—Using AML variables and functions

- Setting AML variables
- Using AML variables
- Local and global variables
- Reserved variables
- AML functions
- Nesting functions
- Interpreting special characters
- Order of interpretation

Lesson 1.1—Introducing AML

In this lesson

AML provides you with programming capabilities and a set of tools to tailor the user interface of your ARC/INFO application. You can use AML throughout the ARC/INFO system to perform a variety of tasks and increase your productivity. This lesson provides an overview of AML and describes how you might use it.

A brief history of AML

AML is part of the ARC/INFO geographic information system (GIS) software. AML provides information about ARC/INFO objects (e.g., coverages) and specific ARC environments. When it was introduced with ARC/INFO Version 4.0, AML unified applications programming in ARC/INFO across all hardware platforms.

AML is modeled after Prime Computer Inc.'s Command Procedure Language (CPL). If you're familiar with the Bourne or C shells in the UNIX® operating system, you'll find that AML is similar in principle and functionality to these languages in its ability to execute sophisticated actions such as branching, variable manipulation, and argument transfer.

Don't worry if you don't know a programming language. This workbook includes the introductory programming concepts, principles, and techniques you'll need to successfully use AML.

How you can use AML

AML is an *interpreted language*. Interpreted languages translate each command line to machine language and execute it before moving to the next line. In addition to grouping frequently used command sequences as one task, you can use AML to prompt the user for information, validate user input, provide meaningful error messages, create your own commands, and create a menu-driven interface to ARC/INFO operations.

Here are some examples:

Customize ARC/INFO for your applications

- Combine ARC/INFO tools to create your own specialized functions
- Build interactive menus
- Create your own commands
- Enhance native ARC capabilities
- Implement complex applications
- Perform modeling

Increase productivity

- Create a user interface for your application
- Standardize procedures
- Reduce typing by storing command sequences in a file
- Help less experienced users become productive more quickly
- Provide users with an alternative to memorizing ARC/INFO commands
- Help reduce errors by presenting users with only valid choices
- Provide managers with display and query capabilities
- Automate frequently performed and repetitive tasks
- Automate database access
- Create quick keyboard macros

In summary

AML provides a wide variety of operations to complement ARC/INFO. The best way to learn what AML can do is to use it, so move on to lesson 1.2 and get started!

Lesson 1.2—Getting started with AML

In this lesson

Directives are AML commands that instruct AML to perform specific tasks. This lesson introduces some basic AML directives and shows how to use them in AML programs. In this lesson directives record command input to a file, convert the file to an AML program, and execute the AML program. You'll also learn how to get usage lines or a list of available directives.

Creating AML programs without programming

A *program* is a set of instructions that a computer executes in a defined sequence to accomplish a task. An *AML program* is an American Standard Code for Information Interchange (ASCII) text file that contains instructions for the computer. There are two ways to create one: by typing a command sequence into a file opened with your system's text editor or by automatically capturing the commands typed during an ARC/INFO session in a text file called a *watch file*.

Using a text editor

Every day you type many ARC/INFO commands to implement such well-defined tasks as creating plot files, updating attributes, adding features, and building topology. A *macro language* allows you to execute a group of commands under one name. You can create an AML program, or *macro*, by opening a text file and typing an ARC/INFO command sequence. The following lines make up an AML program that creates a display in ARCPLOT:

```
ARCPLOT
DISPLAY 9999 1 POSITION UL
MAPEXTENT INDEX
LINESYMBOL 5
POLYGONS CENSUS
RESELECT ROADS LINE DESC CN 'MAJOR'
LINESYMBOL 12
ARCS ROADS
POINTMARKERS SCHOOLS 24
```

An AML program that runs in ARCEDIT and draws arcs and node errors to the screen is entered like this:

```
ARCEDIT
EDIT VEGETATION
DRAWENVIRONMENT ARC NODE ERROR
EDITFEATURE ARC
NODESIZE DANGLE .2
NODECOLOR DANGLE 2
NODESIZE PSEUDO .15
NODECOLOR PSEUDO 5
DRAW
```

Although it's not required, you should always name text files containing AML programs with the .AML extension (for example, FILENAME.AML). You can only omit the extension when running a program if it has the .AML extension. For instance, if you name an AML stuff.txt, &RUN *stuff* will not execute the program.

FYI

On the NT platform, be careful of more complex text editors like WRITE. These text editors store files in a format that AML can't understand. Use text editors like DOS EDIT or NOTEPAD.

Finishing your program

Every AML program you create should include the &RETURN directive as the last line. When you run an AML program, control passes to the program. &RETURN indicates that the program has executed and returns control to the calling source, which in this example is the keyboard. The final ARCEDIT™ program looks like this:

```
ARCEDIT
EDIT VEGETATION ARC
DRAWENVIRONMENT ARC NODE ERROR
EDITFEATURE ARC
NODESIZE DANGLE .2
NODECOLOR DANGLE 2
NODESIZE PSEUDO .15
NODECOLOR PSEUDO 5
DRAW
&RETURN
```

Running AML programs

The &RUN directive executes a specified AML program and passes control from the keyboard to the AML program. Suppose you're asked to recreate your ARCEDIT display. Instead of typing all the commands again, you can save time and energy by running your AML program.

You can abbreviate &RUN (&R) and/or omit the .AML extension on the file name. To run an AML program, type any of the following command lines from the Arc: prompt:

```
Arc: &run <textfile_name>.aml
Arc: &run <textfile_name>
Arc: &r <textfile_name>.aml
Arc: &r <textfile_name>
```

Commenting your programs

Comments are a vital part of any program because they provide such information as the author's name, name of the program, date written, variables used, and an outline of the main tasks performed. When written at the beginning of the AML program, this information is often referred to as the *program header.* Detailed comments are essential to maintaining your code. It is difficult to edit an uncommented program you wrote months ago.

Comments are denoted by placing a *comment symbol* before the descriptive text. The comment symbol is a combination of two characters, a forward slash and an asterisk (/*). This symbol can appear anywhere on a line; all text that follows it is ignored. The following example shows comments in a program header:

```
/* Author: Josephine Programmer
/* Name: school.aml
/* Date: 07/16/96
/*
/* Creates a display in ARCPLOT that shows census tracts,
/* roads, and schools
```

FYI

Many programs in this book aren't commented as thoroughly as you should comment yours. Our examples are often short code fragments and are explained in the text instead of with comments. We also use comments for instructional purposes; yours should be more design and task oriented.

Using watch files

While typing commands in a text file isn't difficult, it can be time-consuming. In addition, your text editor can't give you feedback if you type a command incorrectly. A *watch file,* on the other hand, captures commands entered during an ARC/INFO session. As you type commands at the keyboard, AML writes them to the watch file. This method is easy, fast, and allows you to see the results as you go. As you'll see, a watch file can be the predecessor to an AML program.

To open a watch file, type the AML directive &WATCH, followed by a file name. Naming watch files with a .WAT file name extension is optional but helps you identify and organize these files. &WATCH creates or opens a text file with the specified name. &WATCH &OFF closes the file. The following example, which creates an ARCPLOT™ display of school locations, initiates a watch file, captures commands, and closes the watch file:

```
Arc: &watch school.wat
Arc: ARCPLOT
Copyright (C) Environmental Systems Research Institute, Inc.
All rights reserved.
Arcplot: DISPLAY 9999 1 POSITION UL
Arcplot: MAPEXTENT INDEX
Arcplot: LINESYMBOL 5
Arcplot: POLYGONS CENSUS
Arcplot: RESELECT ROADS  LINE DESC = 'MAJOR'
ROADS lines: 21 of 153 selected
Arcplot: LINESYMBOL 12
Arcplot: ARCS ROADS
Arcplot: POINTMARKERS SCHOOLS 24
Arcplot: &watch &off
```

The resulting text file, SCHOOL.WAT, is created and stored in the local workspace and looks like this:

```
Arc: |> ARCPLOT <|
Copyright (C) Environmental Systems Research Institute, Inc.
All rights reserved.
Arcplot: |> DISPLAY 9999 1 POSITION UL <|
Arcplot: |> MAPEXTENT INDEX <|
Arcplot: |> LINESYMBOL 5 <|
Arcplot: |> POLYGONS CENSUS <|
Arcplot: |> RESELECT ROADS LINE DESC = 'MAJOR' <|
ROADS lines: 21 of 153 selected
Arcplot: |> LINESYMBOL 12 <|
Arcplot: |> ARCS ROADS <|
Arcplot: |> POINTMARKERS SCHOOLS 24 <|
Arcplot: |> &watch &off <|
```

In addition to commands you typed at the keyboard, the file includes ARC/INFO system messages and special brackets (i.e., |> <|). Notice that only the ARC/INFO commands are enclosed in brackets. The brackets allow AML to distinguish between system output and what was typed.

You can append commands at the bottom of a closed watch file by opening it using the &WATCH directive with the &APPEND option. For example, you can open the SCHOOL.WAT file as follows:

```
&watch school.wat &append
```

If the &APPEND option isn't used, the contents of SCHOOL.WAT are overwritten.

FYI

&WATCH with the &COMMANDS option captures commands executed in another AML program or a menu selection as well as those typed at the keyboard. See the &WATCH directive in the online help for more information on the &COMMANDS option.

Converting watch files to AML programs

The AML directive &CONV_WATCH_TO_AML (&CWTA) converts a watch file to an AML program. &CWTA copies the text that's enclosed in the special brackets (|> <|) from the watch file and writes it to another file as an AML program.

It's valid for an AML program to contain operating system commands, ARC/INFO commands, and other AML elements (i.e., directives, variables, and functions), but not system messages and prompts. &CWTA copies the user input from the watch file to the AML while removing any system output.

The following example shows how to specify the input file (watch file) and output file (AML program). The following statement converts the watch file named SCHOOL.WAT to an AML program named SCHOOL.AML.

```
&cwta school.wat school.aml
```

SCHOOL.AML is shown below. It contains only the commands input by the user because &CWTA eliminated the system output.

```
ARCPLOT
DISPLAY 9999 1 POSITION UL
MAPEXTENT INDEX
LINESYMBOL 5
POLYGONS CENSUS
RESELECT ROADS LINE DESC CN 'MAJOR'
LINESYMBOL 12
ARCS ROADS
POINTMARKERS SCHOOLS 24
&watch &off
```

When using commands like MOVE, IDENTIFY, or KEYPOSITION, the asterisk (*) is often used to locate the cursor position. To capture this x,y coordinate in a watch file, use &WATCH with the &COORDINATES option. You must then use &CWTA with the &COORDINATES option to copy the coordinate value from the watch file to the AML program. The coordinates are placed in a set of brackets just as if they were typed at the keyboard. Notice that the coordinate brackets (i.e., |>* *<|) use asterisks (*) to differentiate them from the standard command brackets (i.e., |> <|).

```
Arcplot: |> IDENTIFY SOILS POLY * <|
Enter point |>* 6801983.345829 1984920.987536 *<|
```

To recreate your ARCPLOT display, run SCHOOL.AML instead of typing all the commands again. To run SCHOOL.AML, type the following at the Arc: prompt:

```
Arc: &run school.
```

FYI

You can use the &RUNWATCH directive to run a watch file without first converting it to an AML program. For more information, see the &RUNWATCH command reference in the online help.

Editing an AML program converted from a watch file

After a watch file is converted to an AML program, you need to use a text editor to add some finishing touches. Replace the &WATCH &OFF directive at the end of your converted program with &RETURN to return control to the keyboard after the program runs. In addition, add a program header and descriptive comments.

The program that results from converting a watch file might not be exactly what you want. There might be commands that you typed wrong, commands that you no longer want, or other information you want to change or add.

Another reason for editing a converted watch file could include changing the subsystem from which the AML is executed. For example, if you want SCHOOL.AML to run from the Arcplot: prompt instead of the Arc: prompt, delete the ARCPLOT command from the AML program.

The edited version of SCHOOL.AML looks like this:

```
/* Author: Josephine Programmer
/* Name: school.aml
/* Date: 07/16/96
/*
/* Creates a display in ARCPLOT which shows census tracts,
/* roads, and schools
DISPLAY 9999 1 POSITION UL
MAPEXTENT INDEX
LINESYMBOL 5
POLYGONS CENSUS
/* Draw only major roads
RESELECT ROADS LINE DESC CN 'MAJOR'
LINESYMBOL 12
ARCS ROADS
/* Draw the schools
POINTMARKERS SCHOOLS 24
&RETURN
```

AML directives

Directives provide the framework for all AML operations; they are AML commands that perform specific tasks. The arguments appended to a directive indicate the operation to perform.

Directives always begin with an ampersand (&). This distinguishes them from ARC/INFO commands. You've already seen some AML directives in this lesson: &WATCH, &CWTA, &RUN, and &RETURN. These directives are used to create, execute, and complete AML programs.

Some directives set terminal characteristics (&TERMINAL), manage screen listings (&FULLSCREEN), type messages to the screen (&TYPE), change workspaces (&WORKSPACE), and send commands to the operating system (&SYSTEM).

The following example incorporates these directives in a program that establishes an ARC/INFO working environment, types a "good morning" message, and lists the contents of the current workspace:

```
/* start.aml
&terminal 9999
DISPLAY 9999 SIZE 900 700
&fullscreen &popup
&workspace /laguna1/vicki/project5
&type Good morning Vicki
&system ls
&return
```

Unlike ARC/INFO commands, AML directives work in any ARC/INFO subsystem. For example, the ARC/INFO command DESCRIBE, which provides a detailed description of a geographic data set (e.g., a coverage) and stores the information in AML reserved variables, only works in the ARC environment. The AML directive &DESCRIBE, however, will extract the same information from ARC, ARCEDIT, ARCPLOT, or any ARC/INFO subsystem.

Similarly, the &WORKSPACE and &SYSTEM directives are useful when you want to change workspaces or issue operating system commands from prompts other than Arc:. The WORKSPACE command is only valid from the Arc: prompt; the &WORKSPACE directive performs at any subsystem prompt. When you type operating system commands at the Arc: prompt, they are automatically passed to the operating system, but the &SYSTEM directive passes operating system commands from any subsystem prompt.

AML directives also provide sophisticated programming capabilities, such as:

- Testing and choosing between alternative instructions (branching)
- Repeating an action (looping)
- Handling errors encountered during program execution (error checking)
- Finding errors in a program that cause it not to execute (debugging)
- Changing information in a program dynamically (variable substitution)

These programming techniques, and the directives that implement them, are the focus of subsequent chapters in this workbook.

Getting help

There are several ways to get on-screen help in AML. You can access an argument list (syntax), a detailed help file, or a listing of available AML directives. The &USAGE directive provides the structure, called *syntax*, for a specific AML directive. For example, examine how issuing &USAGE invokes the syntax of the &TERMINAL directive:

```
Arc: &usage &terminal
Usage: &TERMINAL <device> {&CURSOR|&TABLET|&MOUSE|&KEYPAD}
```

Some directive arguments are a choice of *keywords*, (e.g., &CURSOR, &MOUSE) and must be typed exactly as presented, even though they are not case sensitive. Arguments that are keywords are always preceded by an ampersand (&) just like directives. Arguments such as <device> are not keywords and therefore don't start with an &.

The &COMMANDS directive can list all AML directives and functions. Typing &COMMANDS &C lists all directives that start with the letter C, as shown below:

```
Arc: &commands &c
Directives:
&CALL              &CODEPAGE              &COMMANDS
&Conv_Watch_To_Aml
```

Notice the mixture of uppercase and lowercase letters in &Conv_Watch_To_Aml. The uppercase letters indicate the abbreviation for the &CWTA directive. If the search string is not preceded by an ampersand then &COMMANDS will return a list of functions. Functions are another element of AML that we will discuss in the next section.

```
Arc: &commands c
Functions:
CALC              CLOSE         COPY         COS
CVTDISTANCE
```

In summary

Writing an AML program can be as easy as capturing commands in a watch file and then converting the watch file to an AML program. Running the AML program to replay the commands saves time when you need to repeat a task.

Directives are the foundation of AML. They perform a wide range of tasks in addition to creating, converting, and running programs. Online help is available for all AML directives and their arguments. Directive names and their keyword arguments begin with an ampersand (&) to differentiate them from ARC/INFO commands. Unlike ARC/INFO commands, AML directives work in every ARC/INFO subsystem.

The directives introduced in this lesson appear in many examples in this workbook. Each chapter presents directives that manage the sophisticated operations of AML.

Exercises

Exercise 1.2.1

Write the AML program that results from converting the following watch file:

```
Arcedit: |> ARCEDIT <|
Copyright (C) Environmental Systems Research Institute, Inc.
All rights reserved.
Arcedit: |> EDITCOVER VEGETATION <|
Arcedit: |> DRAWENVIRONMENT ARC NODE ERROR <|
Arcedit: |> EDITFEATURE ARC <|
xxxFeatures selected
Arcedit: |> NODESIZE DANGLE .2 <|
Arcedit: |> NODECOLOR DANGLE 2 <|
Arcedit: |> NODESIZE PSEUDO .15<|
Arcedit: |> NODECOLOR PSEUDO 5 <|
Arcedit: |> DRAW <|
Arcedit: |> &watch &off <|
```

Exercise 1.2.2

(a) When you execute an AML program with &RUN, control passes from the _Keyboard_ to the AML _program_.

(b) When does control return to where it originated?

Exercise 1.2.3

Can you use AML directives during an interactive session of ARC/INFO or only in an AML program?

Exercise 1.2.4

(a) Name two things all AML programs should include.

(b) Other than the two things named in part (a), list two reasons you may want to edit an AML program that was converted from a watch file.

Exercise 1.2.5

Which changes should be made to this file before running it from ARCPLOT?

```
ARCPLOT
MAPEXTENT SOILS
RESELECT SOILS POLY SOIL_CODE = 33
POLYGONSHADES SOILS 14
LINESMBOL 5
LINESYMBOL 5
POLYGONS SOILS
RESELECT ROADS LINE DESC = 'MAJOR'
LINESYMBOL 12
ARCS ROADS
&WATCH &OFF
```

Exercise 1.2.6

How can you capture the coordinate location of the graphical cursor in the watch file and ensure that it appears in the AML program generated by &CWTA?

Lesson 1.3—Using AML variables and functions

In this lesson

Programs like SCHOOL.AML created in the previous lesson are useful when you need to perform a task repeatedly. SCHOOL.AML creates a display in ARCPLOT to show census tracts, roads, and schools. The functionality of this AML program is limited because the elements are always the same—major roads are always drawn with line symbol 12, schools are always symbolized with point marker 24, and so on.

This lesson explores how AML variables and functions allow you to substitute new values in a program without rewriting it. *Variables* are named storage entities whose value can be used and changed as a program executes. AML *directives* manage AML variables by performing such tasks as assigning, listing, and deleting variable values. *Functions* perform a defined operation that returns a value to the program. The value of a function can be used in a command or stored in a variable to use later.

This lesson also presents the special characters AML recognizes and discusses how they affect program interpretation. Special characters allow the AML processor to distinguish between program components. The ampersand (&), paired square brackets ([ ]), and paired percent signs (% %) are some of the special characters that have a defined order in command-line evaluation.

Setting AML variables

Variables have two parts: a name and an assigned value. Variables are defined, or set, by the user with the &SETVAR directive. Here's an example of setting a variable to store a coverage name:

```
&setvar covname = parcels
```

The variable named COVNAME is assigned the value PARCELS. Because variable names aren't case sensitive, the same variable is set by this statement:

```
&setvar COVNAME = parcels
```

The *value* of a variable, however, *is* case sensitive. This is particularly important to remember when storing text strings. In the following examples, the value of the variable COMPANY is different because one is in uppercase and the other is in lowercase.

```
&setvar company = ESRI
&setvar company = esri
```

Text strings assigned to a variable can include blanks:

```
&setvar coname = Environmental Systems Research Institute
```

You can assign variables using no operator or one of the two operators shown here:

```
:=
 =
```

The variable assignment resulting from the three statements that follow is the same:

```
&setvar scale := 1200
&setvar scale = 1200
&setvar scale 1200
```

Choose the style you prefer and stick with it—mixing styles makes it difficult to find errors.

Abbreviate the &SETVAR directive as &SET, &SV, or &S. This workbook uses the &SV abbreviation with the equal sign (=) operator. Any abbreviation works with either operator or with no operator. The following statements are equivalent:

```
&setvar scale := 1200
&sv scale = 1200
&s scale 1200
```

Whenever we set a variable we don't tell the system what type of data we're storing in that variable. Even though we enter characters as the value of the variable, internally the AML processor can differentiate between different types of data. These different data types are: character strings, integer numbers, real numbers, and Boolean values. The following table lists examples of each.

Data type	Examples
Character string	ESRI; Environmental Systems Research Institute
Integer number	1; 1200; 15000000
Real number	.75; 245.331
Boolean value	.TRUE.; .FALSE. (only)

FYI See chapter 4 for a discussion of Boolean variables and how to use them in AML programs.

The directive &LISTVAR (abbreviated &LV or &L) displays a list of currently assigned variables. Given the variables set in previous examples, the following listing appears when you issue &LISTVAR:

```
Arc: &listvar
Local:
    COVNAME        parcels
    COMPANY        esri
    CONAME         Environmental Systems Research Institute
    SCALE          1200
Global:
    No global variables are defined
Program:
    :PROGRAM               ARC
```

Notice the three kinds of variables in the list: local, global, and program. Local and global variables are discussed later in this lesson. Program variables are covered in chapter 11.

Variables can also be written to an output file using &LISTVAR. Each variable name and its value are written to a separate line of the output file. This could be nice for storing variables that are no longer needed within the current session. Chapter 10 will show how to read a file and use its contents.

Using AML variables

You can substitute variables anywhere in ARC/INFO: in commands, in directives, or in functions. When a variable name appears enclosed in percent signs, the variable is being *referenced*. When a variable is referenced, the value of that variable is substituted for the variable name.

```
&sv covname = CENSUS
BUILD %COVNAME% POLY
```

The value CENSUS is substituted for %COVNAME%, resulting in the following command line:

```
BUILD CENSUS POLY
```

Only enclose a variable name in percent signs to reference the variable. No percents signs are used when variables are assigned (&SETVAR), listed (&LISTVAR), or deleted (&DELVAR).

You can combine as many variables as are needed to execute a command. Consider the following:

```
&sv covname = CENSUS
&sv feature = POLY
BUILD %covname% %feature%
```

You can also join variables like this:

```
&sv covname = CENSUS
&sv feattable = PAT
LIST %covname%.%feattable%
```

which results in this:

```
LIST CENSUS.PAT
```

Question 1-1: Which command line results from the following three lines of code?

```
&sv covname = CENSUS
&sv region = 4
EDITCOVER %covname%%region%
```

(Answer on page 1-39)

Now that you know how to assign variables, you're ready to use them to provide information to an AML program. First consider the following program, which uses no variables and creates the same display each time it executes:

```
/*drawparzone.aml
ARCPLOT
DISPLAY 9999 3
SHADESET COLOR
MAPEXTENT PARCELS
POLYGONSHADE PARCELS ZONING ZONING.LUT
POLYGONS PARCELS
&return
```

Incorporating variables in the program makes it more flexible:

```
/*drawmap.aml
&sv cov = parcels
&sv item = zoning
ARCPLOT
DISPLAY 9999 3
SHADESET COLOR
MAPEXTENT %cov%
POLYGONSHADE %cov% %item% %item%.LUT
POLYGONS %cov%
&return
```

The COV and ITEM variables are assigned to store a coverage name and item name, respectively. The variable reference %COV% replaces all occurrences of the coverage name PARCELS in the program. %ITEM% replaces all occurrences of ZONING.

Assigning a variable and referencing it throughout your AML program makes your programs easier to update and manage. Suppose you want to change a coverage name that appears twenty-five times in a program. If you used a variable to store the coverage name, you only need to change one variable assignment (&SV), instead of changing twenty-five entries.

Local and global variables

As your AML programs get larger and more complex, it's a good idea to divide them into several smaller programs, each program doing one part of the larger task. Because these smaller programs must interact to perform the larger task, variables assigned in one program may be needed in another program.

Local variables are only valid in the AML program where they're set; their value is unknown to other programs. *Global* variables, however, are accessible to every AML program executed during the ARC/INFO session in which they're assigned. You can set both local and global variables using the &SETVAR directive. Global variable names begin with a period (.) to distinguish them from local variables. Similar to the &LISTVAR directive, which lists all variables, &LISTLOCAL (&LL) lists local variables, and &LISTGLOBAL (&LG) lists global variables.

Continuing the previous example, you might write one AML program that assigns the variables and another that draws the map. Written as two AML programs, it looks like this:

```
/* subs.aml
&sv .cov = parcels
&sv .item = zoning
&run drawmap2.aml
&return

/*drawmap2.aml
ARCPLOT
DISPLAY 9999 3
SHADSET COLOR
MAPEXTENT %.cov%
POLYGONSHADE %.cov% %.item% %.item%.LUT
POLYGONS %.cov%
&return
```

In SUBS.AML the variable assignments are global, and the variable references in DRAWMAP2.AML reflect this by beginning with a period.

Chapter 8 explores writing large applications as a group of smaller programs and managing local and global variables.

Reserved variables

Reserved variables provide a wealth of information to the AML programmer. Reserved variables are set by ARC/INFO; they store information held internally by the system, including information about coverages, the graphical cursor, and current status of AML programs. These variables have strict naming conventions, and the names are reserved (i.e., you should *never* set these variables).

About coverages: DSC$ and PRJ$

Approximately seventy-seven reserved variables containing coverage information and projection information are assigned when you issue either the &DESCRIBE directive or the DESCRIBE command. The names of these reserved variables always begin with DSC$ (for general coverage information) and PRJ$ (for projection information). DSC$ contains coverage information such as the processing tolerances, number of features, whether there's current topology, and so on. PRJ$ contains projection information such as the projection name and the units of the projection. The table below lists some examples:

Reserved variable	Value
DSC$POLYS	Number of polygon features
DSC$QEDIT	.TRUE. or .FALSE.—whether or not the coverage has been edited since last BUILD or CLEAN
PRJ$NAME	Projection name
PRJ$DATUM	Datum of the projection

For a complete listing of DSC$ and PRJ$ variables, see the &DESCRIBE command reference in the online help. You can find examples of how to use the DSC$ variables in chapters 4 and 9.

 FYI
Other reserved variables set by &DESCRIBE and DESCRIBE: ARC/INFO GRID™ (GRD$, PRJ$), ARC/INFO GRID STACKS (STK$, PRJ$), ARC/INFO TIN™ (TIN$, PRJ$), and IMAGE data sets (IMG$).

About the graphical cursor: PNT$

Other reserved variables store three important pieces of information about the location of the graphic cursor on the screen: the mouse key used to select a screen location and the x- and y-coordinates of the location. This group of variables (PNTKEY, PNTX, and PNT$Y) are set by issuing the directives &GETPOINT or &GETLASTPOINT. See chapter 6, "Getting and using coordinates," for more information about the PNT$ variables.

About AML program status: AML$

Reserved variables also store information regarding the status of AML programs. These reserved variable names begin with AML$ and can report which line is being executed, what the last error message was, and so on. You don't need to issue any commands to initialize AML$ variables—they're automatically created whenever an AML is executed. The table below gives some examples of AML$ variables:

Reserved variable	Value
AML$MESSAGE	Holds the last message issued by AML. This variable is set whenever an error or warning message is generated.
AML$SEV	The severity status of the last line processed by AML or the current ARC program.
AML$LINE	The line number of the current AML being executed.

See chapter 9, "Expecting the unexpected," or search the online help for "variables" to find a thorough discussion of AML$ variables.

AML functions

Like variables, *functions* provide information to your program. Functions perform a defined operation that returns a value, which can be substituted directly into any command or directive or stored in a variable to use later in a program. AML functions are always enclosed in square brackets (i.e., [ ]).

The following example illustrates how the [USERNAME] function retrieves the current user's name (i.e., the person who logged in) and substitutes this value in the &TYPE statement.

```
&type Good morning [username]
```

If Judy is the current user, &TYPE prints this message on the screen:

```
Good morning judy
```

The [DATE] function returns the current date, which has several formatting options. The -CAL option returns a calendar-style date as follows:

```
&type Today's date is [date -cal]
Today's date is November 8, 1996
```

See the [DATE] function command reference in the online help for a complete listing of all formats. Variables can store the value returned by a function. Examine the following code:

```
&sv today =  [date -cal]
&type Today's date is %today%
Today's date is November 8, 1996
```

&SETVAR (&SV) assigns the value of the [DATE] function to the variable TODAY. The variable is referenced in the &TYPE statement and produces the same result as the previous example.

You can use more than one function on a command line. Consider this example:

```
&type Good morning [username], today's date is [date -cal]
Good morning judy, today's date is November 8, 1996
```

Functions perform the following kinds of operations:

- Prompting for user input from the keyboard or a menu
- Calculating mathematical values
- Modifying and extracting values from character strings
- Reporting current conditions
- Reading and writing files

It's convenient to group functions based on the kinds of operations they perform. The following discussions present the most commonly used functions in each group and list the chapter in this workbook devoted to describing their operations.

Prompting for user input from the keyboard or a menu

The AML functions shown in the following table prompt the user to input information by typing a response or selecting from alternatives presented on a menu.

Function	Method of retrieving input
[QUERY]	Asks the user a yes/no question and retrieves the value typed
[RESPONSE]	Asks the user for information and retrieves the value typed
[GETCOVER]	Presents a list of coverages and retrieves the coverage chosen
[GETSYMBOL]	Presents a list of symbols and retrieves the symbol chosen

Suppose you write an AML program to shade a set of selected polygons. You want the user to specify the number of the shade symbol to use and then shade the polygons accordingly. The following code performs this task:

```
POLYGONSHADES LANDUSE [response 'Enter a symbol number']
```

The user sees the following prompt in the dialog window:

```
Enter a symbol number:
```

The value of the [RESPONSE] function is the number the user types in response to the prompt. The value is substituted for the function in the command line. If, for example, the user types a 6, the following command line executes:

```
POLYGONSHADES LANDUSE 6
```

The [GETSYMBOL] function is an alternative to [RESPONSE] for soliciting user input for symbology. As shown below, [GETSYMBOL] displays a list of symbol choices to the user.

```
POLYGONSHADES LANDUSE [getsymbol -shade]
```

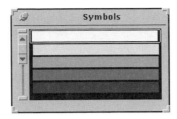

[GETSYMBOL -SHADE] returns the number of the shade symbol the user selects and substitutes that value in the POLYGONSHADES command line.

[QUERY] and [RESPONSE] appear throughout this workbook in examples that require user input. Chapter 2 discusses AML's suite of [GET...] functions.

Calculating mathematical values

There are more than twenty-five AML functions that perform mathematical calculations. This group of functions handles everything from simple addition to trigonometry to measurement conversion. An often-used mathematical function, [CALC], is introduced and used below.

Function	Mathematical operation
[CALC]	Performs the five basic mathematical operations: addition, subtraction, multiplication, division, exponentiation

Continuing the previous example, suppose you're using a customized shadeset with symbols numbered 101 through 199. For the sake of simplicity, you want the user to be able to type the numbers 1 through 99 instead. The following code calculates the correct symbol number:

```
&sv shade = [response 'Enter a color number']
POLYGONSHADES LANDUSE [calc %shade% + 100]
```

The [RESPONSE] function is used to set the variable SHADE, which in turn is used in the [CALC] function. If the user enters symbol number 6, the final command line evaluates to:

```
POLYGONSHADES LANDUSE 106
```

You'll find the [CALC] function explained in chapter 6 and used in examples throughout the workbook.

Modifying and extracting values from character strings

Character strings contain all kinds of information. Usually, character strings contain more than one piece of information. Extracting and modifying individual parts of a character string is essential to many AML programs.

Function	Character string manipulation
[EXTRACT]	Finds and extracts a specific element from a character string
[UPCASE]	Converts a string from lowercase to uppercase letters
[LENGTH]	Returns the number of characters in a character string

Suppose you need an AML program to update a parcel database with new owners and all names need to be in uppercase. The [UPCASE] function ensures that the names are converted to uppercase as shown in the following code:

```
&sv owner = [response 'Enter new owner']
MOVEITEM PAR_OWNER = [upcase %owner%]
```

If the new owner is typed Brenneman or brenneman, [UPCASE] converts the input to BRENNEMAN and the final command line looks like this:

```
MOVEITEM PAR_OWNER = BRENNEMAN
```

Chapter 7 includes a complete discussion on handling character strings.

Reporting current environments

Numerous AML functions report on current environments that exist in ARC/INFO. The [SHOW] function is worth noting because of the variety and detail of the information it returns.

Among other things, [SHOW] returns values such as information about the operating system, the current display device, the number of selected features, the current symbol number, the edit cover, and the number of vertices in an arc. See the [SHOW] function command reference in the online help for a listing of information that [SHOW] can return.

FYI

The &SHOW directive displays information established through other AML directives. For example, &SHOW returns the hardware device set with the AML &TERMINAL directive but not the device set with the ARC/INFO command DISPLAY.

In ARCEDIT, it's common to change the map extent for viewing and editing different areas. [SHOW MAPEXTENT] returns the corner coordinate values of the map extent. Using [SHOW MAPEXTENT] to capture the current map extent in a variable enables you to return to the original after changing the map extent. This procedure is shown here:

```
MAPEXTENT *    (user locates the corners of desired region)
&type [show mapextent]
1560.89475 3456.72839 1798.84955 4432.67849
&sv NWREGION = [show mapextent]
MAPEXTENT LANDUSE
DRAW
MAPEXTENT %NWREGION%
DRAW
```

The [SHOW] *function* can also return any information that the SHOW *command* can produce. For instance [SHOW MAPEXTENT] is used in the last example. The documentation for [SHOW MAPEXTENT] is found in online help references under ARCEDIT for the command SHOW MAPEXTENT, not under the Command reference for the [SHOW] function. The only difference in syntax between the [SHOW] function and the SHOW command is that the SHOW function has brackets.

```
Arcplot: SHOW MAPE
6811199.847125,1829647.747925,6823179.992775,1836516.177075
Arcplot: &TYPE [SHOW MAPEXTENT]
6811199.847125,1829647.747925,6823179.992775,1836516.177075
```

You could use the [SHOW MAPEXTENT] function to store previously set map extents for quick display when ARCEDIT is initialized. The macro on the previous page is an example of how this might be done.

The [SHOW] function often returns a character string containing several elements or words. Character strings are discussed in chapter 7.

Reading and writing files

A small group of functions performs file input and output operations. These functions return a value indicating the success of an operation, a record of information, or the reference number of the file. The following table lists some of the functions for opening and reading files:

Function	Operation
[OPEN]	Opens a system file for reading or writing
[READ]	Reads a record from an open file

The following code opens a file containing a list of coverages, reads a coverage name, and copies it to preserve a backup copy:

```
&sv openfile = [open cover.list stat -read]
&sv cover = [read %openfile% stat]
copy %cover% %cover%_old
```

Chapter 10, "Reading and writing files," covers functions that perform file input and output.

Nesting functions

Now that you understand how a single function is evaluated, consider how combining, or nesting, functions can reduce the amount of code needed to accomplish a task.

A *nested function* occurs when one function is placed inside another function. Look at how to rewrite some of the previous examples using nested functions.

You've seen a variable set by a function and then used inside another function like this:

```
&sv answer = [RESPONSE 'Enter new owner']
&sv owner = [UPCASE %answer%]
MOVEITEM [QUOTE %owner%] TO PAR_OWNER
```

Here's how to perform the same task using nested functions:

```
MOVEITEM [QUOTE [UPCASE [RESPONSE 'Enter new owner']]] TO
PAR_OWNER
```

When functions are nested, the inside function is always evaluated first and its value returned to the outside function. The outside function is then evaluated. You can nest any number of functions. Evaluation always moves outward from the inside function, regardless of how many functions are nested.

Assume the user enters burrows as the new owner in the example above. First, the [RESPONSE] function is replaced by burrows, yielding:

```
MOVEITEM [QUOTE [UPCASE burrows]] TO PAR_OWNER
```

Then [UPCASE] converts the value to uppercase, yielding the following:

```
MOVEITEM [QUOTE BURROWS] TO PAR_OWNER
```

Finally, [QUOTE] puts single quotes around the string so that the final ARCEDIT command looks like this:

```
MOVEITEM 'BURROWS' TO PAR_OWNER
```

Interpreting special characters

In addition to knowing about functions, directives, and variables, understanding the order in which AML interprets these components is vital to understanding how an entire AML program executes.

Special characters direct the AML processor to interpret a line in a unique way. The table on the next page lists the special characters recognized by AML and describes how they're interpreted. You can display most of these special characters on the screen by issuing the &LISTCHAR directive.

Characters	AML interpretation
[ ]	An AML function
&	An AML directive
% %	An AML variable
~ (at end of line)	A single command is continued on the next line
~ (at beginning of line)	Suppress AML interpretation of the command line
;	Separate the multiple commands typed on the same line
/*	Disregard descriptive information during execution
(! !)	Repeat a command for each element in a delimited list
{! !}	Expand one command with all elements in a delimited list

Command continuation

Sometimes commands are too long to fit on one line and, although ARC/INFO will wrap the command onto the next line, you may wish to type a tilde (~) before pressing the carriage return. This tells ARC/INFO that you want to continue typing a command on the next line. Otherwise, when you enter a carriage return, the command line executes. The tilde instructs AML to ignore the carriage return and interpret both typed lines as one command line.

```
Arc: BUILD ~
Arc: WELLS P~
Arc: OINT
```

evaluates as:

```
Arc: BUILD WELLS POINT
```

The tilde will allow you to write cleaner AMLs that don't have lines of text scrolling off the side of the text editor. You would probably never use command continuation with the above example, but when we get to the &MENU directive we'll see many uses for command continuation.

Command separation

You can string together commands on one line by separating the commands with a semicolon (;). AML interprets a semicolon as a carriage return. The following example shows how to do this:

```
Arcedit: EDIT PARCELS; EF POLY; DE POLY; DRAW
```

Comments

When AML reads a comment symbol (/*), it ignores the text that follows. Comment symbols, introduced in lesson 1, are the way to include descriptive comments in your programs.

Command repetition

When the AML processor encounters a command followed by a list that's surrounded by exclamation points and enclosed in parentheses (i.e., (! !)), it repeats the command with each item in the list.

Issuing this command:

```
Arc: COPY /GIS/DATA/(!PARCELS LANDUSE STREETS!)
```

is equivalent to typing:

```
Arc: COPY /GIS/DATA/PARCELS
Arc: COPY /GIS/DATA/LANDUSE
Arc: COPY /GIS/DATA/STREETS
```

Command expansion

When a command line includes a list enclosed by exclamation points and brackets (i.e., {! !}), the command is expanded using all elements in the list. Issuing this line:

```
Arcplot: MAPEXTENT QUAD{!15 17 23 24!
```

results in this command:

```
Arcplot: MAPEXTENT QUAD15 QUAD17 QUAD23 QUAD24
```

 FYI

Operating system or application-specific character definitions may conflict with the AML special characters definitions. In this case, use the &SETCHAR directive to modify AML interpretation. Document any changes you make and be sure to maintain consistency.

Suppressing AML interpretation

Using a tilde (~) at the beginning of a command line suppresses AML interpretation of special characters. This is often needed when you want to use these characters in text (i.e., with no special meaning attached). For example, consider the problem of creating text containing a percent sign (%) on a map. Issuing the following TEXT command line in ARCPLOT produces the following error message:

```
Arcplot: TEXT 25% of the area is affected by flooding.
AML  ERROR - Unmatched variable delimiter
```

Pairs of percent signs delimit a variable, so AML interprets 25 as a variable with one percent sign missing. Placing the tilde at the beginning of the line allows the TEXT command to read the text string correctly:

```
Arcplot: ~TEXT 25% of the area is affected by flooding.
```

Quotation marks also suppress AML interpretation, so quoting the text as follows produces the same result as the tilde used in the previous example:

```
Arcplot: TEXT '25% of the area is affected by flooding.'
```

Order of interpretation

Now look at the order in which the AML interpreter evaluates special characters and the other components of AML programs, such as ARC/INFO commands, operating system commands, directives, functions, and variables. AML interprets variables first, then functions, and finally directives or commands.

Knowing the priority AML assigns to the program components allows you to determine how AML will evaluate a line code. As your knowledge of AML increases, your programs get more sophisticated, but the order of evaluation never changes. Use the diagram on the following page as a reference throughout this workbook.

The following diagram indicates AML's order of evaluation:

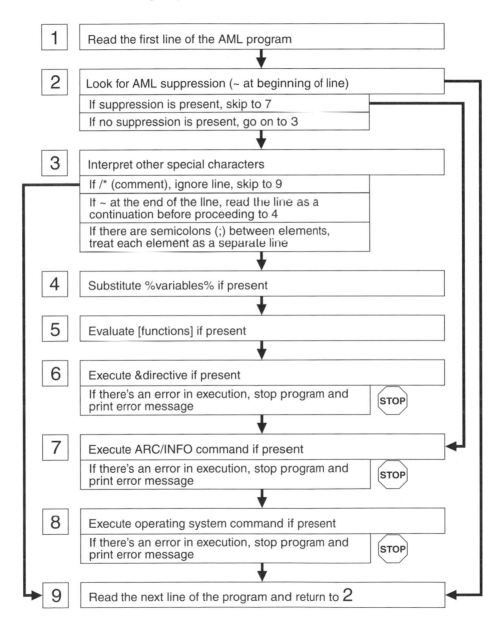

Examine the following lines of AML code and how AML evaluates them, assuming that the variable TYPE contains the value SHADE:

```
&sv symbol = [getsymbol -%type%]
POLYGONSHADES [getcover * -poly] %symbol%
```

Read the first line
```
&sv symbol = [getsymbol -%type%]
```

Substitute the variable
```
&sv symbol = [getsymbol -shade]
```

Evaluate the function
```
&sv symbol = 7
```

Execute the directive
```
Assigns a variable named symbol a value of 7
```

Read the second line
```
POLYGONSHADES [getcover * -poly] %symbol%
```

Substitute the variable
```
POLYGONSHADES [getcover * -poly] 7
```

Evaluate the function
```
POLYGONSHADES landuse 7
```

Execute the ARC/INFO command
```
Creates a display on the screen
```

In summary

Including variables and functions in an AML program enables it to perform under changing conditions. Variables are assigned once and then referenced as many times as needed throughout the program. Variables must always be enclosed in percent signs (i.e., % %) when they are referenced.

User-assigned variables can be either local or global. Local variables can be used only in the program in which they're assigned. Global variables are available to any program run during the ARC/INFO session in which they're assigned. Global variables are distinguishable from local variables because their names begin with a period (.). Reserved variables are assigned by ARC/INFO—their names are reserved and should never be overwritten.

Functions yield a value resulting from some defined operation (e.g., prompting the user, mathematical operations, or extracting part of a character string). The value of a function can be stored in a variable or used as a command argument. Nested functions can perform complex operations.

Special characters allow AML to differentiate between the components of a program, provide you with some shortcuts, and enable you to document your programs. AML adheres to a hierarchical order of interpretation for every element that occurs in the program. AML is an interpreted language, so the interpretation of AML code defines what is finally executed by the computer.

Answer 1-1: The resulting command line is as follows:

```
EDITCOVER CENSUS4
```

Exercises

Exercise 1.3.1

Suppose two variables are set as shown here:

```
&sv name = michael
&sv NAME = mike
```

What is returned by the following command and why?

```
&type %name%
```

Exercise 1.3.2

A variable named %SNAPTOL% is set in an AML program named SETEDIT.AML.

(a) Why can't %SNAPTOL% be referenced successfully in an AML named SNAPPIT.AML?

(b) Which change could you make so SNAPPIT.AML can use this variable?

Exercise 1.3.3

DSC$ variables provide information about _____ .

PNT$ variables provide information about _____ .

Exercise 1.3.4

Match the definitions with the option it best describes.

(a) Directives perform an operation and return a value.

(b) Variables execute an operation.

(c) Functions store and return values.

Exercise 1.3.5

What does the following line of AML code do?

```
&workspace /gis/project5/[username]
```

Exercise 1.3.6

How can you decide whether to use a function in a command line or to set its value to a variable first?

Exercise 1.3.7

Create a nested function from the following two lines of code:

```
&sv shade = [response 'Enter a shade symbol number']
POLYGONSHADES LANDUSE [calc %shade% + 100]
```

Exercise 1.3.8

Determine the resulting command line in the following example. Show any intermediate steps.

```
&sv letter = [upcase [response 'Enter  A, B, C, or D']]
```

Suppose the user types: b

Exercise 1.3.9

Which commands are executed as a result of the interpretation of this line?

```
Arc: KILL (!SOILS PARCELS FIREDIST!)_OLD ALL
```

Exercise 1.3.10

Write a single command that executes the following three commands:

```
Arcplot: ARCS ZONE
Arcplot: ARCS LANDUSE
Arcplot: ARCS ROAD
```

Exercise 1.3.11

Determine how AML evaluates the following command line. Show all intermediate steps.

```
CALCULATE BLD_SPACE = [calc %side% - [response 'Enter front offset']]
```
Assume that %side% = 250 and the user types an offset of 35.

2 Menu-driven ARC/INFO

In ARC/INFO terms, a menu is a graphical list of options from which the user can select an action such as drawing a map, overlaying geographic data sets, or relating tables. AML allows you to create menus that provide users with an easy-to-learn, easy-to-use alternative to memorizing ARC/INFO commands. Menus make it easier for new users to learn ARC/INFO and help all users accomplish GIS tasks more efficiently.

Consider the following reasons for designing and creating a menu interface for your ARC/INFO application:

- Help less experienced ARC/INFO users become more productive with less training.
  A well-designed, visually appealing menu interface is usually easier for beginners to learn than a complex series of commands.
- Eliminate user errors by displaying only valid choices.
  Throughout this workbook you'll see the advantage—from the programmer's view—of ensuring correct user input. Presenting only valid choices reduces errors and also helps keep users from becoming confused by too many options.
- Represent user-defined tasks as menu choices.
  You can place a series of ARC/INFO commands that accomplish a series of tasks in an AML program and execute it from a menu.
- Customize terminology and standards.
  Your menus can contain terminology specific to your organization or application.
- Create a visually appealing, easy-to-use interface for demonstrating the capabilities of GIS technology.
  The usefulness of menus, combined with their inherent visual appeal, allows tasks to be accomplished in a way that makes the system inviting to new users.

The task of writing AML menus is easier than you might think. This chapter looks at the various types of menus, how they're created, and how they appear to the user. Some AML functions automatically create menus with a set style and screen layout. Other types of menus allow you to customize their look, positioning, and mode of interaction.

This chapter covers the following topics:

Lesson 2.1—Creating simple menus

- [GET...] me a menu!

Lesson 2.2—Creating a menu interface

- Menu basics
- AML programs vs. AML menus
- Menu types and files
- The &MENU directive: Beyond the basics

Lesson 2.3—Bringing AML programs and menus together

- Introducing menu design
- Starting a menu from an AML program
- Running AML programs from a menu
- Starting a second menu

Lesson 2.1—Creating simple menus

In this lesson

In ARC/INFO, you can display a list of available files in a number of ways, including using the LISTCOVERAGES command in ARC; the DIRECTORY command in ARCEDIT, ARCPLOT, TABLES™, and INFO™; or the operating system's command for listing directories (e.g., ls, dir). What if you don't really want to look at the list, but, rather, you want to select one of its elements from a menu? Some functions display a menu containing a list from which you can interactively select an element. This lesson discusses the AML functions that create this kind of interactive interface.

Functions that begin with [GET...] automatically create a menu of choices from which the user selects. You'll see how easy it is to construct menus to perform some basic ARC/INFO operations. Put your keyboard away and put running shoes on your mouse—operations like selecting coverages, files, and items are as easy as point and click.

FYI

This workbook reflects how menus appear and act when generated on a UNIX workstation running the Common Desktop Environment (CDE) user interface. The Microsoft® Windows NT® window interface looks similar to CDE. On ANSI terminals (i.e., those without a windowing system), AML menus appear and act differently. Consult the online help for a more detailed discussion of menu operation on ANSI terminals.

(GET...) me a menu!

Many operations in ARC/INFO require the user to provide a coverage name, INFO file name, text file name, item name, item value, and so on. AML provides the [GET...] functions to assist the user in creating a graphical user interface (GUI) for ARC/INFO. The [GET...] functions allow you to easily generate scrolling list menus with a sophisticated appearance. Common operations, like choosing a coverage or item name, become as easy as scrolling through the list and clicking the mouse on the desired selection.

The following is a list of the [GET...] functions and the choices they display on the menu.

Function name	Menu choices offered
[GETCHOICE]	User-defined choices (e.g., names, project numbers, map sheets)
[GETCOVER]	Coverage names
[GETDATABASE]	ArcStorm™ databases, connections, tables, layers, libraries, or historical views
[GETDATALAYER]	Layers of a Spatial Database Engine™ (SDE™) data set
[GETDEFLAYERS]	Defined layers of an SDE data set
[GETFILE]	Directory or file names (e.g., workspaces or INFO files)
[GETGRID]	Grid names
[GETIMAGE]	Image names
[GETITEM]	Item names contained in a feature attribute table or other INFO file
[GETLAYERCOLS]	Columns of a defined SDE layer
[GETLIBRARY]	Map libraries, tiles, or layers
[GETSTACK]	Grid stacks
[GETSYMBOL]	Symbols of the specified type (e.g., marker, shade, text, or line)
[GETTIN]	TIN names
[GETUNIQUE]	Unique item values in a feature attribute table or other INFO file

FYI

[GETCHAR] doesn't appear in the table because it doesn't create a menu. Instead, it prompts the user to type one character using the keyboard to initiate an action. See the online help Command references for more information.

For menus to correctly position, scale, and display text, the &TERMINAL directive must be issued before the [GET...] function. &TERMINAL sets up screen characteristics and specifies the pointing device used for picking from the menu (e.g., the mouse). You can use what's called a startup AML to set &TERMINAL (see page 2-35) or type it on the command line before using a [GET...] function. The &TERMINAL directive only needs to be issued once during an ARC/INFO session—it remains set until you quit from ARC/INFO.

To tell [GETCOVER] to search the current workspace for coverages, use the wildcard character, an asterisk (*), within the function. The following lines of AML code create the menu shown below:

```
&terminal 9999
&sv covername = [getcover *]
```

Menus display up to five menu choices at a time. Whenever there are more than five choices available, the scroll bar on the side of the menu allows you to scroll through the list. ARC/INFO defines the menu's size and location on the screen; this can't be modified.

All [GET...] functions can access other workspaces—perhaps that of a coworker or centralized database—by changing the asterisk (*) to the workspace pathname:

```
&sv covername = [getcover /meling1/vicki]
```

Many times, *system variables* can be set to store path names. This can make changing the location of an application or its data very simple. [GET...] functions can also access Windows NT or UNIX system variables.

```
&sv covername = [getcover $PROJDATA]
```

Use [GET...] functions to set variables or incorporate them directly into ARC/INFO commands. In the following examples from ARCPLOT, both uses of the [GETCOVER] function are valid:

```
&sv covername = [getcover *]
MAPEXTENT %covername%
```

or

```
MAPEXTENT [getcover *]
```

In both cases, the function evaluates to the coverage name picked from the menu, and the map extent is set to that coverage. Although the format shown in the second example is shorter, the variable %COVERNAME% isn't set and consequently isn't available to use with other commands (e.g., to draw coverage features). To determine which format is appropriate, consider how you'll use the information returned by the function.

The [GETCOVER] function returns the full pathname to a coverage, not just the coverage name. This is also true for [GETTIN], [GETGRID], and [GETFILE].

For example, given this code:

```
&sv covername = [getcover /meling1/vicki]
```

%COVERNAME% equals:

```
/meling1/vicki/soils
```

If you want the coverage name without the pathname, use the [ENTRYNAME] function described in chapter 7.

Many ARC/INFO commands require an item name. The [GETITEM] function displays a list of item names and retrieves any item the user picks. Assume a polygon coverage is selected and stored in the variable %COVERNAME%.

The line of code shown below displays a menu of all items in the coverage's polygon attribute table (PAT) and sets the variable %POLYITEM% to the item the user selects.

```
&sv polyitem = [getitem %covername% -poly]
```

The following menu was created using the [GETITEM] function with a coverage named SOILS:

The following AML program incorporates [GETCOVER] and [GETITEM]. The program prompts the user for a polygon coverage and an item name, shades the coverage based on the item values, lists out the feature attribute table, and labels each polygon with text representing the value of the item. Substituting the choices selected from the menu into the command arguments makes this task easy to accomplish.

```
/* viewdata.aml
ARCPLOT
&sv covername = [getcover *]
&sv polyitem = [getitem %covername% -poly]
MAPEXTENT %covername%
POLYGONSHADES %covername% %polyitem%
LIST %covername% POLY
POLYGONTEXT %covername% %polyitem%
```

Errors could arise from the way the [GETCOVER] and [GETITEM] functions are used in this AML program. The [GETCOVER] menu includes all coverages, regardless of their feature type (point, line, poly); however, the ARC/INFO commands that follow require a polygon coverage. If the user picks a point coverage, the task can't be completed. Similarly, [GETITEM] displays the values of all items, but only integer values are valid for the POLYGONSHADES command. In this example, the choices on the menu should be restricted to only polygon coverages and integer-valued items.

Presenting only valid choices to the user reduces program errors. [GET...] functions provide options for limiting menu choices. Each [GET...] function has slightly different options depending on the type of data displayed. The following table outlines some of the important options. The online help includes a complete description of the syntax for these and other [GET...] functions.

Function name	Option	Description
All [GET...] functions	-SORT	Sorts menu choices in ascending order
	-NONE	Includes _NONE_ as a menu choice
	-OTHER	Includes _OTHER_ as a menu choice; prompts user for a value
[GETCOVER]	wild_card (*)	Coverages with common names (e.g., soils*)
	-coverage_type	Feature types (e.g., poly, line, route.subclass)
[GETFILE]	wild_card (*)	Files/directories with common names (e.g., *.gra, *.aml, *.pat)
	-type-	File or directory types (e.g., INFO files, system files, workspaces)
	NOEXTENSION	No extension displayed on file name
[GETGRID]	wild_card (*)	Grids with common names (e.g., soils*)
[GETITEM]	-item_class	Item class (ordinary, redefined, or both)
	-item_type	Item type (character, numeric, integer)
[GETLIBRARY]	-LIBRARY; -TILE; -LAYER	Map libraries; map tiles; map layers
[GETSYMBOL]	-LINE; -SHADE; -MARKER; -TEXT	Line symbols; shade symbols; marker symbols; text symbols
[GETTIN]	wild_card (*)	TINs with a common name (e.g., quadelev*)

Look at some scenarios that illustrate how the [GET...] options can tailor menus to display only valid choices.

[GETCOVER] with the feature type option

Suppose your AML program draws and lists information for only polygon coverages (as in VIEWDATA.AML on page 2-7). Using the [GETCOVER] function with the -POLY option as follows displays a scrolling list of valid polygon coverages:

```
&sv covername = [getcover * -poly]
```

[GETCOVER] with the wildcard option

The coverages you want reside in another workspace. All the coverage names use the word ZONE as a prefix. Use [GETCOVER] as follows:

```
&sv covername = [getcover /gis/sally/quad5/zone*]
```

[GETFILE] with the wildcard option

Use [GETFILE] to select a lookup table for creating a display in ARCPLOT. All the files named with a .LUT extension are listed.

```
&sv filename = [getfile *.lut -info]
```

Issuing a pathname that includes a wildcard can create the same character pair (/ *) that denotes a comment symbol to AML (e.g., /project100/quad5/*.gra). You must quote the entire string (i.e., '/project100/quad5/*.gra') to keep AML from interpreting these characters as a comment.

[GETFILE] with an instructional prompt and the -NONE option

Suppose you want to add an instructional prompt across the top of the menu to help the user understand what's expected. You also decide to include the _NONE_ choice in case the user doesn't want to select a lookup table. Use [GETFILE] with a text string and the -NONE option as shown below. Notice that the instructional prompt needs quotes when it includes more than one word.

```
&sv filename = [getfile *.lut -info 'Please select a lookup ~
   table' -none]
```

[GETITEM] with the -INTEGER option

You want the user to pick an item from which to shade polygons. Only items defined as integer type are appropriate for this operation and are displayed using:

```
&sv polyitem = [getitem %covername% -poly -integer]
POLYGONSHADE %covername% %polyitem%
```

[GETSYMBOL] with the -SHADE and -RANGE options

Use [GETSYMBOL] as shown here to create a menu of shade symbols in the range of 1 to 25:

```
&sv symbol = [getsymbol -shade -range 1-25]
```

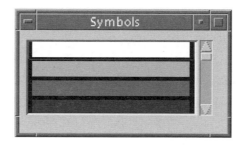

[GETUNIQUE] with the -OTHER option

Suppose you want an application that automates data input. The data input is "USE" codes for a parcel base. All new codes input should be consistent with the preexisting codes. Use [GETUNIQUE] as follows to allow the user to pick a unique value from a list for the item USE and for the polygon coverage PARCELS. This value can then be used for the parcels being input:

```
&sv value = [getunique parcels -poly use 'Pick a use' -other]
```

Notice that if the user wants to type in the value they can choose _other_. If _other_ is chosen, the user is prompted to input the value at the keyboard. The prompt on the command line is the same as the prompt indicated within [GETUNIQUE]. The value input must still be consistent with the values returned by [GETUNIQUE].

```
Pick a use: 510
```

[GETCHOICE] with the -SORT option

Use [GETCHOICE] to create a menu of user names that are sorted in alphabetical order and supply a prompt:

```
&sv user = [getchoice vicki nick judy bill andy -prompt ~
'Who are you?' -sort]
```

[GETCHOICE] is the only [GET...] function that needs -PROMPT specified to display a prompt or message. -PROMPT distinguishes the message from the choices in the code.

You have the ability to substitute AML variables for any of the options in the [GET...] functions. This allows you to change what the functions do in an AML program at run time. For example, suppose you choose a coverage from a [GETCOVER] function. You can determine the correct symbol type based on topology and store this data in a variable called TYPE. You can use this variable with the [GETSYMBOL] function to choose symbology relevant to the coverage type chosen:

```
&sv symbol = [getsymbol -%type%]
```

FYI

It's always important to think about coverage- and file-naming conventions. Standardized names allow you to take advantage of the wildcard option (*) provided with the [GET...] functions.

Q

Question 2-1: What is the line of code that would create a menu of all items in PARCELS.PAT and set a variable PARCELITEM equal to the item name the user selects? Include a message that instructs the user to select an item from the menu.

(Answer on page 2-13)

In summary

A graphical user interface (GUI) can simplify many operations in ARC/INFO for users. [GET...] functions generate menus that display choices from which the user selects the one containing the appropriate data. The user selects a menu choice by pointing and clicking with the mouse.

You can reduce errors in your AML programs by using the optional parameters provided with the [GET...] functions. These options limit the choices offered on the menu to only those that are valid for the task.

A

Answer 2-1:

```
&sv parcelitem = [getitem parcels.pat -info 'Please select an item']
```

 or

```
&sv parcelitem = [getitem parcels -polygon 'Please select an item']
```

Exercises

Exercise 2.1.1

Use [GETCOVER] and [GETITEM] to obtain data for the POLYGONSHADES command.

Exercise 2.1.2

Write a [GET...] function to display a menu of textset files (.TXT) in a workspace named
/gis/symbolsets. Use the function as the argument for the TEXTSET command in ARCPLOT.

Exercise 2.1.3

Write the AML code that creates the following menu:

Exercise 2.1.4

Write the code that creates the following menu, which allows the user to define an EDITFEATURE:

Exercise 2.1.5

Write the code that creates the following menu, which allows the user to define an EDITCOVERAGE:

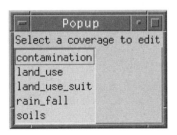

Lesson 2.2—Creating a menu interface

In this lesson

The [GET...] functions are powerful tools for creating menus. You can't, however, control how they look and interact, or where they're located on the screen. Menu interfaces need versatility to support many types of users and situations. AML supports eight types of menus that you can customize to support your ARC/INFO application. These encompass a variety of graphical displays, screen layouts, and methods of interaction. This lesson introduces the eight menu types. Three types—pulldown, sidebar, and matrix—are discussed in detail.

Menu basics

Like AML programs, AML menus are ASCII text files created with any text editor. AML menus aren't AML programs because they differ in how they're named, formatted, and executed.

The naming convention for menu files is a descriptive file name followed by a .MENU extension. For example, if you write a menu interface to display information for a project in Salt Lake City, you might call the menu file SLC.MENU. Menu files are normally stored in your current directory. Accessing menus stored in a centralized location is discussed in chapter 3.

The &MENU directive executes a menu file and invokes the graphical user interface (GUI). The menu described above is invoked by typing &menu slc.menu on the command line. It's alright to leave off the .MENU extension (e.g., &menu slc). The menu remains on the screen until the user dismisses it.

The &TERMINAL directive must be set before issuing the &MENU directive. &TERMINAL defines the characteristics of the terminal used to display the menu, such as device size and pointing device.

Menu choices and the actions they execute are defined in the menu file. The format of each menu file uniquely identifies its type and appearance on the screen. The indentation of each line in the file dictates where each choice appears when the menu is displayed.

AML programs vs. AML menus

AML supports two types of files. AML *programs* are text files named with the .AML extension and are executed with the &RUN directive. AML *menus* are also text files, but are named with the .MENU extension and are invoked with the &MENU directive. Either file can include AML directives, functions, and variables.

Programs and menus have different purposes. An AML program should accomplish a task. It may require user input and use a menu interface (e.g., a [GET...] function). On the other hand, an AML menu is designed as a GUI. You can customize the visual appearance and performance of an AML menu.

Often the two files are used in conjunction. A selection from a menu can execute a program, and programs can invoke menus. Menu files are discussed further in the next lesson and in chapter 12.

Menu types and files

You can create eight types of menus with AML. As outlined in the following table, a unique number in the menu file specifies each type:

Menu type	Number
Pulldown	1
Sidebar	2
Matrix	3
Key	4
Tablet	5
Digitizer	6
Form	7
Enhanced Pulldown	8

Menu types differ in how they look, where they're positioned on the screen, and how the user interacts with them. The pulldown, sidebar, and matrix menus appear on the screen in various positions, and the user makes a selection by pointing to a choice and clicking the mouse.

Key menus appear in the dialog area or window where commands are typed. You can use this style on terminals that don't support windows. Key menus prompt the user to respond by typing a single key. The [GETCHAR] function creates a key menu.

Tablet and digitizer menus are paper menus placed on a tablet or digitizing board. The user interacts with this menu using the digitizing cursor. Locations on the paper correspond to defined operations. Selecting a location on the paper menu executes the operation.

Form menus are the most powerful type of menu. Not only do they enable the user to make choices, such as pulldown and sidebar menus, but they also accept input from the keyboard and can incorporate scrolling lists within the menus themselves. You'll learn more about form menus in chapter 12, "Form menus."

Enhanced pulldown menus allow you to have sidebar choices within each pulldown. Menu choices can also be grayed out to limit the user's choice to only those options that apply to a particular circumstance. Enhanced pulldown menus are discussed further in chapter 13.

The remainder of this chapter covers pulldown, sidebar, and matrix menus in detail. Search the online help index under menus in AML for additional information on creating key, tablet, and digitizer menus.

Pulldown menus

Pulldown and sidebar menus are similar in file format, appearance, and operation. Both styles display a main menu and, optionally, one or more submenus. First, look at a pulldown menu and the menu file that generates it.

The following pulldown menu draws coverages and lists attribute tables. The *main menu* is oriented horizontally across the terminal screen. The choices appear as buttons and the user makes a selection by pointing to one with the screen cursor and clicking with the mouse. A main menu is displayed below:

A menu selection can directly execute an operation, but more typically it displays additional choices on a submenu. These submenus appear to *pull down* from the main menu, hence the name of this style.

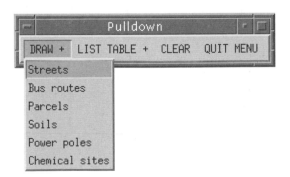

Notice in the last example that there are plus signs (+) attached to two of the menu options. This is a user-added symbol put into the name of the options that have pulldown choices. When programming your own applications, pick the symbol that best conveys the correct message. Enhanced pulldown menus have arrows automatically built into the menu type.

The pulldown main menu remains on the screen while the submenu is present. The submenu remains on the screen only until a choice is made. The menu file that generates this menu is listed below.

```
[1]     1 expull.menu
[2]     /* Draws coverages and lists attribute tables
[3]     'DRAW +'
[4]        Streets            ARCLINES STREET TYPE
[5]       'Bus routes'        ROUTELINES STREET BUS RT_NO
[6]        Parcels            POLYGONS PARCELS
[7]        Soils              POLYGONSHADES SOILS SOIL_TYPE SOIL.LUT
[8]       'Power poles'       POINTMARKERS POLES 10
[9]       'Chemical sites'    POINTMARKERS CHEMSITES CLASS
[10]    'LIST TABLE +'
[11]       Streets            LIST STREET.AAT INFO
[12]      'Bus routes'        LIST STREET.RATBUS INFO
[13]       Parcels            LIST PARCELS.PAT INFO
[14]       Soils              LIST SOILS.PAT INFO
[15]      'Power poles'       LIST POLES.PAT INFO
[16]      'Chemical sites'    LIST CHEMSITES.PAT INFO
[17]    CLEAR
[18]    'QUIT MENU'          &return
```

The number on line [1] indicates the menu type (refer to the table on page 2-18). Text appearing on the first line after the type is read as a comment. This is often used to record the menu file name, a short explanation, or the author and date. Write additional comments anywhere in the menu using the comment symbol (/*) shown here on line [2].

Each line in a menu file defines two things: (1) the *choice* that's graphically displayed to the user, and (2) the *action* that's executed when a choice is selected. Choices appear on the left side of the line, either in the first column or indented a few spaces.

Menu choices

Indentation determines how choices appear on a menu. A choice that's *not indented* (i.e., it starts in the first column) appears on the main menu. You can see how lines [3], [10], [17], and [18] in the file EXPULL.MENU are displayed:

Indented choices appear on a submenu. You must indent at least one space; after that, it's personal preference. Each submenu corresponds to one choice in the main menu. When DRAW is selected from the main menu, lines [4] through [9] create a submenu. The submenu for LIST TABLE is created by lines [11] through [16].

> Always indent menus and AML programs using spaces instead of tabs. AML interprets tabs as special characters rather than as spaces. If a menu fails to initialize, set the text editors you're using to identify spaces. In vi use *:set list* to mark the tabs with the ^I. Be careful of texteditors, like vi, that automatically insert tabs. In vi, adjust your tabstop to control tab insertion, or do a global search and replace on existing tabs.

Whenever a choice contains two or more words, the choice must be enclosed in single quotes (e.g., 'Bus routes' and 'LIST TABLE +'). Sometimes you may want to separate the choices in a pulldown menu. To do this with spaces, put a quoted blank space (' ') as one of the choices in the menu.

Sometimes the choice displayed on the menu is also the command that performs the action, like the CLEAR command in line [17]. In this case, you only need to write the choice in the file once. You may, for clarity, prefer to specify commands, like CLEAR, as both a choice and an action in your menu files.

The maximum number of displayed choices you can define in the menu file (including the main menu and all submenus) is 150. EXPULL.MENU contains sixteen choices. The main menu is limited to twenty choices, but may have fewer depending on the width of your terminal. Each choice is limited to thirty-two characters (e.g., 'Chemical sites' is fourteen characters).

Menu actions

The action defines what happens when a choice is selected and is written to the right of the choice on the same line. Separate actions from choices using several spaces and visually align them to make your program more readable.

An action executed by a menu choice can include more than one command as long as the commands are separated by semicolons. One menu choice could, for example, clear the screen, set a new map extent, and draw a coverage. This line in the menu file might look like this:

```
Streets        CLEAR; MAPEXTENT STREET; ARCLINES STREET TYPE
```

If you need more than one line of text for a command, use the tilde (~) before you hit the carriage return and continue typing on the next line. Remember that a carriage return indicates a new command to AML. You can write long commands or long command sequences in an AML program that's executed by the menu choice. AML programs that run from a menu are discussed in lesson 2.3.

FYI

The [MENU] function can return a single value from a menu. Once a selection is made, the menu is erased from the screen and the selected value is returned. See the online help Command references for additional information on the [MENU] function.

Question 2-2: Describe how EXPULL.MENU would look if all the indented lines were left-justified (i.e., moved to the first column).

(Answer on page 2-30)

Sidebar menus

After you understand the basics of pulldown menus, it's easy to understand the modification needed to create a sidebar menu. The main difference between sidebar and pulldown menus is their orientation. In sidebar menus, choices appear vertically down the terminal screen and the submenus pull out to the right.

The menu files for the sidebar and pulldown menus are almost identical. Line indentation for the main menu and submenus is defined the same for sidebar menus as for pulldown menus. The only difference between the two menu files is the menu type specified on the first line of the file. Sidebar menus are specified with the number 2 instead of the number 1, which indicates a pulldown menu—it's that easy.

Substituting a 2 (sidebar menu type) for the 1 (pulldown menu type) in the first line of EXPULL.MENU produces this sidebar menu:

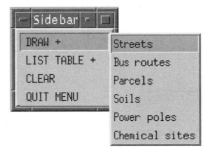

FYI

You can force a sidebar display from a pulldown file (and vice-versa) by invoking the menus like this:

```
&menu expull.menu &sidebar
&menu exside.menu &pulldown
```

Matrix menus

Like pulldown and sidebar menus, matrix menus provide the user with a
point-and-click interface, but matrix menus present all possible choices on a
single menu—there are no submenus. The menu automatically sizes to
accommodate all the choices that are listed in rows and columns—hence the
name, "matrix menu."

A matrix menu designed to draw coverages might look like this:

The following file generates the menu displayed above:

```
[1]     /* exmatrix.menu
[2]     3
[3]     0 0
[4]     Select a map to draw
[5]     Streets          ARCLINES STREET TYPE
[6]     'Bus routes'     ROUTELINES STREET BUS RT_NO
[7]     Parcels          POLYGONS PARCELS
[8]     Soils            POLYGONSHADES SOILS SOIL_TYPE SOIL.LUT
[9]     'Power poles'    LIST POLES.PAT INFO
[10]    'Quit menu'      &RETURN
```

Notice that all lines start in the first column. There is no indentation because there
are no submenus. The first line in the example above is a comment indicating the
title of the menu file. You can include as many comment lines as you need.

The first *executable* line in the file, [2], specifies the menu type (3 for a matrix menu). The next executable line, [3], consists of two numbers that define the orientation and position of the menu on the screen. The third executable line, [4], places a title or an instruction along the top of the menu. These lines must always be the first three executable lines in the file, but you can insert comments between them.

The maximum number of displayed choices that you can define in a matrix menu file is 100. AML limits your choices to thirty-two characters. The instruction line is limited to eighty characters.

Matrix menus can be positioned in one of six possible locations on the screen. The two numbers in the second executable line generate the six positions. The first number defines a menu's *orientation* as follows:

Value of first digit	Orientation
0	Horizontal
1	Vertical

The second number defines the menu's *position* on the screen as follows:

Value of second digit	Position
-1	Along the left side
0	In the center
1	Along the right side

These numbers specify the six possible locations shown in the following diagrams:

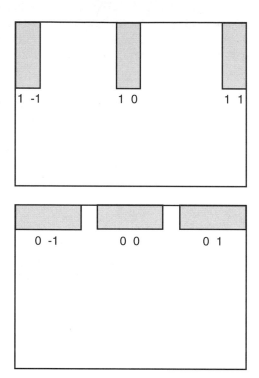

Question 2-3: What's wrong with the beginning of this matrix menu file?

```
3
Select a coverage
0 0
/* displays Seattle coverages
```

(Answer on page 2-30)

The &MENU directive: Beyond the basics

Previously, you learned that the &MENU directive executes a menu file that, in turn, invokes a graphical menu. You can also use this directive to modify certain characteristics of the menu. &MENU's optional parameters, &POSITION, &SIZE, and &STRIPE, modify the position, size, and window title of the invoked menu, respectively.

The placement of matrix menus is defined on the second line of the menu file. Pulldown and sidebar menus have default screen positions. Subsequent displays of pulldown and sidebar menus are positioned slightly down and over from where the upper left corner of the previous menu appeared. Although you would only see one menu at a time, the positioning of subsequent menus follows the pattern shown here:

Pulldown and sidebar menus have default positioning based on the location of the display window (i.e., ARCPLOT or ARCEDIT windows). When the display is set to 9999, pulldown menus are positioned above your display, and sidebar menus appear to the left of your display. This holds true unless the position of your display would cause the menu to be generated off-screen. In this case, AML ensures that the whole menu fits on the screen.

You can use &MENU to position menus precisely by providing explicit placement parameters. These include specifying an absolute x,y location (in pixels) or one of the nine relative screen positions displayed here:

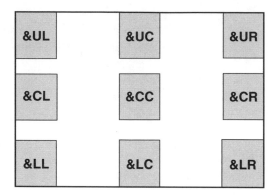

Suppose you want to create a menu interface that displays every pulldown menu starting in the upper right corner of the screen. The command to invoke one of these menus is:

```
&menu listatt.menu &pulldown &position &ur
```

Because of the &MENU directive's arguments, this is the same as the following:

```
&menu listatt.menu &pulldown &position &ur &screen &ur
```

Although the menu type is already indicated in the menu file, pulldown and sidebar menus are interchangeable. A pulldown menu can be made to look like a sidebar by simply changing the first argument. Outside of pulldown and sidebar menus, the first argument is only used for documentation.

To center a matrix menu on your screen, use the following command:

```
&menu drawcov.menu &position &cc
```

Chapter 12, "Form menus," discusses other positioning techniques that you might want to incorporate into your user interface. Also consult the online help Command references and look at usage for the &MENU directive for a complete list of positioning options.

Question 2-4: Which AML code displays EXPULL.MENU as a sidebar menu located in the upper middle of the screen?

(Answer on page 2-30)

Use the &STRIPE option to place descriptive text on the title bar of the window. Earlier, you learned to place descriptive text at the top of a matrix menu; however, the window title bar remained "Matrix." If, for example, you want to change the title bar of the window to read "Coverages for Project Mohican," invoke the menu as follows:

```
&menu mohican.menu &position &cc &stripe ~
  'Coverages for Project Mohican'
```

In the example, the length of the window title defined by &STRIPE exceeds the window width, so the title gets truncated. You can use &SIZE to enlarge the window to accommodate the title.

The &SIZE option defines the size of a menu. By default, size is controlled by the number of choices in the menu file. Use the &SIZE option to create a consistent look for menus that would otherwise vary in size. Standardize on the largest menu because making menus smaller truncates some of the choices. To make a menu smaller without truncating it, use the operating system to reduce the font size.

&SIZE sets width and height in pixels. The menu in the following example is positioned in the center of the screen and its size is 350 pixels by 155 pixels; widen the menu to accommodate for the length of the title:

```
&menu mohican.menu &position &cc ~
  &size 350 155 &stripe 'Coverages for Project Mohican'
```

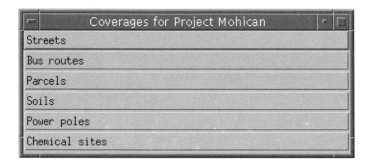

In summary

A menu interface consists of two parts: a menu file and the graphical user interface that's displayed on the screen. Menu files are ASCII text files containing menu choices and their subsequent actions. Menus graphically display the user's choices.

AML supports eight types of menus: pulldown, sidebar, matrix, key, tablet, digitizer, form, and enhanced pulldown. These menus are invoked using the &MENU directive. Menus have a default screen position, size, and window title. You can modify the default settings by using the &MENU directive's optional arguments.

Answer 2-2: All choices would appear on the main menu.

Answer 2-3: The second and third lines are out of order.

Answer 2-4: &menu expull.menu &sidebar &position &uc

Exercises

Exercise 2.2.1

Are other menus interchangeable like pulldown and sidebar menus? Why, or why not?

Exercise 2.2.2

Write the menu file that generates the following menu. Don't worry about the actions, just format the choices.

Exercise 2.2.3

Create a menu file for a pulldown menu with the following choices:

```
Page Size
Page Units
Map Limits
```

Possible page sizes that are set should include the following: 8.5 x 11, 11 x 17, 17 x 32, 22 x 34, 34 x 46. Page units should be set to INCHES or CM, and the map limits should be set interactively with the mouse.

Exercise 2.2.4

Draw a picture of the menu that the following menu file creates. In addition, show the choices that appear on the submenus.

```
1
EDITFEATURE
  ARC            EDITFEATURE ARC
  NODE           EDITFEATURE NODE
  LABEL          EDITFEATURE LABEL
ARCSNAP
  ON             ARCSNAP ON *
  OFF            ARCSNAP OFF
NODESNAP
  FIRST          NODESNAP FIRST *
  CLOSEST        NODESNAP CLOSEST *
MOVE             MOVE *
EXTEND
  BOTH           EXTEND BOTH *
  FROM           EXTEND FROM *
  TO             EXTEND TO *
```

Exercise 2.2.5

What's executed when the choice Landuse is picked?

```
3 matrix
0 0

Streets LINESET STREET.LIN; ARCLINES STREET ST_CODE
Landuse SHADESET LANDUSE.SHD; POLYGONSHADES LANDUSE LU_CODE;~
  POLYGONS LANDUSE
Planimetric POLYGONS PLANIMETRIC
Gas LINESET GAS.LIN; ARCLINES GAS GAS_LN_TYP GAS_LN_TYP.LUT
```

Lesson 2.3—Bringing AML programs and menus together

In this lesson

Menu designers often want to keep the end user from having to touch the keyboard or interact with ARC/INFO at the command line. You can achieve this by invoking a menu from an AML program. Your interface may require that a menu selection execute an AML program or another menu. This hierarchical structure of AML programs and menus increases the range of tasks you can handle. This lesson discusses how multiple menus and AML programs are managed, as well as some design considerations.

Introducing menu design

Up to this point, the discussion of menus was limited to the process of creating a stand-alone menu. Now you're ready to tackle a more flexible interface that accesses many menus and AML programs.

Building a complete menu interface should be approached as a design project. Before you sit at the keyboard and start to write an interface, you need to interview the end users to find out what they need. Put yourself in the users' shoes; don't assume that something that's intuitive to you is straightforward to users. A good menu interface accommodates users' needs and is easy to use, but is also sophisticated enough to get the job done. Consider the following issues during the design process:

- Which menu types are your users comfortable using?
- Which tasks/operations need to be choices in the menu?
- Which terminology should be used for the choices?
- Where should menus, graphics, and listings appear on the screen?
- On which hardware will the menu interface run?
- Does anyone need access to the operating system or ARC/INFO command line?

The user may interact with several menus, so a consistent look throughout the interface is important. Here are some ways to make your menus more consistent:

- Place groups of commands in the same position on each menu.
  Example: If drawing operations are first on the list in one menu, place them first on all other menus on which they appear.
- Use the same terminology for the same operation.
  Example: If the choice CLEAR means to clear the screen on one menu, don't use it to clear the selected set in another menu.
- Use menus of a consistent style.
  Example: If you use a matrix menu to present file names, don't also use [GETFILE], which creates a scrolling list.
- Present informational messages in a consistent format.

Modifications are easy to incorporate if you create a prototype of your menu before it's fully implemented. You can use a prototype to show the user the layout of the menu and how the choices are displayed. This can be done before writing the functionality. The example below illustrates this idea:

```
QUERY
    'Select coverage'        &type Not yet implemented
    'Select feature type'    &type Not yet implemented
    'Identify features'      &type Not yet implemented
```

After the users approve of this interface, you can write the functionality. Soliciting feedback from the user periodically during implementation helps ensure that the final menu interface is easy to use and expedites getting work done.

Starting a menu from an AML program

Instead of having the user invoke a menu from the command line, as you saw earlier, your AML programs can contain the &MENU directive to invoke a menu. The next example demonstrates how this works.

It's useful to have an AML program like this that contains the commands necessary to set the required and optional parameters for an ARC/INFO session:

```
/*startredlands.aml
ARCPLOT
DISPLAY 9999 3
&terminal 9999
&fullscreen &popup
SYMBOLSET COLOR
MAPPOSITION UR UR
&menu redlands.menu
&return
```

Typing &run startredlands.aml passes control from the keyboard to the AML program. ARCPLOT starts, and the commands up to and including &MENU are executed. &MENU invokes REDLANDS.MENU and transfers control from the AML program to the menu. Control remains with the menu until the user quits; all user interaction is with the menu, not the keyboard. As soon as the user quits the menu, control returns to the AML program at the &RETURN line. The AML program stops as soon as the &RETURN line executes. Control has come full-circle, back to the keyboard.

This is a convenient way of establishing a working environment for a novice user who doesn't want to interact with ARC/INFO at the command line. However, the user still needs to know enough to type &run startredlands.aml.

If you want the user to see the menu interface immediately upon entering ARC/INFO and ensure that the environment is set, name your file ARC.AML. This file is executed every time arc is typed from the operating system and is often referred to as a *startup AML program*. ARC.AML must exist in the workspace from which ARC/INFO is started. In this case, users can see and interact with REDLANDS.MENU as soon as they type arc.

```
/*arc.aml
ARCPLOT
DISPLAY 9999 3
&terminal 9999
&fullscreen &popup
SYMBOLSET COLOR
MAPPOSITION UR UR
&menu redlands.menu
&return
```

FYI
Like ARC.AML, files named ARCEDIT.AML and ARCPLOT.AML run automatically when the user types `arcedit` and `arcplot`, respectively. See chapter 3 in this workbook for additional information on startup AML programs.

In chapter 1, you learned that local variables are known only to the AML program that sets them. Local variables, however, are known to all menus started by the AML program. The calling program is still current, but is inactive while the menu is displayed.

In the following example, the variable %DATAPATH% is set in the startup program to indicate where coverages are located:

```
/*arc.aml
ARCPLOT
DISPLAY 9999 3
&terminal 9999 &mouse
&fullscreen &popup
SYMBOLSET COLOR
MAPPOSITION UR UR
&sv datapath = /meling1/vicki
&menu redlands.menu
&return
```

%DATAPATH% can be used in REDLANDS.MENU to perform such tasks as listing coverage attributes, as shown below.

```
/*redlands.menu
1
DISPLAY
  'Street attributes'    list %datapath%/streets arc
  'Slope attributes'     list %datapath%/slope poly
  'Zoning attributes'    list %datapath%/zone poly
```

Question 2-5: When SHADEPOLYS.MENU is invoked, can %COLOR% be referenced from the menu? Why, or why not?

```
/*shadepolys.aml
&sv color = 13
&menu shadepolys.menu
&return
```

(Answer on page 2-40)

Running AML programs from a menu

When you need a series of commands to accomplish one task, it's best to write these commands into an AML program. You can execute this program from a selection off a menu. A portion of the menu might look like this:

```
/* redlands1.menu
1
"DISPLAY +'
    Streets        &run drawstreets.aml
    Slopes         &run drawslope.aml
    Zoning         &run drawzone.aml
```

DRAWSTREETS.AML executes when you choose Streets from the menu, DRAWSLOPE.AML, when you choose Slopes, and so on. The menu choice runs the AML program, so while the AML program is running the menu is inactive. Different window managers indicate that a window is inactive in different ways: Windows NT shows an hourglass, while CDE shows a watch in place of the cursor. When the AML program finishes, the cursor turns into a pointer, and the menu is once again active.

Consider the AML programs DRAWSTREETS, DRAWSLOPE, and DRAWZONE. Each program executes the same commands using a different coverage and different line symbol number.

```
/* drawstreets.aml
mapextent streets
arclines streets 3
&return

/* drawslope.aml
mapextent slope
arclines slope 5
&return

/* drawzone.aml
mapextent zone
arclines zone 2
&return
```

You could write one AML program to draw all three if the coverage name and line symbol are stored in variables. These can be global variables defined by the menu and used by the AML program or local variables defined in the menu and passed as arguments with the &RUN statement. Use the &ARGS directive to define variables by accepting arguments passed with the &RUN statement. Chapter 8 covers the &ARGS directive in detail.

Starting a second menu

A menu choice can also invoke other menus. This is useful when the number of choices becomes so large that the menu becomes confusing and difficult to use.

Note that menus invoked with the &MENU directive are referred to as single menus—only *one* menu appears on the screen at a time.

In the next example, when Select streets is chosen, SELSTREET.MENU displays and REDLANDS2.MENU disappears from the screen. When &RETURN executes, SELSTREET.MENU disappears and REDLANDS2.MENU reappears.

```
/* redlands2.menu
1
CONDITIONS
    'Select streets...'        &menu selstreet.menu ~
                               &stripe 'Street selection'
    'Select landuse...'        &menu sellu.menu ~
                               &stripe 'Landuse selection'
    'Select parcels...'        &menu selpar.menu ~
                               &stripe 'Parcel selection'
```

Both local and global variables known in the first menu are also available in the second menu.

Chapter 13 discusses threads, which allow multiple menus to be displayed simultaneously.

Question 2-6: Given the following code, which menus are displayed (a) while SETCOLOR.AML is running, and (b) when SETCOLOR.AML terminates?

```
/*editpoly.menu
DRAW
    Outlines        &run setcolor.aml; &menu drawpoly.menu
```

(Answers on page 2-40)

In summary

A successful menu interface needs thoughtful design. User feedback and prototypes are tools for creating an interface that's easy to use and increases productivity.

A menu interface may include one or more AML programs and menus. When the &MENU directive is used to invoke menus, only one menu can be displayed at a time.

Programs can invoke menus, and menus can execute programs. Menus can also invoke other menus. Variables and program control must be managed throughout this process. Program control is maintained by either the keyboard, a program, or a menu.

Answer 2-5: Yes, %COLOR% can be used as a variable by any menu invoked from SHADEPOLYS.AML.

Answer 2-6:
(a) EDITPOLY.MENU
(b) DRAWPOLY.MENU

Exercises

Exercise 2.3.1

Below is the code for a choice in a pulldown or sidebar menu. As you can see, this choice contains a lot of AML code. Is there a more efficient way to write this code using an AML program and a menu?

```
'Draw a coverage'   &sv cov = [getcover]; MAPE %cov%; ~
&sv feat = [getchoice ARC POLY POINT NODE -prompt ~
'Select a feature class to draw']; %feat%s %cov%
```

Exercise 2.3.2

A municipal application begins with a program called START.AML which confirms environments and launches a menu with the following four choices:

- Planning
- Engineering
- Assessment
- Zoning

A choice from the menu should run an AML program (e.g., PLANNING.AML) that, in turn, should execute the application's menu (e.g., PLANNING.MENU). In the spaces provided on the diagram below, fill in the following items for each choice. The answers for the PLANNING choice are provided:

- The command executed by the menu choice
- The name of the AML program run
- The executable statement to launch the menus
- The name of the menu executed

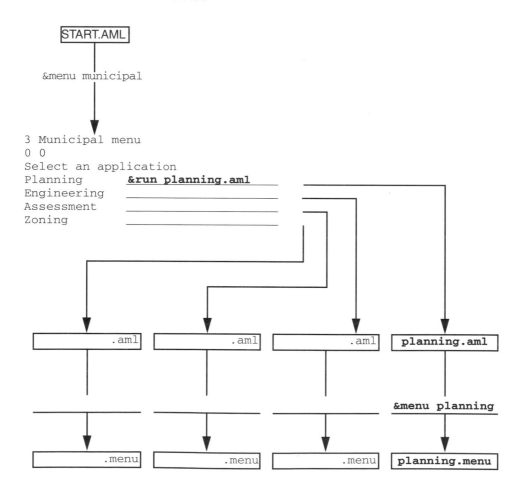

Looking for code in all the right places

3

Large applications need to access many programs. Usually the programs your application uses aren't stored in a single directory. AML enables you to access programs and menus no matter where they're stored.

AML can also access and automatically run programs stored in special directories whenever ARC or one of its subsystems starts. Use these startup programs to establish the environments needed for an ARC/INFO session. Additionally, AML can find and execute customized ARC/INFO commands called ATOOLS.

Your application programs and menus must be accessible to anyone who uses the application, so, for security reasons, you may want to protect them. AML enables you to alter your programs so that others can execute but not read them.

This chapter covers the following topics:

Lesson 3.1—Accessing programs in other directories

- Accessing AML programs
- Accessing AML menus
- Startup programs that run automatically

Lesson 3.2—Creating your own commands

- AML abbreviations
- ATOOL commands

Lesson 3.3—Protecting your programs

- Encoding AML programs and menus

Lesson 3.1—Accessing programs in other directories

In this lesson

In this lesson, you'll learn how to access AML programs stored in other directories without having to specify pathnames each time the program runs. You'll also learn how AML programs associated with an ARC/INFO subsystem such as ARC, ARCPLOT, or ARCEDIT can run automatically each time the subsystem starts.

Accessing AML programs

AML can search multiple directories to locate AML programs. This allows you to organize your programs to meet your organizational needs. For example, the individual programs needed to run an application might be managed by different departments.

&AMLPATH directs the AML processor to search one or more directories for an AML program if it can't be found in the current working directory. Consider the following example:

```
Arc: &amlpath /usr/utilities/data   /usr/transportation/data
Arc: &run delivery
```

In this case, the AML processor looks first in the current directory for the AML program, DELIVERY.AML. If it doesn't find the program in the current directory, AML searches the directories /usr/utilities/data and /usr/transportation/data. If AML can't find the program file in the current directory or in directories specified with &AMLPATH, it returns an error message like this one:

```
AML ERROR - Unable to run file delivery
```

&AMLPATH can set up to thirty pathnames. The list of pathnames to search resets each time you issue &AMLPATH, so you must specify each directory you want searched every time you issue &AMLPATH. Consider this example:

```
Arc: &amlpath /usr/utilities/data  /usr/transportation/data
Arc: &show &amlpath
/usr/utilities/data  /usr/transportation/data
Arc: &amlpath /usr/planning/data
Arc: &show &amlpath
/usr/planning/data
```

The directories set by the first &AMLPATH (i.e., /usr/utilities/data /usr/transportation/data) don't show in the current &AMLPATH (i.e., /usr/planning/data) because they're not specified the second time you issue the &AMLPATH directive.

FYI

> Drive letters should preface each one of these pathnames if you're using Windows NT. A Windows NT &AMLPATH should look something like this:
>
> ```
> Arc: &amlpath D:\usr\utilities\data E:\usr\planning\data
> ```
>
> Even though the Windows NT operating system uses back slashes, ARC/INFO on Windows NT also allows forward slashes.

You can append a pathname to the current list using the [SHOW] function as follows:

```
Arc: &show &amlpath
/usr/utilities/data  /usr/transportation/data
Arc: &amlpath [show &amlpath] /usr/planning/data
Arc: &show &amlpath
/usr/utilities/data  /usr/transportation/data
/usr/planning/data
```

[SHOW &AMLPATH] returns the existing pathnames set by a previous &AMLPATH. The additional path is specified after the function, so the entire list consists of three pathnames.

AML always searches the current directory before searching the pathnames specified with &AMLPATH. You don't need to use &AMLPATH if your application programs are in the current directory. If, however, your application allows the user to change directories, then specify &AMLPATH to enable AML to search all directories containing the required program files.

&AMLPATH is only valid for the current ARC/INFO session, so it's a good idea to include it in an AML startup program.

Accessing AML menus

AML can also search multiple directories to locate AML menus or icons. &MENUPATH directs the AML processor to search a specified directory or directories for an AML menu file or icon if it can't find it in the current working directory. &MENUPATH works like &AMLPATH and can also set up to thirty pathnames.

Question 3-1: Examine the following directory structure and then write the &AMLPATH and &MENUPATH directive lines to make AML search the two directories AMLDIR and MENUDIR. Assume the application is running from the PROJECT1 directory.

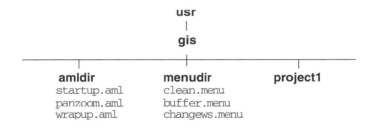

Question 3-2: Examine the following directory structure, which now includes a startup AML program in the PROJECT1 directory. Study the following commands and then answer the question below:

```
Arc: &workspace /usr/gis/project1
Arc: &amlpath /usr/gis/amldir
Arc: &run startup
```

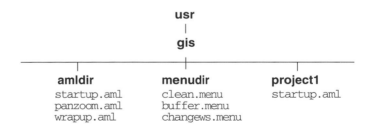

Which STARTUP.AML program would AML run, the one in AMLDIR or PROJECT1?

(Answers on page 3-8)

Startup programs that run automatically

ARC/INFO software provides AML system startup programs stored in a subdirectory called STARTUP under the ARC executable directory (e.g., $ARCHOME/startup). Individual system startup programs are associated with ARC and each of the subsystems. The appropriate startup program runs automatically when a subsystem starts. You can modify these programs, but be careful—anyone using the same copy of ARC/INFO will also launch your modified program. You can also create personal startup programs, but they must be in the same directory in which the subsystem is launched. &AMLPATH isn't used to search for these startup programs, so they must be in the current directory. These programs are most useful for setting up environments for a particular project. The following table summarizes the naming convention used for these system startup programs:

Startup program file name	When the program runs
arc.aml	Runs every time ARC starts
arcedit.aml	Runs every time ARCEDIT starts
arcplot.aml	Runs every time ARCPLOT starts
grid.aml	Runs every time ARC/INFO GRID starts
librarian.aml	Runs every time ARC/INFO LIBRARIAN starts
etc.	

By default, the system searches the user's home directory for a specially named, user-defined program. The home directory is where the system places the user after logging in. If such a program exists, it's executed after the system startup program finishes. The user-defined program must be named after its associated subsystem and begin with a period. (Note: on the UNIX platform, these programs are hidden; on Windows NT, they aren't.) The following table summarizes this naming convention:

Program file name	When the program runs
.arc	Runs every time ARC starts
.arcedit	Runs every time ARCEDIT starts
.arcplot	Runs every time ARCPLOT starts
etc.	

Some text editors on Windows NT don't allow you to save a file as `.arc`. The text editor doesn't recognize this as a valid file name. For instance, DOS EDIT doesn't allow you to save a file as `.arc`, but NOTEPAD does.

After looking in the user's home directory, the system startup programs search the current workspace for another specially named, user-defined program. If one exists, it too executes. The naming convention for these startup programs is the same as the system startup programs (i.e., ARC.AML, ARCEDIT.AML, etc.).

The following diagram illustrates how you might store AML startup programs:

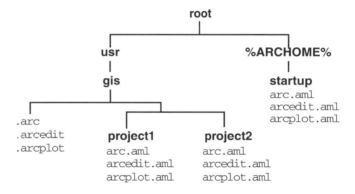

Assume a user's login directory is named GIS. If the user changes directories to PROJECT2 and starts ARC, three ARC startup programs run. The first is ARC.AML in the $ARCHOME/STARTUP directory, followed by .ARC in the GIS directory, then ARC.AML in the PROJECT2 directory. If the user now starts ARCEDIT, three ARCEDIT startup programs run in the order described for the ARC programs.

The &STATION directive was an older version of the .ARC and ARC.AML files. &STATION runs a station file that contains information on hardware setup, like DISPLAY and &TERMINAL. The ARC.AML and .ARC files have a great deal more flexibility than the old station files because they can contain parameters other than just system setup.

Question 3-3: Use the startup program diagram shown on page 3-7 to answer the following question: Suppose the user starts ARC from the PROJECT2 directory. The .ARC program sets DISPLAY to 9999 and the ARC.AML in the PROJECT2 directory sets DISPLAY to 9999 3.

What's the current DISPLAY setting?

_____

(Answer on page 3-8)

In summary

&AMLPATH and &MENUPATH direct AML to search other directories for AML programs and menus. You must specify all pathnames to search each time you issue the &AMLPATH or &MENUPATH directives. [SHOW &AMLPATH] and [SHOW &MENUPATH] allow you append paths to the current path.

You can name and store AML programs so that they run automatically each time ARC or one of its subsystems starts. System startup programs that come with ARC/INFO software always run when ARC or one of its subsystems starts. You can also store startup programs in your login directory and working directories. Up to three programs can run automatically when ARC or one of its subsystems starts (i.e., a program in ARC/INFO software's STARTUP directory, one in the user's login directory, and one in the user's working directory).

Answer 3-1:
```
&amlpath /usr/gis/amldir
&menupath /usr/gis/menudir
```

Answer 3-2: The STARTUP.AML program in the PROJECT1 directory would run.

Answer 3-3: The current setting for DISPLAY is 9999 3.

Exercises

Use the following diagram to complete exercises 3.1.1 to 3.1.4:

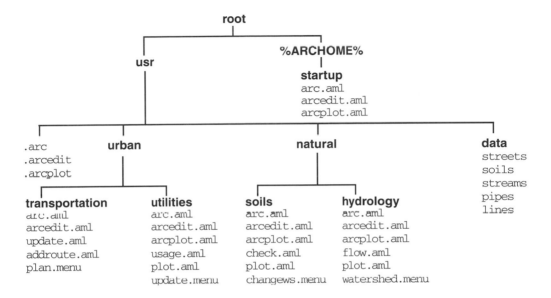

Exercise 3.1.1

An application needs to use the programs and menus stored in the directories located under URBAN. The directories should be searched in the following order: TRANSPORTATION, then UTILITIES. Write AML statements to direct AML to search the directories for programs and menus in the order specified.

Exercise 3.1.2

Suppose the application in exercise 3.1.1 expands in scope and needs to access programs and menus in the subdirectories stored in the directories below NATURAL. Assume SOILS and HYDROLOGY need to be searched in that order. Write the AML statements that would append the pathnames of these directories to the current set of pathnames set in the previous exercise.

Exercise 3.1.3

Suppose the order of searching pathnames is TRANSPORTATION, UTILITIES, SOILS, and HYDROLOGY. From which directory would an AML program named PLOT.AML be used if you issue the AML directive &RUN PLOT?

Exercise 3.1.4

The GIS project manager wants the AML startup programs to execute the following actions when users start ARCPLOT. Match the letter designating an action to the corresponding startup program file.

ACTIONS:

    (a) PAGESIZE 17 11 (for all users whose login directory is usr and will work in workspaces below usr)

    (b) MAPUNITS FEET; MAPSCALE 24000 (for all users creating maps in the soils workspace)

    (c) DISPLAY 9999 3; &TERMINAL 9999 (for all users on the system who use ARC/INFO)

FILES:

    ____ $ARCHOME/startup/arcplot.aml

    ____ /usr/.arcplot

    ____ /usr/natural/soils/arcplot.aml

Lesson 3.2—Creating your own commands

In this lesson

This lesson teaches you to use AML variables to execute such actions as
ARC/INFO commands. You'll also learn to run an AML program without the
&RUN directive by typing its name, just as you would a command.

AML abbreviations

An AML abbreviation is an AML variable. Abbreviations differ from other
variables only in the way they're evaluated. The most common way of evaluating
an AML variable is to enclose it in percent signs. An AML abbreviation can be
evaluated without the percent signs if it's the first word on the command line and
you activate the &ABBREVIATIONS setting. The &ABBREVIATIONS
directive activates abbreviations in an ARC/INFO session (i.e.,
&ABBREVIATIONS &ON).

AML variables can store values that AML or ARC/INFO can use to perform an
action. When you use a variable as an AML abbreviation, the stored actions are
performed whenever the variable appears as the first word on the command
line—just like typing an ARC/INFO command. The percent signs aren't needed
in this case.

Consider the following example:

```
Arc: &sv b = build
Arc: b soils poly
Submitting command to Operating System ...
b: Command not found.
Arc: &abbreviations &on
Arc: b soils poly
Building polygons...
```

The variable b is set to the BUILD command. The first attempt to use b as an AML abbreviation isn't successful because &ABBREVIATIONS is set to &OFF (the default). After AML abbreviations are activated by typing &abbreviations &on, the variable evaluates to the command BUILD because it's the first character on the command line.

Consider another example:

```
Arc: ARCPLOT
Arcplot: &sv .ps = POLYGONSHADES SOILS SOIL_TYPE
Arcplot: ps
```

The global variable .ps is set to an ARCPLOT command and its required arguments. It's not necessary to type the period in the variable name when it's used as an AML abbreviation. The entire command line POLYGONSHADES SOILS SOIL_TYPE executes when ps is typed at the program prompt.

FYI

Both local and global variables that appear as the first word on a line are evaluated as AML abbreviations when &ABBREVIATIONS is set to &ON.

The next example illustrates how to set a variable to an AML directive and then execute it as a command:

```
Arcplot: &abbreviations &on
Arcplot: &sv dc = &run drawcov soil poly
Arcplot: dc
```

The variable dc is set to the AML directive &run drawcov soil poly. Typing dc at the Arcplot: prompt executes the directive like a command.

AML evaluates abbreviations before evaluating directives, functions, or ARC/INFO commands. Refer to the online help index under *AML, processing order* for more information on AML interpretation.

An AML abbreviation can be a small program of instructions. Consider this example:

```
Arcedit: &abbreviations &on
Arcedit: &sv cid = [unquote 'select;calc $id = [response ~
'Enter new user-id']']
Arcedit: cid
Point to the feature to select
Enter point
     .
     .
     .
```

The variable cid contains multiple commands. These commands make a small program that could be used to update User-IDs. The commands are delimited by semicolons, which act as line separators. The quotes enclosing the commands temporarily suppress AML so that the commands don't execute when the variable is set. The UNQUOTE function strips the sets of single quotes, leaving cid equal to select; calc $id = [response 'Enter new user-id']. When cid is typed as the first word on the line, the actions stored in the variable execute. This is one way to create customized commands.

You're probably thinking that AML abbreviations are great and how you'll use them to perform some AML tricks. Beware! When abbreviations are activated, every variable in your program is evaluated as a potential abbreviation when it appears as the first word on a line. Abbreviations reduce the amount of typing needed to execute long command sequences, but use caution whenever &ABBREVIATIONS is set to &ON and deactivate it when you no longer need it.

Examine the following code fragment. Assume the user has set &ABBREVIATIONS to &ON in a startup AML program.

```
/* Enter a map title
&sv text = [response 'Enter the map title']
move 5 8 /* Location in page units for the map title
text %text% /* Draw the title on the map
Unrecognized command
```

Because abbreviations are activated, ARCPLOT evaluates the `text` variable before the TEXT command. ARCPLOT issues an error message because it can't find a command that equals the value of a map title entered by the user. Obviously, a map title can't be accepted as a valid command.

Never give your AML abbreviations the same names as commands. This can only cause trouble and confusion, though some have found it an interesting joke to set the following abbreviation:

```
&sv build = kill
```

which yields the following message and a good bit of rage:

```
Arc: BUILD GEOLOGY
Killed GEOLOGY with the ARC option
```

Needless to say, this isn't a great way to win friends.

AML abbreviations can also cause typing mistakes to have unwanted results. Though most often used interactively, AML abbreviations in your programs can make your code difficult to read.

Question 3-4: Suppose you start ARC and want to create an AML abbreviation named id that executes the ARCPLOT command line IDENTIFY PARCELS POLY *. Write the two AML directive lines that set the variable and activate the AML abbreviation.

_____

_____

(Answer on page 3-20)

ATOOL commands

AML gives you the ability to create your own commands. To do so, store command sequences in an AML program and execute them by typing the program file name at the command line—just as you would an ARC/INFO command.

AML programs executed in this manner are called ATOOL commands. If you enter a command that ARC/INFO doesn't recognize, AML searches for an ATOOL command with the same name. AML programs used as ATOOL commands are given a .aml suffix and, like ARC/INFO commands, are specific to the ARC/INFO subsystem for which they're created (e.g., ARCEDIT).

An AML program must be stored in a program-specific directory to function as an ATOOL command. This directory must be named the same as the subsystem for which the ATOOLs apply (e.g., ARC, ARCEDIT, ARCPLOT).

The following diagram illustrates how AML programs provided with ARC/INFO software are stored as ATOOL commands in program-specific directories:

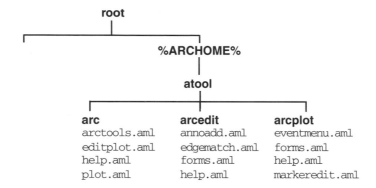

The diagram doesn't list all of the AML programs provided by ARC/INFO, but serves as an example. Suppose you want to perform edgematching in ARCEDIT. From any workspace, type `edgematch` at the `Arcedit:` prompt. Following AML's order of interpretation, ARCEDIT gets the command first. ARCEDIT can't find EDGEMATCH in its library of commands, so AML searches the `$ARCHOME/atool/arcedit` directory for an AML program named `edgematch.aml`. AML locates and runs the program of that name just as if a command was entered.

Notice that all three program directories contain an AML program named HELP.AML. Thus, you can type `help` at any program prompt to initiate the help tool.

You can store AML programs in multiple directories and execute them as ATOOL commands. The &ATOOL directive can specify pathnames to directories that store ATOOL commands. &ATOOL can set up to twenty pathnames.

The ARCEXE ATOOL pathname doesn't need to be set in the &ATOOL list— AML automatically searches the system ATOOL directory after searching user-specified pathnames.

Examine the following diagram:

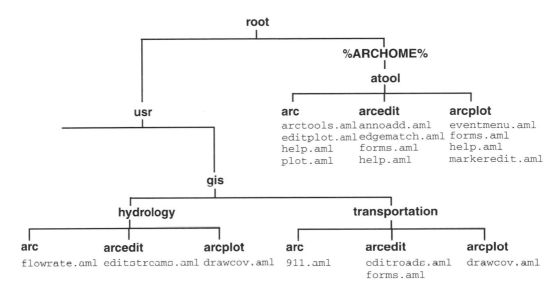

Assume that you want to access the programs stored in the HYDROLOGY and TRANSPORTATION directories as ATOOL commands. Use &ATOOL, as follows, to specify the pathnames to the directories where AML should search for the programs:

```
Arc: &atool /usr/gis/hydrology  /usr/gis/transportation
```

Notice that the directory doesn't have to be named ATOOL. The program directories, however, must have the same name as the ARC/INFO subsystems they represent.

Given the pathnames specified in the &ATOOL example, suppose you type EDITROADS at the Arcedit: prompt (from any directory). EDITROADS isn't an ARCEDIT command, so AML searches for ATOOL commands. AML first searches the ARCEDIT directory under /usr/gis/hydrology because it's the first pathname in the &ATOOL list. When AML can't locate the AML program file there, it searches the ARCEDIT directory under /usr/gis/transportation. AML finds EDITROADS.AML and runs it as an ATOOL command.

If EDITROADS.AML isn't found in the directories specified with &ATOOL, AML makes one last search in $ARCHOME/atool/arcedit before returning an error message. Remember that the $ARCHOME/atool pathname doesn't need to be set with &ATOOL; AML automatically searches there for ATOOL commands.

The directory $ARCHOME/atool is a good place to install programs that your entire organization uses, but don't forget to copy them out of the directory whenever you install a new version of ARC/INFO or they'll be overwritten.

Question 3-5:
Study the commands below and then use the diagram on page 3-17 to answer the questions that follow:

(a) Arc: ARCEDIT
    Arcedit: &atool /usr/gis/hydrology /usr/gis/transportation
    Arcedit: FORMS

From which directory will the ATOOL command FORMS be used?

_____

(b) Arcedit: QUIT.
    Arc: ARCPLOT
    Arcplot: DRAWCOV

From which directory will the ATOOL command DRAWCOV be used?

_____

(Answers on page 3-20)

The following table summarizes the differences between &ATOOL and &AMLPATH:

&ATOOL	&AMLPATH
Can set 20 paths.	Can set thirty paths.
Program-specific—the AML program won't run in the wrong program environment (e.g., ARC, ARCEDIT, etc.).	Not program-specific—the AML can run in any program environment.
&RUN isn't required. The AML program becomes a command in a program environment.	The AML is executed with &RUN.

If your ATOOL paths are set, any ATOOLs that you add are shown whenever you issue the COMMANDS command. Your ATOOLs show up as separate subsections with their directory path as the header.

```
Arc: COMMANDS
    .
    .
    .
ATOOL directory /tavarua2/tom/atool/arc
bindex        d_to_u        dos_conv        flip

ATOOL directory /san1/esri71.slrs/arcexe71/atool/arc
addcogoatt            addindexatt            airequest
arctools             asciihelp              atusage
etc.
```

Use the ATUSAGE command to get a usage line on any ATOOLs. In order for ATUSAGE to work on ATOOLs that you create, copy over the ATUSAGE.AML from the program directory under $ARCHOME to your local ATOOL program directory.

```
Arc: &type [copy $ARCHOME/atool/arc/atusage.aml ~
/tavarua2/tom/atool/arc/atusage.aml]
```

Once ATUSAGE.AML is local, you can make whatever changes you want to incorporate new ATOOLs. This ATUSAGE.AML in your local ATOOL directory should contain all of the usage lines for the pre-existing ATOOLs, as well as all of the usage lines for any ATOOLS you created. The local copy of the ATUSAGE.AML is now read, while the $ARCHOME copy isn't.

In summary

You can create your own commands using AML abbreviations and ATOOL commands. An AML abbreviation is an AML variable containing one or more actions. AML abbreviations are executed when they are active (&ABBREVIATIONS &ON) and the name of the variable appears as the first word in the command line. ATOOL commands are AML programs that can run as commands in the appropriate program environment. ATOOL programs must be stored in user-specified ATOOL directories or the ARC/INFO system ATOOL directory.

Answer 3-4:
```
&abbreviations &on
&sv id = IDENTIFY PARCELS POLY *
```

Answer 3-5:
(a) /usr/gis/transportation
    (This overrides the ARCEXE ATOOL FORMS command.)
(b) /usr/gis/hydrology

Exercises

Exercise 3.2.1

Write the AML statements that activate AML abbreviations and set an AML variable so that the text string `al` executes as the ARCPLOT command `arclines street street_code`.

Exercise 3.2.2

Explain what's wrong with the following procedure that uses AML abbreviations:

```
Arcedit: USAGE EDIT
Usage: EDIT <cover> {feature_class}
Usage: EDIT <info_file> INFO
Arcedit: EDIT streets arc
Arcedit: &abbreviations &on
Arcedit: &sv edit = [unquote 'select;calc street_code = [response ~
Arcedit: ''Enter new street code'']']
Arcedit: edit
Point to the feature to select
Enter point
Arc 237 User-ID: 2119 with 2 points
1 element(s) now selected
Enter new street code: 23
Arcedit: EDIT streets.exp info
Point to the feature to select
Enter point
```

Exercise 3.2.3

Explain why the following directory structure doesn't support ATOOL commands:

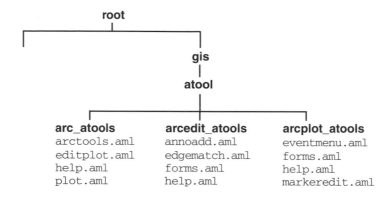

Exercise 3.2.4

Use the following diagram to complete exercises 3.2.4 and 3.2.5:

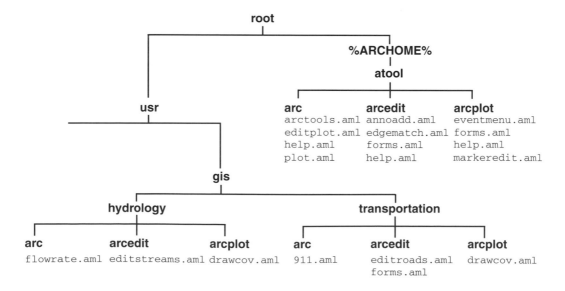

Write the AML statement that would set the proper environment to access programs in the HYDROLOGY and TRANSPORTATION directories as ATOOL commands.

Exercise 3.2.5

Refer to the graphic in exercise 3.2.4. Suppose &ATOOL hasn't been issued in a current ARC/INFO session. Explain what happens if the user types forms at the Arcedit: prompt.

Lesson 3.3—Protecting your programs

In this lesson

As you know by now, AML programs and menus are ASCII files, which anyone can read. If a user makes a copy of an AML program or menu and then modifies it, your application may not perform according to standards, or it may not work at all. You can avoid unwanted modifications to your programs and menus as well as protect a proprietary software product written in AML. In this lesson, you'll learn how to encode your AML programs and menus. *Encoding* is a data security measure that makes your code unreadable—users can run your programs, but can't read them.

Encoding AML programs and menus

When you want your programs to be unreadable, use the &ENCODE directive to encode your AML programs and menus. Encoded program names have a .eaf extension; encoded menu names have a .emf extension.

If both an encoded version and the original AML program exist at the same location, the encoded version is used by default when the program is executed. You can override this priority by specifying the extension of the program name you want to execute. Examine the following example:

```
Arc: &encode &encrypt update.aml
Arc: ls update* /* '&sys dir update*' on Windows NT.
Submitting command to Operating System ...
update.aml update.eaf
Arc: more update.eaf
Submitting command to Operating System ...
bp|snS]9b7|5p**z?g{{gFn(nw&;OloH+Zkoynbf0lmtv©+eiDM74zS
&Z(.zp~ (unjet51{~[&D]^n.xcpkk{|.
&("DM
&b{(oMcSD¥ah"
   --More--(63%)
Arc: &run update
Arc: &run update.aml
```

&RUN UPDATE runs the encoded version of the program UPDATE.EAF, which is unreadable to the user. &RUN UPDATE.AML runs the original version of the program.

If you modify the original AML program, be sure to use &ENCODE again to create an updated encoded version.

> The encoding process is *not* reversible. Keep a copy of your original AML programs and menus in case you need to update the code.

Move your original AML programs and menus to a location where only approved users have read access. The encoded versions should be stored where they're accessible to the application by using &AMLPATH, &MENUPATH, and &ATOOL.

You can put a single-line header in the encoded version of a program file that anyone can read. Examine the following example:

```
Arc: &encode &encrypt update.aml
Arc: &encode &listheader update.eaf
/usr/gis/project1/update.aml 16 July 96 12:58:34 Monday
Arc: &encode &encrypt update.aml This AML program updates~
  utility line codes. Coded by [username]. Last modified on~
  [date -full].
Arc: &encode &listheader update.eaf
This AML program updates utility line codes. Coded by jesse. last
modified on 96-07-16.15:27:41.Mon
```

The header information wasn't specified the first time the program file was encoded, so AML provides the path to the program or menu and the date in the header by default. The second time the program file is encoded, the user specifies the header information as an argument to the &ENCODE directive. You could write an AML program that encodes AML program files with standard header information.

Question 3-6: Suppose a user initiates an application by running the startup AML program shown in the directory structure given below. The application, which can use all of the programs and menus listed, keeps the user in the project1 directory. Study the commands and diagram shown below and then answer the question.

```
Arc: &workspace /usr/gis/project1
Arc: &amlpath /usr/gis/amldir
Arc: &menupath /usr/gis/menudir
Arc: &run startup
```

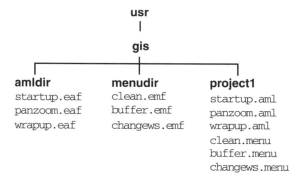

If the project manager wants the encoded versions of the AML programs and menus to be used when the application runs, what's wrong with the directory organization?

_____

_____

(Answer on page 3-27)

In summary

Encoding protects your AML code so that others can't read or modify it. Remember to keep a copy of your original AML programs and menus in case you need to modify the code later. Protect the original programs by placing them in a location where only selected users have read access.

Answer 3-6: When AML programs or menus are called, AML always searches the current directory for them. Although &AMLPATH and &MENUPATH establish pathnames to where the encoded programs are stored, the original AML programs are used because they're located in the current directory when the application runs.

Exercises

Exercise 3.3.1

Use the following diagram to complete exercises 3.3.1 and 3.3.2:

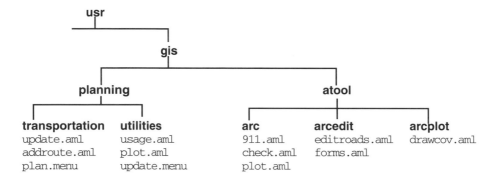

Write the AML statements that fulfill the following requirements:

(a) Direct AML to search the TRANSPORTATION and UTILITIES directories for AML programs and menus when &RUN and &MENU are issued.

(b) Direct AML to search the ATOOL directory for ATOOL commands.

Exercise 3.3.2

Write the AML statements that would encode UPDATE.AML and PLAN.MENU in the TRANSPORTATION directory.

Exercise 3.3.3

What are the names of the encoded programs resulting from exercise 3.3.2?

Exercise 3.3.4

Explain what should be done if the GIS manager doesn't want to leave the readable versions of UPDATE.AML and PLAN.MENU in the current workspace.

4 Branching out

In programs you've written so far, the order of execution never varies. Each line executes sequentially and there are no choices presented that significantly alter the program's course. In this chapter, you'll learn a technique for writing *dynamic* AML programs, that is, programs that react to new or changing information. This technique, called *branching*, creates programs that allow the user to make choices that affect a program's execution as it runs.

This chapter presents the AML directives, variables, and functions commonly used in branching. You'll use branching to test conditions and choose from alternative courses of action. Testing may include validating current ARC/INFO parameters (e.g., is the display device set?), checking coverage status (e.g., what's the type of topology?), and verifying user input (e.g., was the correct response entered?). By executing different actions based on the outcome of the test, you can prevent errors and provide a friendly interface for performing ARC/INFO tasks.

This chapter covers the following topics:

Lesson 4.1—Branching and logical expressions

- The true/false test: Logical expressions
- Getting at the truth: The &IF &THEN structure
- Truth or consequences: The &ELSE directive
- Grouping multiple actions: The &DO block
- Testing coverage parameters
- Testing user input

Lesson 4.2—Choosing from many alternatives

- &IF more alternatives are needed
- &SELECTing from many alternatives

Lesson 4.3—Compound expressions and more testing techniques

- Constructing and evaluating compound expressions
- Hierarchical testing using nested &IF statements
- Testing for the correct [TYPE]

Lesson 4.1—Branching and logical expressions

In this lesson

A branching structure is composed of a condition and a set of alternatives. Based on the condition, one of the alternatives is executed. The condition for branching is written in the form of a *logical expression*—a statement that is either true or false.

This lesson introduces the components, evaluation, and usage of logical expressions. You'll practice using these expressions with AML directives to control program flow (i.e., parts of a program execute while other parts are skipped). You'll learn about the AML functions used to extract information from and test conditions in ARC/INFO.

The true/false test: Logical expressions

You've constructed true/false tests all your life by posing questions that have only two possible answers: yes or no. The question, "Dad, can I borrow the car Friday night?" is one that almost always had "NO!" for an answer, though there was a chance that the answer would be "Yes." If you express this question as the statement, "I can borrow the car Friday night," then the statement is either true or false (based on your father's mood at the time). This kind of statement is called a logical expression. A logical expression always evaluates to either true or false.

It's often beneficial to perform this kind of testing before executing ARC/INFO operations. For example, if you start with the question, "Does the coverage I want to draw exist?" you can convert it to the statement, "The coverage I want to draw exists" and have the AML program evaluate whether it's true or false. Similarly, you can express the question, "Do I have an item called DISTRICT in my PAT file?" as the logical expression "The item called DISTRICT exists in my PAT file," which your program can evaluate.

FYI

You'll find logical expressions are also referred to as *Boolean expressions* after George Boole, who devised this system of testing logic. Likewise, a variable that can be set to a true or false value is often referred to as a *Boolean variable*.

AML represents the values true and false as .TRUE. and .FALSE., respectively. There are several ways to generate these values in an AML program. The easiest way is to directly assign these values to a variable. When a variable is assigned either a .TRUE. or .FALSE. value, it's called a *logical variable*. For example:

```
&sv okdraw = .TRUE.
&sv nodraw = .FALSE.
```

Like logical expressions, logical variables always evaluate to .TRUE. or .FALSE.. AML substitutes variables and functions in a logical expression and makes the comparison based on a *relational operator*. The following table lists the AML relational operators. AML accepts either the mathematical symbol or the character representation of a relational operator.

Mathematical symbol	Character representation	Meaning
=	EQ	Equal to
<>; ^=	NE	Not equal to
>	GT	Greater than
>=	GE	Greater than or equal to
<	LT	Less than
<=	LE	Less than or equal to
	CN	Contained in
	NC	Not contained in
	IN	Included in (a set or range)
	LK	Like (a wildcard (*) match)

The following logical expression uses the [DATE] function with the -DOW (i.e., day of week) option.

```
[date -dow] = Thursday
```

If [DATE -DOW] evaluates to Friday, then, after substitution, the expression yields `Friday = Thursday`, which evaluates to .FALSE..

Given the expression:

`%linesymnum% > 15`

If the variable %LINESYMNUM% evaluates to 16, then, after substitution, the expression is:

`16 > 15`

and the logical expression yields .TRUE.

Question 4-1:
(a) If `%newname% = judy` and `[username] = mike`, what's the value of the expression `%newname% NE [username]`?
(b) If `%day% = FRIDAY` and `[date -dow] = Friday`, what's the value of the expression `%day% = [date -dow]`?

(Answers on page 4-18)

Getting at the truth: The &IF &THEN structure

Now that you know how to construct and evaluate logical expressions, you need a structure that tests for .TRUE. or .FALSE. and directs the program to take the appropriate action based on the result. In AML, the &IF &THEN directive provides the structure for managing a program's execution, or the *flow of control*. &IF &THEN allows you to execute or skip lines in your program based on the outcome of the logical expression. &IF tests the logical expression and &THEN indicates which action to take when the expression evaluates to .TRUE..

AML syntax dictates that &THEN must *always* follow the &IF directive on the same line. Follow this rule or your AML program won't run.

A simple &IF test looks like this:

```
&if <expression> &then
  <true action>
<next statement>
```

The flow of the program depends on whether the expression is true or false. If the expression is true, then perform the `<true action>` and continue to the `<next statement>`. If it's false, ignore the `<true action>` and continue with the `<next statement>`.

In the previous example, only one statement executes when the expression is true. Later in this lesson, you'll learn how to execute multiple statements when the logical expression is true.

Here are some expressions written in this format:

```
&if [username] = jorge &then
  &type Good morning jorge, have a good day

&if %linesymnum% > 15 &then
  &sv %linesymnum% = 1

&if [date -dow] = Thursday &then &type Today is Thursday, ~
  weekly backups tonight!
```

FYI

Unlike most other directives, you're not allowed to use the &IF directive on the command line. The &IF directive can only be used in an AML program.

Question 4-2: Write an &IF &THEN statement that completes the following task: All valid project numbers are greater than or equal to 1,000. Use the [RESPONSE] function to get a project number from the user and set a variable named PROJECTNUM to the result. Test the variable %PROJECTNUM% to see if it's valid and return a message that indicates the result.

(Answer on page 4-18)

Truth or consequences: The &ELSE directive

In a simple &IF structure, no actions are defined for the case when the logical expression is false; control simply passes to the statement following the true action. Now consider how to execute an alternate action when the expression is false. In this case, use the &ELSE directive with the &IF &THEN statement to define the alternate action as follows:

```
&if <expression> &then
  <true action>
&else
  <false action>
<next statement>
```

When true, &THEN executes the `<true action>`, then skips to the `<next statement>`. When false, the `<true action>` is skipped and &ELSE executes the `<false action>` and continues to the `<next statement>`.

STYLE

Notice how indentation is used with the &IF &THEN &ELSE structure. Indenting the lines after the &IF &THEN makes your code easier to follow and debug. Although it may not be obvious in a simple example, complex structures are almost impossible to follow without line indentation. How many spaces to indent is up to you, but usually three spaces are sufficient to organize your AML program without using up too much room. Other styles work equally well to organize your code. Whichever style you decide on, use it consistently.

Question 4-3: Modify the code you wrote for question 4-2 to type a message telling the user that the project number is invalid.

(Answer on page 4-18)

Grouping multiple actions: The &DO block

The &IF &THEN &ELSE structure supports the execution of only one statement in either the true or false condition. Most of the time, however, you want to perform several actions, not just one. The &DO directive groups a series of statements in one block. An &DO block begins with the &DO directive and *always* ends with the &END directive. AML interprets all statements between the &DO and &END and executes the block as a single statement.

The following example expands on previous examples by adding actions and grouping them with an &DO block. Reference the line numbers (in brackets) to help you follow the flow of the program. The program first checks for the user name [1], and if it's Jorge, types a good morning message [3], moves to the appropriate workspace [4], and starts ARCPLOT [5]. If the user isn't Jorge [7], a message is typed [9] and the program quits ARC/INFO [10]. &END ends the &DO block [11] and &RETURN ends the program [12].

```
        /*namecheck.aml
[1]     &if [username] = jorge &then
[2]       &do
[3]         &type Good morning Jorge, have a good day
[4]         &workspace /project23/jorge
[5]         ARCPLOT
[6]       &end /*if user is jorge
[7]     &else
[8]       &do
[9]           &type You are not authorized to use this computer!
[10]          QUIT
[11]      &end /*if user is not jorge
[12]    &return
```

STYLE

Instead of placing the &DO on a separate line as seen in this chapter, many people place the &DO on the line with the &IF...&THEN and &ELSE directives, and line up the &END with the &IF or &ELSE rather than with the &DO:

```
&if <expression> &then &do
  <true action>
&end
&else &do
  <false action>
&end
```

Question 4-4: If the program includes lines [2] and [8] shown in
NAMECHECK.AML (on the preceding page), what other lines are absolutely
necessary? Why?

(Answer on page 4-18)

Testing coverage parameters and user input

Now that you understand the structure of the &IF &THEN &ELSE statement,
take a look at techniques designed to trap errors that users typically encounter.
For example, you can test coverage parameters set by ARC/INFO for a true or
false value with the &IF directive and AML reserved variables. Testing these
variables before executing commands makes your programs more flexible and
allows you to send the user descriptive error messages when problems occur.

FYI

Some ARC/INFO commands require many conditions to be met to execute
properly, so testing can be extensive. The more your programs allow user
interaction, the more you need to check for errors. The time initially spent
writing error-checking code is compensated for by programs that don't fail
and code you can use with other programs.

Chapter 9 shows you how to anticipate errors in a more systematic way than
checking for every possible error.

You can test coverage information using the reserved variables AML provides.
The AML &DESCRIBE directive is similar to the ARC/INFO DESCRIBE
command—both store coverage information in AML reserved variables named
with the DSC$ prefix. Using &DESCRIBE in your programs has two advantages
over using DESCRIBE. &DESCRIBE stores coverage information without
producing the long screen listing that DESCRIBE generates. In addition, because
&DESCRIBE is an AML directive, it executes from any ARC subsystem (e.g.,
ARCPLOT, ARCEDIT, etc.). The table on the next page gives a few examples of
DSC$ variables.

Reserved variable	Value
DSC$QEDIT	Set to .TRUE. if the coverage has been edited since topology was last constructed
DSC$QTOPOLOGY	Set to .TRUE. if there is polygon topology
DSC$REGIONNAMES	A list of names of region subclasses in a coverage
DSC$ARCS	Set to the number of arcs in a coverage
DSC$PRECISION	Set to the precision of the coverage (i.e., single or double)

FYI   See &DESCRIBE in the online help Command references for a complete listing of the reserved variables set by this directive.

STYLE   Naming variables with a consistent scheme allows you to quickly identify variables and delete those you no longer need. Notice that the names of the logical reserved variables in the table begin with the letter Q after the DSC$ prefix. This particular naming scheme makes it easy to see which of the reserved variables are logical variables. Choose a consistent naming convention for your variables when you write AML programs.

Many processes in ARC/INFO require a coverage to have current topology before an operation can be performed. You've probably encountered the following error message, which indicates that polygon topology either isn't present or isn't current:

```
Cannot open PAL file
```

Before trying to execute a command, it's useful to ascertain whether a coverage has current topology. DSC$QEDIT stores information on whether a coverage was edited since the last CLEAN or BUILD, allowing you to test for this condition. Consider the next example, which returns an informative message to the user regarding coverage topology. This could be executed before an overlay operation to ensure current topology for both of the coverages involved:

```
[1]    &describe streetcov
[2]    &if %dsc$qedit% &then
[3]       &return Update topology using BUILD or CLEAN.
[4]    &else
[5]       &type Coverage has current topology.
[6]    &return
```

[1] &DESCRIBE initializes the DSC$ variables.

[2] %DSC$QEDIT% is evaluated. Its value determines which action to take.

[3] If %DSC$QEDIT% is .TRUE., then the coverage has been edited. The program ends and informs the user to update topology.

[4] If %DSC$QEDIT% is .FALSE., then the coverage hasn't been edited and topology is current.

[5] The user is sent the appropriate message.

[6] The program ends.

If you bother to check for current topology, why not go one step further? In addition to returning an error message to the user, also execute the CLEAN command. Although not required by ARC/INFO command syntax, the AML program should create an output coverage for backup purposes when the CLEAN command executes. In the ARC example that follows, the action indicated by &THEN renames the coverage and sends a message to the user. This step makes it easier for the user because the resulting coverage has the same name as the original.

```
/* checktopology.aml
&describe streetcov
&if %dsc$qedit% &then
  &do
      &type /&Topology must be reconstructed
      &type /&Your original coverage is renamed to streetcovold
      RENAME STREETCOV STREETCOVOLD
      CLEAN STREETCOVOLD STREETCOV # # LINE
  &end
&return
```

Notice the use of the /& with the &TYPE directives. This acts as a carriage return and inserts a blank line on the screen before the message displays. Those who use your AML programs appreciate such programming considerations as informative and easy-to-understand messages.

FYI

Prior to ARC/INFO Version 7.1, the backslash was used to insert blank lines on an &TYPE line. The backslash still works on the UNIX platform, but not on Windows NT. /& should replace all backslashes in older AML programs in order to make them compatible with different platforms. This is only true for the &TYPE directive; the backslash still acts as a line feed for text stacking in many ARCPLOT commands.

Q

Question 4-5: What happens if %DSC$QEDIT% is false in the previous example?

(Answer on page 4-18)

Think for a minute about what happens if the previous AML program runs a second time. The line RENAME STREETCOV STREETCOVOLD causes the program to fail because STREETCOVOLD already exists and ARC/INFO doesn't allow you to overwrite existing coverages. A better way to write this program is to check if the coverage name already exists before using RENAME. This is important for any command that creates a new coverage.

There are other potential problems with the program. The CLEAN command is hard coded with the LINE option, and the program is written for a coverage named STREETCOV. This kind of specific coding doesn't consider what would happen if, for example, a polygon coverage is used with this program, or if someone wants to use the program with a different line coverage.

Testing user input

Instead of hard-coded options, consider generalized AML programs that adapt to a variety of situations. The user supplies the specific information to generalized programs. A well-written program verifies all user input before continuing, never assuming the user entered the correct information. If your program doesn't test user input, it will frequently fail, making users unhappy. Many AML functions support the capture and testing of user input. The following table lists six of these functions:

Function	Return value
[EXISTS]	Returns a value of .TRUE. if the object (e.g., coverage, file, etc.) exists and .FALSE. if it doesn't
[NULL]	Returns a value of .TRUE. if the string being evaluated is blank and .FALSE. if otherwise
[QUERY]	Returns a value of .TRUE. or .FALSE. based on the user's response to a yes/no question
[RESPONSE]	Returns the value input by the user
[UPCASE]	Returns an uppercase version of an input string
[LOCASE]	Returns a lowercase version of an input string

The [RESPONSE] and [QUERY] functions capture user input, and the [EXISTS] and [NULL] functions verify user input. The [UPCASE] and [LOCASE] functions convert user input to either uppercase or lowercase, respectively. These functions are often used together in programs requiring user input. The following examples set variables to these functions:

```
&sv qcover = [exists soilscov -cover]
```

%QCOVER% is set to .TRUE. if SOILSCOV exists and .FALSE. if it doesn't.

```
&sv qdraw = [query 'Do you want to draw the coverage']
```

%QDRAW% is set to .TRUE. if the user types YES or Y and .FALSE. if NO or N.

```
&sv qcov = [null %newcover%]
```

%QCOV% is set to .TRUE. if %NEWCOVER% doesn't contain a value and .FALSE. if it does.

Now look at how you can use these functions in the &IF structure. Users often encounter errors that result from entering an incorrect coverage name. Either they type the coverage name incorrectly, try to create a coverage with the same name as one that already exists, or try to access a coverage that doesn't exist in the current workspace. These situations produce ARC/INFO error messages like these:

```
Coverage ... not found.
Cannot open ARC file
Cannot open PAL file
Coverage already exists
```

Examine the next example, which runs from ARCPLOT. This program avoids these errors by checking to see whether the desired coverage already exists. If the coverage exists, the feature attribute table is listed; if it doesn't exist, a message is sent to the user.

```
&sv qcover = [exists soilscov -cover]
&if %qcover% &then
  LIST SOILSCOV.PAT INFO
&else
  &type Soilscov does not exist
```

After listing the table, suppose you want to draw the coverage. Add an &DO block to handle more commands. Then, in the &ELSE statement, add another &DO block to type a message and display a list of coverages to the user.

```
/* showsoils.aml
&sv covername = soilscov
&sv qcover = [exists %covername% -cover]
&if %qcover% &then
  &do
    MAPEXTENT %covername%
    LIST %covername%.pat
    POLYGONS %covername%
  &end /*if
&else
  &do
    &type %covername% doesn't exist
    &type /&Use a name from the following list of coverages
    DIRECTORY COVER
  &end /*else
&return
```

This program works fine for the SOILSCOV coverage, but, because it lists a polygon attribute table and draws polygons, it fails with a point or line coverage. Look at SHOWCOV.AML, which does the same thing as SHOWSOILS.AML, but for any coverage the user specifies. Notice below how NO_RESPONSE is used within the [RESPONSE] function as the default coverage name when nothing is entered. The [EXISTS] function with the -POLY option tests user input.

```
/* showcov.aml
&sv covername = [response 'Enter a coverage name' no_response]
&sv qcover = [exists %covername% -poly]
&if %qcover% &then
```

```
&do
   MAPEXTENT %covername%
   LIST %covername%.PAT INFO
   POLYGONS %covername%
&end /*  if polygon cover exists
&else
   &do
      &type %covername% doesn't exist
      DIRECTORY COVER      /*  list coverages to user
   &end  /*  if polygon cover doesn't exist
&return
```

In the examples you've seen, a variable is set to the value of the function and then the variable is used in the &IF statement. You can save a step by putting the function directly in the &IF statement. [EXISTS] still returns either .TRUE. or .FALSE. and is interpreted by &IF.

To modify the previous example, remove &SV QCOVER and insert [EXISTS] in the &IF statement. The variable %QCOVER% is no longer needed. The modified &IF statement looks like this:

```
&if [exists %covername% -poly] &then
   &do
   .
   .
   .
```

You can also use &IF to verify user input by creating a program that checks with the user before a coverage is deleted with the KILL command. The next example, executed from ARC, is called KILLCOV.AML. It uses the [QUERY] function to verify that the user wants to delete a coverage before the KILL command is executed. The default answer NO, added to the end of the [QUERY] function, acts as a safety net in case the user hits a carriage return. Including a default answer with [QUERY] is a good habit.

```
/*killcov.aml
&sv covername = [response 'Enter a coverage name']
&sv qkill = [query 'DELETE? ARE YOU SURE' .false.]
&if %qkill% &then
   &do
      KILL %covername%
      &type %covername% is deleted from your workspace
   &end
&return
```

In the following example, KILLCOV.AML is modified. Notice how %QKILL% is no longer needed when [QUERY] is placed inside the &IF statement.

```
/*  killcov.aml
/*  modified 12-8-96 by placing [query] in the &if
&sv covername = [response 'Enter a coverage name']
&if [query 'DELETE? ARE YOU SURE?' .false.] &then
  &do
    KILL %covername%
      &type %covername%  is deleted from your workspace
  &end
&return
```

Using the [NULL] function is another way of verifying user input. The program shown below, KILLCOV2.AML, uses the [NULL] function to verify that a coverage name is typed.

```
/* killcov2.aml
&sv covername = [response 'Enter a coverage name']
&if [null %covername%] &then
  &do
    &type You did not enter a coverage name
    &type /&Please run KILLCOV2 again
  &end
&else
  &do
    KILL %covername%
      &type %covername% is deleted from your workspace
  &end
&return
```

[NULL] doesn't actually verify that a coverage name was typed like [EXISTS] does. [NULL] only verifies that the variable %COVERNAME% contains a value. Suppose, for example, it's Monday morning and a tired user enters zzzzz as the coverage. %COVERNAME% is set to zzzzz, so even though no valid coverage name is entered, this input passes the [NULL] test. The program fails when it tries to KILL a coverage with this name.

Notice that the &IF test, [QUERY...ARE YOU SURE?], which appears in KILLCOV.AML, doesn't appear in KILLCOV2.AML. Keeping it requires placing one &IF test inside another, which is discussed in lesson 4.3.

Using the [GETCOVER] function instead of [RESPONSE] is often a better approach for soliciting user input.

Replace:

```
&sv covername = [response 'Enter a coverage name']
```

with:

```
&sv covername = [getcover * -all 'Enter a coverage name' ~
  -none]
```

[GETCOVER] with the -NONE option offers a menu of valid coverages from which to choose plus the additional menu choice, _NONE_. When _NONE_ is selected, %COVERNAME% is null.

> If the user dismisses the popup menu that [GETCOVER] creates, or there are no coverages in the workspace, a null value is returned to the variable.

The [RESPONSE] function requires the most testing because the user is free to type anything at the keyboard. You can avoid testing for [NULL] with [RESPONSE] by putting in a default value. In addition, you can limit the amount of testing you do by using the optional parameters with the [GET...] and [EXISTS] functions.

For example, [exists %cover% -cover] tests whether the variable, COVER, contains a valid coverage, whereas [exists %cover% -poly] tests whether the coverage is a valid polygon coverage. [GETCOVER * -POINT] presents only valid point coverages to the user.

In summary

The &IF &THEN &ELSE structure enables your AML programs to respond to new and changing information. &IF &THEN evaluates a logical expression, executes an action, and handles the flow of the program when the expression is true. The &ELSE directive alters the flow to execute an alternative action when the expression is false. &DO blocks group several statements in one action for more efficient execution.

In the coming lessons, you'll learn to write AML programs that handle more than two alternatives.

Answer 4-1:
(a) .TRUE.
(b) .FALSE.
It's a good idea to use the [UPCASE] or [LOCASE] function to ensure that your comparisons don't return a value of .FALSE. because of a mismatch of uppercase and lowercase letters. Refer to page 4-13 for information about these functions.

Answer 4-2:
```
&sv projectnum = [response 'Enter a project number']
&if %projectnum% >= 1000 &then
   &type You have entered a valid project number
```

Answer 4-3:
```
&sv projectnum = [response 'Enter a project number']
&if %projectnum% >= 1000 &then
   &type You have entered a valid project number
&else
   &type The project number is invalid
```

Answer 4-4: Lines [6] and [11] are also necessary because they end the &DO block—without them, AML would generate an error message.

Answer 4-5: If %DSC$QEDIT% is false in the example, the program skips to &RETURN and terminates.

Exercises

Exercise 4.1.1

Create an AML program that asks if the user wants an output coverage for the BUILD command. The program should prompt the user for a coverage name and the topology type to build, as well as a name if an output coverage is desired.

Exercise 4.1.2

Find the errors in the following AML program:

```
&if [locase [username]] = jorge &then &do
  &type Good morning jorge, have a good day.
  &workspace /project23/jorge
  ARCPLOT
&else &do
  &type You may only work in the guest account on this machine.
  &workspace /guest
&return
```

Exercise 4.1.3

Fix the errors in the AML from exercise 4.1.2 and modify it to return `Good morning` and `Good afternoon` based on the time of day.

Lesson 4.2—Choosing from many alternatives

In this lesson

The &IF directive offers a choice between two actions, but that's not always enough. Your task might be to create an AML program that chooses between three or more actions depending on the value of a single variable or expression. For example, a program might execute one of several actions based on a project number, user name, day of the week, or response to a question.

This lesson introduces two programming structures that choose an appropriate action from several alternatives. One is a sequence of &IF &THEN &ELSE statements, each presenting one alternative. The other, the &SELECT block, presents many alternatives in a streamlined format. With either structure, only one action ultimately executes. You'll learn how to use these structures in your programs and how to decide which structure is appropriate for a particular task.

&IF more alternatives are needed

One way of selecting between three or more alternatives is to append additional &IF directives to the basic &IF &THEN &ELSE structure. For example, suppose you need to check a project number. The initial &IF checks whether the project number equals 1001. If not, another &IF checks whether it's 1002, the next &IF, 1003, and so on. In this case, subsequent &IF statements are evaluated only if the previous &IF statement is .FALSE.. Executing one &IF statement as the result of another is referred to as *nesting*.

A nested &IF structure that presents many alternatives looks like this:

```
&if <expression> &then
  <true action>
&else
  &if <expression> &then
    <true action>
&else
  &if <expression> &then
    <true action>
&else
```

```
<false action>
<next statement>
```

In the structure outlined above, assume that each <expression> is the same variable tested against different values. If the expression is true, the <true action> is executed. If the expression is false, control passes to the associated &ELSE clause. Because the &ELSE clause contains another &IF statement, another value is tested.

In general, as long as the expression is false, control passes to the next &IF statement and another expression is evaluated. As soon as the expression is true, all subsequent statements are skipped and control passes to the <next statement>. If the expression is never true, the <false action> is executed, followed by the <next statement>.

STYLE

You can use any number of nested levels in your code. Remember, however, that you have to maintain your programs. Trying to understand or debug a nested structure that contains ten or fifteen levels can be frustrating. Good code not only works, but is easy to read, understand, and modify.

Look at an example, executed from ARC, that places users in the correct project workspace after they select a project number from the menu [GETCHOICE] creates.

```
/*projwork.aml
&sv projectnum = [getchoice 1001 1002 1003 -prompt 'Project ~
  number?']
&if %projectnum% = 1001 &then
    &workspace /esri/projects/region1
&else
  &if %projectnum% = 1002 &then
    &workspace /esri/projects/region2
&else
    &workspace /esri/projects/region3
&return
```

%PROJECTNUM% is evaluated by each &IF directive. If %PROJECTNUM% is 1001, the first &THEN action executes and the workspace is set to /esri/projects/region1. If %PROJECTNUM% isn't 1001, it's tested again. If %PROJECTNUM% is 1002, the second &THEN action executes (i.e, the workspace is set to /esri/projects/region2). If %PROJECTNUM% is 1003, the final &ELSE executes. After a workspace is set, control passes to &RETURN.

To clean up the format of the nested &IF &THEN &ELSE statements, the &IF and &ELSE statements can be placed on the same line. The order of evaluation is the same and the expression is tested again at every &IF directive. Using this method, the last example could be written like this:

```
&sv projectnum = [getchoice 1001 1002 1003 -prompt 'Project ~
  number?']
&if %projectnum% = 1001 &then
    &workspace /esri/projects/region1
&else &if %projectnum% = 1002 &then
    &workspace /esri/projects/region2
&else
    &workspace /esri/projects/region3
&return
```

This style of nested &IF statement isn't very efficient because the expression must be tested at each branch. Additionally, the flow of the program is hard to follow when the code gets more complex. When the same variable or function is evaluated again and again, you may want to take advantage of the &SELECT directive.

&SELECTing from many alternatives

When an AML program must accommodate many actions based on the test of *one* expression, &SELECT is better suited to the job than &IF. &SELECT is compact, easy-to-follow and, in many cases, faster than nested &IF statements. &SELECT evaluates an expression only once and then executes the action that's defined for that value.

The basic &SELECT structure looks like this:

```
&select <expression>
  &when <value1>
    <action1>
  &when <value2>
    <action2>

         .
         .
         .
&otherwise
    <other action>
&end
```

The evaluation of an &SELECT block flows like this: first evaluate the <expression>. When the expression equals <value1>, execute <action1> and then skip to the &END. When the expression equals <value2>, execute <action2> and then skip to the &END, and so on. If none of the values match, skip to &OTHERWISE, perform the <other action>, and continue to the &END of the block.

The <expression> usually takes the form of a variable or a function and is usually *not* a logical expression (i.e., it can take on values other than simply .TRUE. or .FALSE.). The &WHEN statement is executed only when its <value> matches the value of the <expression>; other &WHEN clauses are ignored. The &SELECT block can include as many &WHEN clauses as you need to support the task. &OTHERWISE serves as a "catchall" whenever the value of the expression isn't matched by any &WHEN values. &END always completes the &SELECT block.

Earlier, you saw code that used a nested &IF structure to place a user in the correct project workspace. The following example implements the same task using &SELECT. In this case, the -NONE option is added to [GETCHOICE], allowing the user to decline to choose a project number. You'll find this example is easier to follow and more efficient because the test of %PROJECTNUM% only happens once.

```
/*  selectproj.aml
&sv projectnum = [getchoice 1001 1002 1003 -prompt 'Project ~
  number?' -none]
&select %projectnum%
  &when 1001
    &workspace /esri/projects/region1
  &when 1002
    &workspace /esri/projects/region2
  &when 1003
    &workspace /esri/projects/region3
  &otherwise
    &return No project number was matched.
&end
&return
```

Question 4-6: How could you modify SELECTPROJ.AML to avoid setting %PROJECTNUM%?

Question 4-7:
(a) What does the user see if the %PROJECTNUM% isn't equal to 1001, 1002, or 1003?
(b) Could this happen with this implementation of [GETCHOICE]? Why or why not?

(Answers on page 4-29)

Now that you understand a simple &SELECT block, look at ways to make your programs more sophisticated and forgiving. The &SELECT block allows you to group different values that execute the same set of instructions (e.g., uppercase and lowercase versions of the same word, a group of users, or correct and incorrect versions of a word or command that's often typed incorrectly).

To incorporate this functionality, place all acceptable values after the &WHEN clause, separated by commas. The following AML program for display in ARCPLOT illustrates this idea of anticipating shortcuts and grouping like values:

```
/* selectfeat.aml
&sv covername = [getcover]
MAPEXTENT %covername%
/* capitalize acceptable shortcuts in [RESPONSE]
&select [response 'Feature type (Arc | Poly | POint)' ARC]
   &when A,ARC,a,arc
      ARCS %covername%
   &when P,POLY,p,poly
      POLYS %covername%
   &when PO,POINT,po,point
      POINTS %covername%
   &otherwise
      &type You didn't pick a valid feature type.
&end /*select
&return
```

To keep from having to test all possible values, use the [UPCASE] function to convert the user input from [RESPONSE] to uppercase. This eliminates all lowercase values from your &WHEN clauses.

```
/* selectfeat2.aml
&sv covername = [getcover]
MAPEXTENT %covername%
/* capitalize acceptable shortcuts in [RESPONSE]
&select [upcase [response 'Feature type ~
  (Arc | Poly | POint)' ARC]]
  &when A,ARC
    ARCS %covername%
  &when P,POLY
    POLYS %covername%
  &when PO,POINT
    POINTS %covername%
  &otherwise
    &type You didn't pick a valid feature type.
&end /*select
&return
```

To group several statements in one action for a single value, add an &DO block to any &WHEN clause. See lesson 4.1 if you're unfamiliar with the structure and purpose of the &DO block.

Instead of placing the &DO on a separate line, as seen in this chapter, many people place the &DO on the same line with the &WHEN or &OTHERWISE. If you choose this indentation style, AML syntax dictates that you place a semicolon before the &DO as follows:

```
&when <value> ; &do
&otherwise ; &do
```

Another approach you can use to match values in an &SELECT block is to establish a list of all possible values, retrieve the position of the desired value in the list, and match the position to the &WHEN clause. The [KEYWORD] function returns the position of a specified element from a list of elements. In the following example, [KEYWORD] returns Position is 3 because point is the third element in the list.

```
&sv feat = point
&type Position is [keyword %feat% arc, poly, point]
Position is 3
```

[KEYWORD] also matches uppercase and lowercase as well as unique abbreviations. This reduces the number of values that need to be tested for each &WHEN clause. Adding the [KEYWORD] function to SELECTFEAT2.AML requires the following modifications to SELECTFEAT2.AML. [KEYWORD] replaces the [UPCASE] function and the values of &WHEN are positions instead of the actual character strings.

```
/* selectfeat3.aml
/*  modified selectfeat.aml to use [keyword] to capture
/*  feature type
&sv covername = [getcover]
MAPEXTENT %covername%
&select [keyword [response 'Feature type ~
  (Arc | Poly | Point)'] Arc Poly Point]
  &when 1
     ARCS %covername%
  &when 2
    POLYS %covername%
  &when 3
    POINTS %covername%
  &otherwise
     &type You didn't pick a valid feature type./&Try again.
&end /*select
&return
```

Another ARC/INFO task that users often perform is setting tolerances for automating data. Part of this task involves setting a dangle length and fuzzy tolerance for the CLEAN command (used in ARC or ARCEDIT) and tolerances for the ARCSNAP, NODESNAP, and INTERSECTARCS commands (used in ARCEDIT).

Many tolerances are based on the scale and accuracy of the source material. A useful AML program using &SELECT would ask the user what the scale of the source material is and, based on the answer, execute a group of commands that set different tolerances.

The following AML program runs from the Arc: prompt and starts ARCEDIT to set the fuzzy tolerance, dangle length, and arc- and node-snapping tolerances based on the scale of the source map. The program saves time in ARCEDIT and simplifies the topology-building process. It also provides consistency to the database automation process. (Note: Tolerances were set assuming the coverage units are feet and using the standard digitizer resolution of .002.)

```
/* settol.aml
&sv covername = [getcover *]
&sv sourcescale = [getchoice 100 400 -prompt ~
  'Enter map scale: 1" = ?' -other]
&select %sourcescale%
  &when 100
    &do
      TOLERANCE %covername% FUZZY .2     /*ARC command
      TOLERANCE %covername% DANGLE 20    /*ARC command
      ARCEDIT                 /*ARCEDIT commands follow
      ARCSNAP ON .3
      NODESNAP CLOSEST .3
      INTERSECTARCS ALL
    &end /* when 100
  &when 400
    &do
      TOLERANCE %covername% FUZZY .8     /*ARC command
      TOLERANCE %covername% DANGLE 30    /*ARC command
      ARCEDIT                 /*ARCEDIT commands follow
      ARCSNAP ON .9
      NODESNAP CLOSEST .9
      INTERSECTARCS ALL
    &end /* when 400
  &otherwise
    &do
      &type Tolerances aren't defined for %sourcescale%
      &type /&Have your project manager define them.
    &end  /* when not 100 or 400
&end  /*select
&return
```

Question 4-8: What happens if SETTOL.AML is executed from ARCEDIT?

(Answer on page 4-29)

This format works for other coverages in a project. Coverages have different tolerances due to density or type of data. For example, soils and street centerlines use different processing tolerances due to differences in scale and accuracy of the data.

STYLE

Notice how comments organize the &DO blocks within an &SELECT block. This is an easy way to track the values associated with each &DO block. Also, because &DO and &END must always be paired, commenting in this manner helps to reveal errors.

Commenting AML programs is essential to well-written code. Comments help communicate what's happening in your program. In the previous chapters, comments were used primarily to explain the basic AML structures. Later, you'll see how comments can communicate a program's purpose, design, and execution. See chapter 1 for more information on commenting.

As the number of alternatives increases, &SELECT offers some advantages over using nested &IF directives. An expression is evaluated only *once* in an &SELECT statement, while it is evaluated *once for each* &IF directive in the nested structure. &SELECT tends to speed up a program because fewer evaluations are performed to get the same result. &SELECT is a compact, efficient format and is as easy to follow with fifty choices as it is with five.

FYI

Lesson 4.3 looks at nested structures that handle complex conditions, such as evaluating compound expressions or several different variables. In these cases, the nested &IF is required and &SELECT can't be substituted.

In summary

This lesson introduced two branching structures that choose an appropriate action from three or more alternatives. Both the nested &IF and &SELECT block test the value of a single expression and match it to an action defined for that value. Too many nested &IF statements make code hard to read. As the number of alternatives increases, &SELECT becomes the preferred structure.

Answer 4-6: Delete the &SV statement and change &SELECT as follows:
```
&select [getchoice 1001 1002 1003 -prompt 'Project number' ~
    -none]
```

Answer 4-7:
(a) The message, "No project number was matched."
(b) Yes, because the user could pick the NONE choice from the menu.

Answer 4-8: ARC/INFO would return an error message indicating an unrecognized command because TOLERANCE isn't a command in ARCEDIT.

Exercises

Exercise 4.2.1

Use the [USERNAME] function in an &SELECT block to move the user to a workspace and return a message reporting the name of that workspace.

Exercise 4.2.2

Use the &IF...&ELSE...&IF structure to create an AML that works in ARCPLOT to display coverage features. Have the user pick the coverage and feature type using [GETCOVER] and [GETCHOICE], respectively. This program should handle the display of points, lines, polygons, or nodes. Don't worry about the symbology; the features will display using the current symbols.

Exercise 4.2.3

Modify exercise 4.2.2 by using the [EXISTS] function to display only the feature types available for the coverage that's selected. In other words, don't display the choice POINT if the coverage doesn't contain point features. If only one feature type is available, display it.

Exercise 4.2.4

Modify exercise 4.2.2 to check the selected feature type. If that type isn't available, inform the user and return.

Exercise 4.2.5

Rewrite exercise 4.2.2 using an &SELECT block to draw the selected features.

Exercise 4.2.6

You saw an example of the &SELECT used with both the [RESPONSE] and [GETCHOICE] functions. Which method do you think is better? Why?

Exercise 4.2.7

Refer to PROJWORK.AML on page 4-21. In this example, why wouldn't you put [GETCHOICE] in the &IF statement, rather than setting the variable at the beginning?

Lesson 4.3—Compound expressions and more testing techniques

In this lesson

This lesson builds on the basic structure of the &IF directive and logical expressions. The &IF structure you saw in the last lesson evaluates a single expression. This lesson shows how the *logical operators* AND, OR, XOR, and NOT can combine two or more expressions in a *compound* expression. Nested &IF directives provide hierarchical testing of multiple expressions, with each level of testing depending on the outcome of the previous test. Additional tools for testing user input are also discussed here.

Constructing and evaluating compound expressions

Many tasks involve asking more than one question (i.e., testing for more than one condition at a time). You can use the &IF directive to evaluate compound expressions by combining the tests with a logical operator. Logical operators used in AML programming include AND, OR, XOR, and NOT. The logical operators AND, OR, and XOR combine two or more expressions into a single compound expression. AML evaluates the individual expressions making up a compound expression independently and then combines the result in a single true or false value. The NOT operator negates or reverses the evaluation of a single expression (i.e., .FALSE. becomes .TRUE. and vice versa).

The AND, OR, and XOR operators

The *truth tables* on page 4-33 indicate all possible outcomes of a compound expression when the AND, OR, and XOR operators are applied to its two expressions, <expression1> and <expression2>. Using the AND operator, both component expressions must be true to yield a true value for the compound expression. All other combinations yield a false value. When joined with the OR operator, if both component expressions are false, the compound expression results in a value of .FALSE. All other combinations yield a value of .TRUE..

For a compound expression to be true, the XOR looks for one, *but not both*, of the component expressions to be true. If the values are the same, the compound expression is false.

<expression1> AND <expression 2>

AND	<expression 1>	
	T	F
<expression 2> T	T	F
<expression 2> F	F	F

<expression1> OR <expression 2>

OR	<expression 1>	
	T	F
<expression 2> T	T	T
<expression 2> F	T	F

<expression1> XOR <expression 2>

XOR	<expression 1>	
	T	F
<expression 2> T	F	T
<expression 2> F	T	F

Suppose your task is to build polygon topology. You need to check to see if the coverage exists as a polygon coverage *and* whether it has been edited. If both are true, build topology. Combining a function and a reserved variable (introduced in lesson 4.1) creates a compound expression that tests for this condition.

The following AML code illustrates how this is done:

```
/* buildcov.aml
&sv cover = [getcover]
&describe %cover%
&if [exists %cover% -poly] AND %dsc$qedit% &then
  BUILD %cover% POLY
&else
  &type Cannot complete the requested operation.
&return
```

[EXISTS] and %DSC$QEDIT% are the two expressions to combine. If the polygon coverage exists *and* has been edited, both expressions are true. As shown in the truth table, compound expressions joined by the AND operator evaluate to .TRUE. when both component expressions are .TRUE.. In this case, the BUILD command is executed. When either or both of the component expressions are false, the final result of the &IF statement is also false, and the message Cannot complete the requested operation displays.

Question 4-9: Which three situations would cause the following compound expression to be false?

```
&if [exists %cover% -poly] AND %dsc$qedit%
```

(Answer on page 4-43)

Suppose the task is to move the user to the correct workspace given either a project number or user's name. You could use the OR operator as shown in the code below to test the project number and the uppercase version of the user's name.

```
&if [upcase [username]] = GISGROUP OR ~
  %projnum% = 2005 &then
   &do
     &workspace /gis/projects/proj2005
     &type The current workspace is /gis/projects/proj2005
   &end /*then
&else
   &do
     &workspace /gis/guest
     &type The current workspace is /gis/guest
   &end /*else
```

In this example, [USERNAME] = GISGROUP and %PROJNUM% = 2005 are the component expressions of the compound statement. Again, checking the truth table, compound expressions joined by the OR operator evaluate to .TRUE. if *either or both* component expressions evaluate to .TRUE.. In this case, the entire &IF statement evaluates to .TRUE., and the &THEN action is executed. Only when both component expressions are false does the &ELSE action execute.

Question 4-10:
(a) Using the previous example, evaluate the expression and trace the program's execution if the user name is Randy and the project number is 2005.
(b) Do the same for user name Julie and project number 2030.

(Answers on page 4-43)

Using these operators to combine logical expressions isn't difficult; the tricky part is figuring out whether to use AND, OR, or XOR. If the action you're trying to perform should be executed only when both conditions are met, the AND operator is appropriate. On the other hand, if you want to execute an action if either of the conditions are met, use the OR operator. Finally, if the action should execute when one but not both of the conditions are met, use XOR.

Note that you can combine more than two expressions in this manner. If you have three or more conditions to test, the expressions are still combined two at a time. You can control the order of evaluation by using parentheses to specify which compound expressions should be evaluated first. Without parentheses, logical expressions evaluate from left to right. Parentheses indicate the compound expressions to be evaluated first. Parentheses require a space on either side; without the spaces, a compound expression won't be evaluated. Parentheses make your code easy to understand and modify. Parentheses also play a vital role in determining the value of an expression. The following examples illustrate how the value of an expression depends on how parentheses are used.

Suppose:
```
a = .FALSE.
b = .FALSE.
c = .TRUE.
```

then: a AND b OR c - .TRUE.

is the same as: (a AND b) OR c = .TRUE.

but not the same as: a AND (b OR c) = .FALSE.

Suppose the task is to test whether the user name is either GISGROUP or MISGROUP and if the project number is 7002. You can state this problem as the following logical expression:

```
%username% = gisgroup OR %username% = misgroup ~
  AND %projnum% = 7002
```

Given:
```
    %username% = gisgroup
    %projnum% = 9238
```

using parentheses like this:

```
( %username% = gisgroup OR %username% = misgroup ) ~
  AND %projnum% = 7002
```

yields a .FALSE. value when the variable values are substituted in the expression. The following diagram demonstrates this:

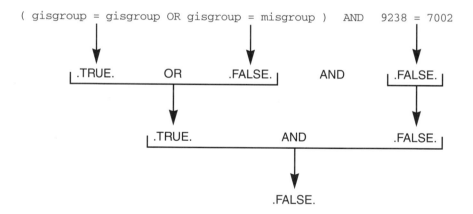

Be careful how you use the parentheses or you might not get the result you intend. Notice how the outcome changes when the components in the expression are grouped differently. When the last two expressions are grouped by parentheses, the expression evaluates to .TRUE..

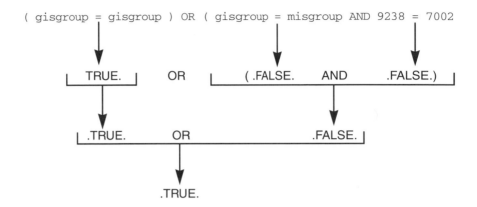

You can see from the preceding examples that parentheses can determine how the evaluation of an expression results.

FYI

There is no change in the way arithmetic operators (+, −, /, etc.) operate when they're combined with logical operators.

The NOT operator

Sometimes the condition may be the reverse of what you want. For example, the coverage does *not* exist, the topology is *not* current, or the variable does *not* contain any information. When this type of condition exists, use the NOT operator. The logical operator NOT is applied after the expression is evaluated. The result of applying NOT is shown in the following table:

	<expression 1>	
	T	F
NOT	F	T

Notice in the examples that follow that you can either type the word or use the symbol (^) to indicate the NOT operator.

Before attempting any action on a coverage, it's good to check whether the particular coverage exists. If it doesn't, you can return a message to the user and display a list of available coverages. This can be coded as follows:

```
&if NOT [exists streams -cover] &then
  &do
    &type The streams coverage doesn't exist
    LISTCOVERAGES
  &end
```

In this example, [EXISTS STREAMS -COVER] is the logical expression. If STREAMS doesn't exist as a coverage, the expression evaluates to .FALSE.. When combined with NOT, this results in a true statement because NOT .FALSE. equals .TRUE.. A true statement executes the &THEN action, which indicates an error and lists the available coverages.

The NOT operator is also useful when testing for current topology. In the following example, if the coverage hasn't been edited, %DSC$QEDIT% is .FALSE.; however, the NOT operator (^) switches the evaluation to true, and an appropriate message is displayed. If the coverage has been edited, the result of the &IF expression is false because NOT .TRUE. equals .FALSE. and the BUILD command is executed.

```
&describe %cover%
&if ^ %dsc$qedit% &then
  &type The coverage has not been edited
&else
  BUILD %cover% POLY
```

The NOT operator is often used with the [NULL] function to test whether a variable has a value (i.e., isn't null or empty). In the following example (run from ARCPLOT), when an INFO file name is chosen from the menu, [NULL %TABLE%] is false. The NOT operator switches the .FALSE. value to .TRUE., and the table is listed.

```
&sv table = [getfile * -info -none]
&if NOT [null %table%] &then
    LIST %table% INFO
&else &type No file selected
```

Question 4-11: Can you rewrite the code in the last example to perform the same actions without using the NOT operator?

(Answer on page 4-43)

Hierarchical testing using nested &IF statements

Many times you test a condition only after another condition has already been satisfied. For example, you test for current topology after making sure the coverage exists. Likewise, you test to see whether a coverage exists after ensuring that the variable holding the coverage name isn't empty. The nested &IF structure effectively handles this method of hierarchical testing.

All the nested &IF statements you saw in lesson 4.2 tested the *same* expression. This lesson takes the nested &IF structure a step further by testing *different* expressions at each &IF directive. This isn't the same as using a compound expression to test two conditions simultaneously. This technique allows you to test conditions separately where the second test depends on the outcome of the first.

Take another look at an example of a compound expression:

```
&describe %cover%
&if [exists %cover% -poly] and %dsc$qedit% &then
   BUILD %cover% POLY
&else &type Cannot complete the requested operation.
```

The returned message doesn't indicate which test failed. You can't tell whether the error occurred because the coverage doesn't exist as a polygon coverage or because it was edited. There is potential for another problem because &DESCRIBE occurs before the test for the existence of a polygon coverage. If the coverage doesn't exist at all, &DESCRIBE fails and generates an error message.

You could enhance this code by testing each condition separately in nested &IF statements as follows:

```
        /* ls3.aml
  [1]   &sv cover = [getcover * -all 'Select a coverage' -none]
  [2]   &if [exists %cover% -poly] &then
  [3]      &do
  [4]         &describe %cover%
  [5]         &if %dsc$qedit% &then
  [6]            BUILD %cover% POLY
  [7]         &else
  [8]            &type The coverage has not been edited
  [9]      &end
 [10]   &else
 [11]      &type %cover% is not a polygon coverage in the ~
            current workspace
 [12]   &return
```

The &IF first tests whether the polygon coverage exists in the current workspace [2].

If the coverage exists, &DESCRIBE executes [4] and the second &IF tests whether the coverage has been edited [5].

If the coverage has been edited, the BUILD command executes [6].

If the coverage hasn't been edited, a message is sent to the user [8].

If the coverage isn't a polygon coverage [2], the entire &THEN block is skipped, and the last &ELSE executes [10] and returns a message to the user [11].

The following graphic illustrates this flow of control:

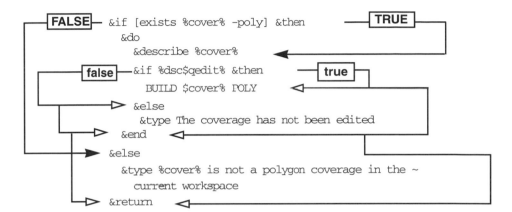

For every &ELSE in a nested structure, there is a corresponding &IF statement. The reverse, however, isn't true. Every &IF doesn't require an &ELSE because the false condition is optional. It's a good idea to establish a visual relationship between these statements by indenting the &ELSE to line up directly under the &IF with which it's associated. This is a visual relationship only; indentation doesn't affect the flow of the program.

In the following example, there is no &ELSE associated with the second &IF. There is an &ELSE after the second &IF, but it's referencing the first &IF. Trying to debug or modify AML programs with no indentation like this could be a full-time job.

```
&if [exists %cover% -poly] &then
&do
&describe %cover%
&if %dsc$qedit% &then
BUILD %cover% POLY
&end
&else &type %cover% is not a polygon coverage in the~
 current workspace
```

Testing for the correct [TYPE]

Testing passed arguments with [NULL] only ensures that they're present, not that they're correct. What if the user types SOILS SOIL_TYPE, when you're expecting a coverage name and map scale? You can't assume that the user will respond as you intend. (Otherwise, there would be no reason for all this testing.) Using the [TYPE] function, you can verify that a value is the correct type of information. [TYPE] returns a value indicating whether the string type is the character, logical, real, or integer. The value returned is shown in the following chart:

Returned value of [TYPE]	Indicates
-2	String type: real
-1	String type: integer
1	String type: character
2	String type: logical

Look at an example executed from ARCPLOT. The task is to capture the number of a census tract and use the RESELECT command to find all the homes in that census tract (i.e., where the item CTNUM has a value of %CENSUSNUM%). If the %CENSUSNUM% isn't an integer, the user receives an error message.

```
&sv censusnum = [response 'Enter the number of the ~
  census tract']
&if [type %censusnum%] = -1 &then
  RESELECT homes points ctnum = %censusnum%
&else &return &inform You did not enter a valid number.
```

In summary

This lesson expanded your knowledge of the nested &IF structure to include compound expressions and advanced tools for capturing and testing user input. Compound expressions create more sophisticated tests needed to support an application. You learned to create these expressions using the logical operators AND, OR, XOR, and NOT.

Truth tables indicate the results of all combinations of true and false values and clarify the evaluation of compound expressions. When many expressions are evaluated, the order of evaluation is critical to the resulting value. Parentheses control the order in which the AML processor evaluates expressions.

You now have the tools to write flexible AML programs that can adapt to varying conditions.

Answer 4-9:
(1) The coverage exists as a polygon coverage, but the coverage hasn't been edited.
(2) The coverage isn't a polygon coverage, but the coverage has been edited.
(3) The coverage isn't a polygon coverage and hasn't been edited.

Answer 4-10:
(a) &IF is true (because the project number matches), so the &THEN &DO block executes.
(b) The &ELSE is skipped and &RETURN is executed to finish the program.

Answer 4-11: Take out the NOT operator and switch the &THEN action with the &ELSE action.
```
&sv table = [getfile * -info -none]
&if [null %table%] &then
  &type No table selected
&else LIST %table% INFO
```

Exercises

Exercise 4.3.1

Modify exercise 4.1.3 to include the messages `Here a little early?` and `Good evening, time to go home.`

Exercise 4.3.2

Given a coverage name, write an AML program that figures out which feature type to draw. The user should choose the coverage from a menu (i.e., [GETCOVER]).

Exercise 4.3.3

Create an AML program that contains nested &IF statements. The program should determine if a LINE coverage stored in the variable COV has been edited. If so, allow the user to choose between the BUILD and CLEAN commands to rebuild topology.

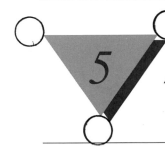

Actions worth repeating

No one likes to perform a boring task over and over. Computers, not people, are designed to execute repetitive tasks efficiently. With AML, you can automate the steps that accomplish a task in a programming structure called a *loop*. A loop organizes these steps and performs them as many times as it is instructed to. No matter how many times the steps are executed, they're written in the program only once. Programming with loops increases your ability to perform many ARC/INFO tasks. Some of these are listed below.

- Set up a series of workspaces and/or coverage names for a project
- Create graphics files from a list of coverages
- List information for each record in a selected set
- Prompt the user until the correct input is given
- Identify features as long as the user wants to continue
- Read and count the number of lines in a file

This chapter shows how loops handle repetitive tasks in programs. You'll learn about the AML directives used to construct basic looping structures, and how they control when a loop starts and stops. You'll also learn how to debug looping programs by tracing the order of command execution through each iteration. The AML functions commonly used in conjunction with loops are also introduced.

This chapter covers the following topics:

Lesson 5.1—Let's &DO loops

- To loop or not to loop
- &DO a given number of times: The counted loop
- Locating looping errors
- &DO for each element in a list: The list loop

Lesson 5.2—&DO loops with logical expressions

- &DO &WHILE a condition is true: The while loop
- &DO &UNTIL a condition is true: The until loop

Lesson 5.3—Compound clauses, nested loops, and indexed variables

- Getting the best of two loops
- Using nested loops
- Creating indexed variables with nested loops

Lesson 5.1—Let's &DO loops

In this lesson

In this lesson, you'll learn how to write simple loops in AML. You'll learn ways to control the number of times a loop executes and how to stop it. You'll expand your knowledge of the &DO block, introduced in chapter 4, by using it to construct loops.

Discussions and examples in this lesson cover &DO loops that execute a specific number of times *(counted loops)* and loops that execute once for each element in a list *(list loops)*. This lesson also presents techniques that help you trace the execution of a loop in order to prevent errors.

To loop or not to loop

Chapter 4 introduced a simple form of the &DO directive called a block. There were no arguments, and all statements between the &DO and &END directives were grouped and interpreted as one task. When &DO is used in this format (i.e., without additional arguments), the task executes once. Review the basic structure of the &DO block:

```
&do
    <action 1>
    <action n>
&end
```

The &DO directive always groups several actions in one task. By using additional arguments to create a *loop control clause*, you can use the &DO directive to perform a task more than once. This structure is referred to as an *&DO loop*.

The &DO loop structure looks like this:

```
&do <loop control clause>
    <action 1>
    <action n>
&end
```

If you want an action to execute again, you must use a loop control clause to specify the conditions under which the loop starts and stops. There are several ways to set these conditions using an &DO loop. The table below lists four types of &DO loops. The type of loop describes the circumstances under which the loop executes.

Type of loop	Number of times the loop executes
Counted loop	Executes a specific number of times
List loop	Executes once for each element in a list
While loop	Executes while a condition is met
Until loop	Executes until a condition is met

&DO a given number of times: The counted loop

The *counted loop* is useful for performing many ARC/INFO operations. You often know the number of records, colors, workspaces, coverages, or features on which you want to perform an operation. Some tasks must be performed at a set increment or interval. The counted loop requires that you specify the number of iterations and optionally the looping increment. The loop control clause is appended to the &DO directive and consists of a *loop control variable* and a range and increment set with the &TO and &BY arguments.

The format for a counted loop is as follows:

```
&do <loop control variable> = <start value> &to ~
  <stop value> &by <increment>
    <action 1>
    <action n>
&end
```

The value of the <loop control variable> controls the processing of the loop. The <start value> is the first value in the range. The <stop value> is the last value in the range. The <increment> is a value added to the <loop control variable> at every iteration. This increment is 1 unless you specify otherwise.

When the loop starts, the <loop control variable> is assigned the <start value>. At each pass through the loop, or *iteration*, the <loop control variable> is incremented and tested against the range of values defined by the <start value> and <stop value>. When &END is encountered, control returns to &DO. When the value of the <loop control variable> is outside the range, the loop terminates.

Thus, adding the loop control clause to the &DO directive instructs the program to return to the &DO, increment the counter, test the range, and either repeat or stop the loop. Look at how a counted loop works in this example:

The following code:

```
&do count = 1 &to 3 &by 1
  &type Count = %count%
&end
```

outputs this:

```
Count = 1
Count = 2
Count = 3
```

and stops when COUNT = 4.

> As a rule, you shouldn't use &SETVAR to change the value of the loop control variable inside the loop. The loop control clause changes the value of the variable and determines whether or not to continue.

The following &DO loop creates nine workspaces for a new project. The workspaces are named TRACT1, TRACT2, TRACT3, etc.

```
/* projwork.aml
&do count = 1 &to 9 &by 1
    CREATEWORKSPACE TRACT%count%
    &type Created workspace TRACT%count%
&end   /* &do count
&return
```

The example starts by initializing the variable COUNT to 1. (*Initialization* is assigning a beginning value to a variable.) The value of COUNT is tested by comparing it to 9. Because 1 is less than 9, the actions in the loop execute once (i.e., the workspace named TRACT1 is created and a message is typed to inform the user). Next, &END is encountered, and control passes back to the &DO loop. Then, COUNT increments by 1, making COUNT = 2. Because the default increment is 1, it doesn't need to be specified, but it's added here for clarity. Again, COUNT is compared to 9; its value is still less, so the loop executes again. This process repeats until COUNT = 10. At this time, COUNT is greater than 9, and the loop is finished.

Question 5-1: Given the code &do count = 0 &to 25 &by 1, how many times does the loop execute? Explain your answer.

(Answer on page 5-16)

Notice when COUNT is used with the &DO directive, it's *not* enclosed in percent signs (%...%). Setting COUNT to the <start value> is functionally the same as using &SETVAR to set a variable. In chapter 1, you learned that the percent signs aren't used to set variables, only to reference them.

For example, when you use the value of COUNT as part of the name of a workspace, you must enclose it in percent signs (i.e., %count%). Note: This variable doesn't have to be named COUNT; it can have any name. As with all variables, descriptive names help document your program and make it easier to follow.

The <start value>, <stop value>, and <increment> depend on your application. They can be set with constants, variables, or functions. It's of particular interest that these values can be negative numbers or fractional parts (e.g., –2, –.1, .25, .5, etc.). Use negative values to define a range in descending order (i.e., highest to lowest). Fractional values are necessary to query databases whose values represent continuous data, such as temperatures, elevations, or chemical concentrations.

FYI

> You can specify a negative increment to cause a loop to count backwards. For example:
>
> ```
> &do i = 50 &to 0 &by -5
> ```
>
> counts downward from 50 to 0 by increments of 5. If you use a negative increment, be sure that the start value is *greater* than the stop value.

Consider the next example, which accesses a coverage of chemical sampling sites in ARCPLOT. The chemical concentrations range between the values 0 and 1 and are recorded to the nearest tenth. You want to select points at each concentration level, list the records, and display each with a different symbol. You'll use the symbol with a number ten times the concentration level.

```
/*   showsites.aml
MAPEXTENT CHEMSITES
&do chemindex = .1 &to .9 &by .1
    RESELECT CHEMSITES POINTS CONCENTRATION = %chemindex%
    LIST CHEMSITES.PAT INFO
    POINTMARKERS CHEMSITES [calc %chemindex% * 10]
    ASELECT CHEMSITES POINTS
    &pause Hit return when ready to continue
&end
&return
```

The [CALC] function returns the value of an arithmetic expression. Here it calculates the values of CHEMINDEX multiplied by ten. (Refer to chapter 6 for a more detailed discussion of [CALC]).

Q

> **Question 5-2:** List the values of CHEMINDEX at each iteration of the loop.
>
> (Answer on page 5-16)

The following AML program runs from the ARC/INFO subsystem, TABLES. It classifies wells according to depth measured at 5-foot intervals. The program uses the command RESELECT to select wells in each interval. A classification number is calculated and assigned to the variable CLASS. For example, wells with depths greater-than-or-equal-to 1 and less-than-or-equal-to 5 are selected and assigned a classification of 2.

```
/* wellclass.aml
SELECT WELL.PAT
&sv maxdepth = 41
&sv mindepth = 1
&sv olddepth = 1
&sv class = 1
&do depth = %mindepth% &to %maxdepth% &by 5
  ASELECT
  RESELECT WELLDEPTH >= %olddepth% and WELLDEPTH <= %depth%
  CALCULATE CLASS = %class%
  &sv class = %class% + 1
  &sv olddepth = %depth%
&end
Q STOP
&return
```

The variables MINDEPTH and MAXDEPTH are initialized with depths of the shallowest and deepest wells, respectively. The variable OLDDEPTH stores the value of the loop variable DEPTH before it increments on the next iteration. The RESELECT statement is built using both of these values. The variable CLASS also increments by one at each iteration.

FYI

You can use the STATISTICS command in conjunction with the [SHOW] function to extract the minimum and maximum depths for initializing these variables. Refer to the online help Command references for a complete discussion of the [SHOW] function.

The next example uses the loop control variable COLORBOX to specify a shade symbol number and to define a text string in ARCPLOT. The task involves creating a palette of color boxes across the bottom of the display screen to reference the available colors. The number of a color is placed in the center of each box. The boxes start at 0,0 and measure .5-inches square. The program also sets text size, the positioning parameters, and the current shadeset.

```
/* colorref.aml
&sv ystart = 0
&sv xstart = 0
&sv ysize = .5
&sv xsize = .5
&sv xtext = %xsize% * .5
&sv ytext = %ysize% * .5
/*   set text characteristics
TEXTSET FONT.TXT
TEXTSYMBOL 2
TEXTSIZE .12
TEXTCOLOR BLACK
TEXTJUSTIFICATION CC
SHADESET COLOR.SHD
/* start creating adjacent solid color boxes for each
/* shade with the shade symbol number centered in the box
&do colorbox = 1 &to 15
  SHADESYM %colorbox%
  PATCH %xstart% %ystart% [calc %xstart% + %xsize%] %ysize%
  /*   now label each box with its shade symbol number
  MOVE [calc %xstart% + %xtext%] %ytext%
  TEXT %colorbox%
  /*   move the start point of the next box
  &sv xstart = %xstart% + %xsize%
&end
&return
```

In this example, the values for the x and y sizes are placed in variables (XSTART, YSTART, XSIZE, YSIZE), even though they remain constant in the program. This makes them easy to modify if, for example, the user wants a different starting point or a box of a different size. If the box size changes, the position of the text changes accordingly.

Question 5-3: In the previous code, notice that all the commands that set text characteristics are placed before the loop. What's the advantage to placing them there instead of in the loop?

(Answer on page 5-16)

Locating looping errors

The key to ensuring that AML loops perform as you expect is to keep track of the loop control variable. To see how the loop is progressing, use &TYPE to print the value of this variable at each iteration. This allows you to see how many times a loop repeats and how the variable increments. Be sure to place the &TYPE statement in the loop so that it executes each time with the new value—a good location is right after the &DO statement where it won't get lost in the rest of the code. If the &TYPE comes before the &DO loop, you'll receive an error that the AML variable isn't defined.

After you're satisfied with the performance of the program, delete the &TYPE statements or change them to comments by inserting a comment symbol (/*). Each option offers an advantage. Deleting the &TYPE statements keeps your code uncluttered. Changing the &TYPE statements to comments allows you to reuse them by removing the comment symbols. Handling errors in AML programs is covered in detail in chapter 9.

The following example adds the &TYPE statement to the program PROJWORK.AML from page 5-6.

```
/* projwork2.aml
&do count = 1 &to 9 &by 1
    &type count = %count%
    CREATEWORKSPACE TRACT%count%
    &type Created workspace tract%count%
&end       /*&do count
&return
```

Question 5-4: What happens if &type count = %count% is placed *after* the &END? Why?

(Answer on page 5-16)

Another error detection technique uses the &ECHO directive. When &ECHO is set to &ON or &BRIEF, the evaluation of each line, including all variables and functions, prints on the screen, making it easy to spot problems in the loop. &ECHO only needs to be activated once, so set it before the loop—it stays on until &ECHO &OFF is issued.

Of the three examples that follow, only one yields the correct result. See if you can find the correct example and determine where the errors are in the other two.

The task is to create a new workspace, move to the workspace, and create new coverages from a master tic coverage. A message is printed for the user when the new coverage is created. The workspace name is TRACTS and the coverage names are SOILS1 through SOILS6.

```
/* example 1
&do covercount = 1 &to 6 &by 1
  CREATEWORKSPACE TRACTS
  WORKSPACE TRACTS
  CREATE SOILS%covercount% /GIS/TEMPLATES/MASTERCOV
  &type Created coverage SOILS%covercount% in TRACTS ~
    workspace
&end


&return
/* example 2
CREATEWORKSPACE TRACTS
WORKSPACE TRACTS
&do covercount = 1 &to 6
  CREATE SOILS%covercount% /GIS/TEMPLATES/MASTERCOV
  &type Created coverage SOILS%covercount% in TRACTS ~
    workspace
&end
&return


/* example 3
CREATEWORKSPACE TRACTS
WORKSPACE TRACTS
&do covercount = 1 &to 6
  CREATE SOILS%covercount% /GIS/TEMPLATES/MASTERCOV
&end
&type Created coverage SOILS%covercount% in TRACTS ~
  workspace
&return
```

In the first example, the &DO loop control clause is correct; however, the workspace is created and moved to within the loop. This means that on each iteration a workspace is created and moved to, and six coverages are created. This creates a hierarchy of six TRACT workspaces inside each other, each containing a coverage.

In the third example, the workspace is created and moved to before the loop starts, so the error in example 1 is eliminated. The loop control clause is also correct even though the &by 1 is absent (1 is the default increment). The error here is that the &TYPE directive is placed outside the loop. Instead of six messages showing the created coverages, the user sees only the following message:

```
Created coverage SOILS7 in TRACTS workspace
```

Example two yields the correct result because the workspace is created and moved to outside the loop while the coverages are created and the message is typed to the user from inside the loop.

&DO for each element in a list: The list loop

Lists help us focus on tasks and track their progress to completion. Whether it's a grocery list, a list of errands, or a list of people to invite to a party, you know the task is completed after each item is checked off the list.

This is the concept behind a *list loop*, which reads a list. Controlled by a list of elements, the list loop repeats once for each element. When the list is exhausted, the loop stops.

The format for the list loop is:

```
&do <loop control variable> &list <list of elements>
    <action 1>
    <action n>
&end
```

A list might include a set of user names, parcel numbers, soil codes, item names, symbol numbers, coverages, or files. The following AML program runs from either ARCPLOT or ARCEDIT and lists all feature classes for a given coverage. This allows you to look at the type of attribute files that exists for any coverage.

```
/* directory.aml
&do covname &list zone util parcel firedist
    DIRECTORY FEATURECLASS %covname%
    &type /&/&
```

```
      &pause Press return to see the next coverage
&end
&return
```

Use this technique when you have a large number of coverages on which to perform the same operation. The following example demonstrates this by using the ADDITEM command to add an item called SOIL_CODE to all soils coverages:

```
/* additem.aml
&do cover &list soilsne, soilsnw, soilsse, soilssw
   ADDITEM %cover%.pat %cover%.pat SOIL_CODE 4 4 i
&end
&return
```

Instead of typing in the list for your &DO loop, use one of the listing functions to create a list for you. The following functions return a list of values:

Function name	Created list
[LISTFILE]	Files of a specific type
[LISTITEM]	Items of a coverage or INFO file
[LISTUNIQUE]	Unique values for a particular item in a coverage or INFO file

[LISTFILE] and [LISTUNIQUE] create a comma-delimited list, while [LISTITEM] creates a list separated by spaces. &DO &LIST accepts lists separated by either spaces or commas. Each of these functions may return a list that can be manipulated in an AML program, but they can also generate a text file. Chapter 10 discusses reading and writing text files.

In the next example, [LISTITEM] creates a list containing all the items in a given feature attribute table. The [ITEMINFO] function retrieves information for a specific item. In the example, the item definition is retrieved, but other options could be used to check to see if the item exists and whether the item is redefined or indexed. Notice how the AML reserved variables set by &DESCRIBE help determine the type of attribute table to search for, accomplished by checking for the existence of records in each of the attribute tables. Allow the AML program to determine the feature type of the coverage so the user doesn't have to know it ahead of time.

```
/*  showitems.aml
&sv cover = [getcover *]
&describe %cover%
/*  set feature type by testing # of bytes in attribute tables
&sv feat
&if %dsc$pat_bytes% > 0 &then &sv feat = poly
   &else &if %dsc$aat_bytes% > 0 &then &sv feat = arc
       &else &sv feat = point
&if [null %feat%] &then
   &return %cover% has no ARC, POINT, or POLYGON attributes.
/*  create a list of all items for the chosen coverage
&sv itemlist = [listitem %cover% -%feat% -all]
&type Item definitions for %cover%
/*  Process through the list,finding name and definition.
&do item &list %itemlist%
    &type /&Name:  %item%
    &type Definition: [iteminfo %cover% -%feat% ~
      %item% -definition]
&end
&return
```

Here is the output from SHOWITEMS.AML:

```
Item definitions for /pathname/amlworkbook/soilsne

Name:  TRACT
Definition:  8,18,F,5

Name:  PERIMETER
Definition:  8,18,F,5

Name:  SOILSNE#
Definition:  4,5,B,0

Name:  SOILSNE-ID
Definition:  4,5,B,0

Name:  SOIL_TYPE
Definition:  2,2,I,0

Name:  SOIL_CODE
Definition:  4,4,I,0
```

The [LISTITEM] and [SHOW COLUMNS] functions do the same thing but [SHOW] can also return items in other databases. The [SHOW] function can create a wide variety of lists that you can incorporate into an AML program.

The following table has some examples of the functionality of the [SHOW] function:

[SHOW] Argument	Created list
ARCSTORM CONNECTS	Active ArcStorm connections
ARCSTORM DATABASES	Available ArcStorm databases
CONNECTS	Each database to which a connection has been established
CURSORS	All declared cursors
DATABASES	All databases available for connection
DATASETLAYERS	SDE layers in the currently connected data set
RELATES	Active relates
LAYERCOLUMNS	Column names for the defined SDE layer
LAYERS	Layers from a specified database library
TABLES	Tables and views of a specified database
TILES	Tiles in a map library

If the selected coverage has both a polygon attribute table (PAT) *and* an arc attribute table (AAT), this AML program lists out only the PAT. If needed, a nested &IF structure can check for the presence of an AAT.

Question 5-5: What's the value of the loop control variable in an &DO &LIST loop after the loop finishes?

(Answer on page 5-16)

In summary

Loops effectively handle repetitive tasks in ARC/INFO. The loop control clause, the most important part of the loop, dictates when the loop starts and stops. Constants, variables, and functions can be used in the loop control clause.

This lesson covered two of the four basic formats of the loop control clause: counted and list loops. In a counted loop, a range of values and an increment control the number of times the loop repeats. A list loop repeats once for each element in the list. When the list is exhausted, the loop is finished. By adding these looping structures to your AML programs, you can efficiently handle repetitive tasks.

Answer 5-1: The loop executes twenty-six times. On the twenty-seventh time, COUNT = 26 and the loop stops.

Answer 5-2: The concentration levels are given at each iteration as follows:
(1) .1, (2) .2, (3) .3, (4) .4, (5) .5, (6) .6, (7) .7, (8) .8, (9) .9
The iterations are in parentheses.

Answer 5-3: The text characteristics remain constant and don't need to be reset each time the loop executes. If these lines are in the loop, the result isn't different, but the code runs slower because these commands execute at each iteration.

Answer 5-4: COUNT = 10 prints. When &DO COUNT = 9, the loop continues one more time and COUNT increments to 10. When COUNT is tested again, the value is greater than 9, so the loop stops. Control passes to the &END, and the statement is typed.

Answer 5-5: After the loop finishes, the value of the loop control variable is equal to the value of the last element in the list.

Exercises

Exercise 5.1.1

Suppose that after you complete a project, your workspace contains thirteen coverages you want to delete. The naming scheme for the coverages is SOILS1 through SOILS13.

The AML code shown below lists the coverages and prompts the user to indicate the base coverage name (SOILS) and the highest number (13), then copies SOILS13 to SOILSFINAL.

Fill in the blank lines with an &DO loop to delete the original thirteen coverages. (The coverages are listed again to verify that they were deleted.)

```
LISTCOVERAGES
&sv covername = [response 'Enter base coverage name']
&sv covernum = [response 'Enter the highest coverage number']
copy %covername%%covernum% %covername%final

_____

_____


_____
LISTCOVERAGES
&return
```

Exercise 5.1.2

Assume that MAXDEPTH is the maximum depth of all wells in a coverage and is stored in the database as a positive number (e.g., 120 for a 120-foot-deep well). The wells need to be categorized at 20-foot intervals. The line of code shown below doesn't set up the loop correctly. What's wrong with it, and how could you correct it?

```
&do welldepth = %maxdepth% &to 0 &by 20
```

Exercise 5.1.3

Given a coverage of the United States named STATES, this partial AML program creates a display of the coverage in a given projection and clears the screen before the next display. Fill in the blanks with an &DO list loop that uses the projection files MERCATOR.PRJ, LAMBERT.PRJ, UTM.PRJ, and PERSPECTIVE.PRJ.

```
          MAPPROJECTION AUTOMATIC %projfile%
          MAPEXTENT STATES
          POLYGONS STATES
          &pause Press return to continue
          CLEAR

        &return
```

Lesson 5.2—&DO loops with logical expressions

In this lesson

Often you can't determine beforehand how many times a loop needs to repeat. You may need a loop that executes while some conditions are true (e.g., while the user enters coverage names), or until a condition becomes true (e.g., until the user presses the 9-button). How a condition in a logical expression evaluates (i.e., to either .TRUE. or .FALSE.) determines the number of times these kinds of loops repeat. In this lesson, you'll use logical expressions to construct loop control clauses.

&DO &WHILE a condition is true: The while loop

In chapter 4, you saw that when a logical expression with an &IF statement is true, an action performs once. Similarly, a loop can repeat the action as long as an expression continues to be true.

In the &DO &WHILE loop, the loop control clause is a logical expression. The loop repeats as long as the expression continues to evaluate to .TRUE.. When the expression evaluates to .FALSE., the loop ends.

The basic structure of the &DO &WHILE loop is shown below:

```
&do &while <logical expression>   /*while expression is true
    <action 1>
    <action n>   /* allows expression to change
&end   /* go to top of the loop and test expression again
```

To begin the loop, the expression is evaluated; if it's true, the actions execute. After all actions in the loop execute, &END is encountered. Control then passes back to the &DO &WHILE statement and the logical expression is evaluated again. If the expression is still true, the loop continues; if not, the loop stops. When the loop stops, control passes to the statement following the &END.

The logical expression in an &DO &WHILE loop is always evaluated at the beginning of the loop, before any action is taken. Because the evaluation is done first, an &DO &WHILE loop might never execute (i.e, when the expression is .FALSE. the first time). Initialization should occur only once, so it should take place outside the loop.

Note that one of the actions within the while loop block must allow the result of the logical expression to change from true to false. If you forget, the loop executes indefinitely because the expression always remains true—this is called an *infinite loop*. An infinite loop hangs up AML processes, using computer resources until it's manually stopped, which usually requires using the workstation interrupt command (e.g., <ctrl>c). To avoid infinite loops, make sure the value of the expression is allowed to change at each iteration. Look at a simple numeric example that illustrates this point:

```
[1]   &sv number = 1
[2]   &do &while %number% < 10
[3]     &type Your number is %number%
[4]     &sv number = [response 'Enter a number less than 10' 1]
[5]   &end
[6]   &return
```

The value of NUMBER is initialized outside the loop [1]. This ensures a true value and avoids an AML error indicating an undefined variable when the logical expression is evaluated [2]. The value of NUMBER can change the loop based on the response from the user [4]. The end of the loop is encountered [5], and control is passed back to the &DO &WHILE [2]. The expression is evaluated again with the new value of NUMBER.

Remember, if NUMBER never changes, staying 1 at every iteration of the loop, the expression is always .TRUE., and an infinite loop results.

The next example, DRAWIT.AML, runs from ARCPLOT. It draws a polygon coverage chosen by the user, then lists its attribute table. The program repeats the process as long as the user wants to continue. The loop control clause contains a logical variable named QDRAW, set with the [QUERY] function. [QUERY] asks the user to indicate whether to continue or quit drawing coverages. If the user presses a carriage return, the default value of .TRUE. is entered as the response.

```
/* drawit.aml
/* Draw polygon coverages and list polygon attribute
/* tables as long as the user wants.
&sv qdraw = .true.
&do &while %qdraw%
  &sv cover = [getcover * -poly]
  CLEAR
  MAPEXTENT %cover%
  POLYGONS %cover%
  LIST %cover%.pat info
  /* asks if the user wants to keep drawing coverages
  &sv qdraw = [query 'Want to draw another (Yes | No)' .true.]
&end   /* &while %qdraw%
&return
```

When testing a logical variable, you might think the looping clause should be as follows:

```
&do &while %qdraw% = .true.
```

However, the variable alone yields the same value as this expression. If QDRAW is true, then QDRAW = .TRUE. is also true, and if QDRAW is false, then QDRAW = .TRUE. is also false. The longer format won't cause the AML program to fail, but it runs slower because of the additional, unnecessary code.

DRAWIT.AML has limited utility because it's designed to work only on polygon coverages. To enhance it so that it works with any coverage, it needs a test for the feature type and a modification to accept any feature type. (In lesson 5.1, you saw a program called SHOWITEMS.AML, which tests for the feature type using DSC$ variables.)

Next, DRAWIT.AML is enhanced using the [EXIST] function to set a variable that indicates the feature type.

```
[1]     /*   drawit2.aml enhanced with feature type checking
[2]     &sv qdraw = .true.
[3]     &do &while %qdraw%
[4]        &sv cov = [getcover * ]
[5]        &if [EXISTS %cov% -poly] &then &sv feat = polygons
[6]           &else &if [EXISTS %cov% -arc] &then &sv feat = arcs
[7]              &else   &sv feat = points
[8]        &if ^ [null %feat%] &then &do
[9]           /*  Draw coverage and list feature attribute table
[10]          CLEAR
[11]          MAPEXTENT %cov%
[12]          %feat% %cov%
[13]          LIST %cov% %feat%
[14]       &end
[15]       &else &type %cover% has no ARCs, POINTS, or POLYGONS.
[16]       /* Update %qdraw% for next iteration of loop
[17]       &sv qdraw = [query 'Want to draw another (Yes | No)' ~
              .true.]
[18]    &end
[19]    &return
```

Changes to DRAWIT.AML include removing the -POLY option from the [GETCOVER] function [4], adding the nested &IF to determine the feature type of the coverage [5-7], substituting FEAT [12] instead of using the POLYS command to draw the features, and modifying the LIST command to use syntax that accepts the type of feature instead of the type of attribute table [13].

In ARC/INFO, you often work with a selected set of features. Loops can step you through each feature in the set.

The following AML program, SHOWSEL.AML, runs from ARCPLOT. It prompts the user to select a group of features from a coverage and lists specific information about them. The loop continues as long as there are selected records, stopping when all are processed. The [GETCOVER] function creates a menu of only those point coverages whose names begin with CHEM. The chosen coverage draws and the user selects a group of features by enclosing them in a box on the screen. The selected features are highlighted with different symbols. While the features are still selected, each feature's User-ID and the value of an item named CONCENTRATION print on the screen.

```
/* showsel.aml
/* Select coverage and set up display parameters
&sv cov = [getcover chem* -point 'Pick a coverage']
MAPEXTENT %cov%
POINTS %cov% IDS
/* Invokes a feature selection
&type Locate two opposite corners of a selection box.
LINECOLOR YELLOW
RESELECT %cov% POINT BOX *
/* Draw selected features
MARKERSET OILGAS.MRK
POINTMARKERS %cov% 202
/*  Initialize loop variable
&sv num = 1
/*  Loop compares %num% against the number of selected
/*  features
&do &while %num% le [extract 1 [show select %cov% point]]
   &type /&Feature number is:
   /* Display User-id for each feature
   &type [show select %cov% point %num% item ~
   [entryname %cov%]-id]
   &type Concentration is:
   /* Display concentration for each feature
   &type [show select %cov% point %num% item concentration]
   /*  Increment %num%
   &sv num = %num% + 1
&end
&return
```

By nesting the [SHOW SELECT] function in the [EXTRACT] function, the number of selected features is returned and used as a component in the loop control clause. The [SHOW SELECT] function returns the number of selected features and the total number of features. [EXTRACT] retrieves the first element in the list. For example, if [SHOW SELECT] returns 45,150, then [EXTRACT 1 45,150] yields 45, which represents the number of selected features. This loop continues as long as the variable NUM is less than or equal to the number of selected features. In other words, it performs the task for each selected feature and then stops. The [EXTRACT] function is discussed in detail in chapter 7.

FYI

You can use the [SHOW] function to retrieve many of the parameters set in ARC/INFO. You should become familiar with its many options. The [SHOW SELECT] function can retrieve attribute information by specifying an item name. The [ENTRYNAME] function used in the previous example returns only the name of the coverage directory from the full pathname of COV to correctly format the name of the User-ID. For example, you need CHEM2-ID, not `/gis/project/chemsites/chem2-id`

&DO &UNTIL a condition is true: The until loop

Like the while loop, the execution of the &DO &UNTIL loop depends on the evaluation of a logical expression. An &DO &UNTIL loop, however, repeats until the expression is true. When the expression changes to true, the loop ends.

There's another important difference between the &DO &UNTIL and the &DO &WHILE loop. The actions within the &DO &UNTIL loop always execute at least one time because the logical expression is evaluated at the end of the loop. Remember that the logical expression in an &DO &WHILE loop is evaluated as soon as the &DO &WHILE statement is encountered; therefore, the loop never executes if the expression is false.

The &DO &UNTIL loop has two advantages. Variables don't need to be initialized prior to encountering the loop, and, because the expression isn't evaluated until the end of the loop, actions within the loop can influence the evaluation of the expression.

ARC/INFO users often want several iterations of a command to be allowed without having to repeat it. As shown in the next example, you can allow users to continue executing a command by using the [QUERY] function to prompt them and an &DO &UNTIL loop to handle their response. The program uses the MEASURE command in ARCPLOT in conjunction with an &DO &UNTIL loop to allow users to continue measuring one object after another until they indicate they want to stop:

```
/*   ruler.aml
&do &until %qdone%
  &sv measurewhat = [getchoice where length tract]
  MEASURE %measurewhat%
```

```
&sv qdone = [query 'Are you finished (NO)' .false.]
&end
&return
```

This loop continues until the expression evaluates to .TRUE., so the question in the [QUERY] function is structured to invoke a .FALSE. response when the user wants to continue. When the question yields a true response, the loop stops. A common mistake is to structure the question incorrectly. Consider the following query:

```
&sv qdone = [query 'Do you want to measure another' .true.]
```

If the user responds YES, the loop stops because QDONE is true. If the user answers NO, the loop continues. &DO WHILE or the NOT (^) operator is needed if the question is structured in this manner because a while loop continues as long as the response is YES (i.e., .TRUE.). If you want to use an &DO &UNTIL loop that continues on a YES response, use the NOT operator to reverse the Boolean value (.TRUE. becomes .FALSE. and .FALSE. becomes .TRUE.).

```
&sv qdone =  NOT [query 'Do you want to measure another'~
    .true.]
```

In this case, the &DO &UNTIL loop continues until the user types NO.

In the [QUERY] functions on the previous page, notice the use of an appropriate default answer allowing the user to continue by pressing the carriage return. Only when the user is ready to quit must a response be explicitly typed.

You might prefer to change [QUERY] to the [GETCHOICE] function to make the choice menu-driven. Keep in mind that displaying YES and NO as choices on the menu is preferable to displaying .TRUE. and .FALSE.. When using YES and NO, the CHOICE variable must be assigned the appropriate value of .TRUE. or .FALSE.. The following code illustrates using [GETCHOICE] in conjunction with an &IF &THEN &ELSE statement to translate YES and NO to .TRUE. and .FALSE.:

```
/*  ruler2.aml
&do &until %qdone%
  &sv measurewhat = [getchoice where length area -prompt ~
    Measure:]
  MEASURE %measurewhat%
  &sv choice = [getchoice YES NO -prompt 'Are you finished?']
```

```
/*   assign the yes/no value to a true/false value
&if %choice% = YES &then &sv qdone = .true.
&else &sv qdone = .false.
&end
&return
```

 FYI

In RULER2.AML, [GETCHOICE] is used to get a YES or NO response and then that choice is tested to see whether the user is ready to stop or not. An alternative way of structuring this is to incorporate the -PAIRS and -VAR options into [GETCHOICE]. The -PAIRS option means that the choices come in pairs, one that's visible to the user and another that's set to a variable. The last [GETCHOICE] could be written like this:

```
&sv choice = [getchoice -pairs  YES .true. NO .false. ~
-var qdone -prompt 'Are you finished?']
```

-PAIRS indicates that the choices are going to be in pairs and -VAR identifies the variable that stores the second value of the pair. In this example, if the user picks YES then %choice% = YES and %qdone% = .true. This code eliminates the need for &IF &THEN &ELSE to test %choice%. In the database, RULER3.AML is an example of how this would work. For more information on [GETCHOICE], look in the online help command references.

When the user gives an invalid response to a prompt, a well-designed program types a gracious error message and continues to ask for input until the user enters the correct response. This sounds like a job for the &DO &UNTIL loop.

In the following example, COVER is tested in a loop expression to verify that it's set. As long as COVER is null, the loop executes and returns an error message. The loop continues until the COVER is not null (i.e, it has a value).

If the coverage isn't provided,

```
[null %cover%]
```

is true,

```
NOT [null %cover%]
```

is false, and the loop continues to ask the user for a coverage.

When the coverage is provided,

```
NOT [null %cover%]
```

is true, the loop is completed, and control goes to the next task.

```
/* do until the coverage is provided (i.e., not null)
&do &until NOT [null %cover%]
   &sv cover = [getcover * -all 'Select a coverage to ~
     continue' -none]
&end
<next task>
```

In this example, &DO &UNTIL is preferred to using &DO &WHILE because the expression isn't evaluated the first time and the user is able to indicate a new coverage inside the loop. If an &DO &WHILE loop is used, the expression is evaluated the first time through the loop so variables must be initialized outside the loop to avoid undefined variables. Undefined variables result in an AML error message and cause a program to fail

In summary

&DO loops use logical expressions to respond to changing conditions. Unlike counted and list loops, which specifically define the number of iterations, &DO &WHILE and &DO &UNTIL loops both use logical expressions to determine when the loop should start and stop. To avoid infinite loops, this logical expression must be changeable inside the loop at each iteration.

Deciding whether to use &DO &WHILE or &DO &UNTIL depends on the application and your personal preference. The type of loop you choose determines when variables are defined and how questions that control the loop are presented (i.e., whether a true or false response is required for the loop to continue). The key points to remember are that the &DO &UNTIL loop always executes at least once, and tests the expression at the end of the loop. This allows variables affecting the loop to be modified within the loop. The until loop looks for a true value to stop the loop. On the other hand, the &DO &WHILE loop may not execute at all because the expression is tested at the beginning of the loop. All variables affecting the &DO &WHILE loop must be initialized before the loop executes. The &DO &WHILE loop looks for a false value to stop.

Exercises

Exercise 5.2.1

Rewrite the following code to use an &DO &WHILE loop:

```
&do &until not [null %cover%]
   &sv cover = [getcover * -all 'Select a coverage to continue' -no
&end
```

Hint: If you don't define COVER before the loop is encountered, you'll receive an AML error.

Exercise 5.2.2

The task is to delete graphics files that the user no longer needs from a workspace and continue deleting graphics files as long as the user desires. The [GETFILE] function creates a menu of all the graphics files. The [DELETE] function deletes the chosen one and returns a zero (0) if successful. Write an &DO &UNTIL statement and supply a question so the user can indicate when to stop. Be sure to add an appropriate default answer in case the user responds by pressing the carriage return.

```
&sv grafile = [getfile *.gra -file]
&if not [null %grafile%] &then &do
  &if [delete %grafile%] = 0 &then &type %grafile% deleted
  &else &type %grafile% not deleted successfully
&end
```

```
&return
```

Hint: See chapter 2 for a discussion on the [GETFILE] function and chapter 10 for more about the [DELETE] function.

Exercise 5.2.3

As in the task outlined in exercise 5.2.2, an AML program needs to delete unwanted INFO files whose names begin with TEMP. Write an &DO &WHILE loop to perform this task. Again, include an appropriate default query response.

```
&sv infofile = [getfile temp* -info]
&if [delete %infofile% -info] = 0 &then &type %infofile% deleted
  &else &type %infofile% not deleted successfully

&return
```

Lesson 5.3—Compound clauses, nested loops, and indexed variables

In this lesson

The previous lessons in this chapter described the four basic types of looping structures. This lesson covers advanced techniques for constructing loops with compound loop clauses and for nesting one loop in another. A compound loop clause allows one of two conditions to control the loop. Nested loops are an efficient way to set and retrieve indexed variables. Indexing variables organizes them in groups so they can be referenced with a common name and unique index number.

Getting the best of two loops

You're probably familiar with automobile companies that warranty a car for 60,000 miles or five years, whichever comes first. This is an example of a compound clause. The car is under warranty until one of the two conditions specified in the compound clause occurs (i.e., 60,000 miles *or* five years). Compound loop control clauses used in AML programs handle similar situations. Consider an AML program that browses through a set of coverages, allowing users to stop whenever they want. In this case, the loop should execute until there are no more coverages or until the user chooses to stop, whichever comes first. Only one of the two conditions must be met for the loop to stop.

You can use compound loop control clauses with both counted and list loops by appending an &WHILE or &UNTIL clause to the &DO &TO &BY and &DO &LIST clauses. A compound loop control clause is like a compound logical expression where two expressions are combined with the AND operator. You can combine two control clauses in a single &DO loop. (If you're unfamiliar with compound expressions and logical operators, refer to chapter 4.)

The loop starts executing when both clauses are true, and stops executing when either clause is false. A list loop stops when either the list of elements is exhausted or the &WHILE expression becomes false. In the example below, the list loop continues until there are no more coverages to browse or the user asks to quit.

```
/*  browsecov.aml
&sv qcont = .true.
&do cov &list soils streets zoning parcels &while %qcont%
  CLEAR
  MAPEXTENT %cov%
  /*  test feature type
  &if [EXISTS %cov% -poly] &then &sv feat = polygons
    &else &if [EXISTS %cov% -arc] &then &sv feat = arcs
        &else &if [EXISTS %cov% -point] &then &sv feat = points
  /*  draw and list feature attribute table
  %feat% %cov%
  LIST %cov% %feat%
  &sv qcont = [query 'Do you wish to continue (Yes | No)' ~
    .truc.]
&end
&return
```

Question 5-6: How will BROWSECOV.AML react if &WHILE is changed to &UNTIL and the user is asked whether or not to continue?

(Answer on page 5-39)

You can also use compound clauses with &DO &WHILE or &DO &UNTIL loops. Earlier, you looked at an &DO &UNTIL loop that returns an error message until the user enters a coverage name. Recall that there was no way to avoid entering a coverage name and letting the loop execute once. Now consider changing the task so the user must pick a coverage with polygon topology. When both conditions are satisfied, the loop executes. See how the code from page 5-27 is revised to include a counter in a compound loop clause:

```
/*  checkcov2.aml
&do &until not [null %cover%] and %dsc$qtopology%
   &sv cover = [getcover * -poly 'Select a coverage to~
     continue' -none]
   &if not [null %cover%] &then &describe %cover%
&end
&return
```

Using nested loops

Once a task is programmed using an &DO loop, you can nest it in another &DO loop. Loops don't have to be of the same type to be nested.

Suppose you write an AML program that creates three coverages. Later, you need to create four project workspaces and place three coverages in each one. You can place the loop that creates the coverages in the loop that creates workspaces (i.e., a list loop inside a counted loop). The code for this task follows:

```
/* creatework.aml
&do wkspace = 1 &to 4
  CREATEWORKSPACE TRACT%wkspace%
  WORKSPACE TRACT%wkspace%
    &do cov &list land road veg
      &messages &off /* no listing as coverages are created
      CREATE %cov%%wkspace%
    &end  /* &do cov
  &type /&
  &messages &on  /* turn on so listcoverages is displayed
  LISTCOVERAGES
  WORKSPACE ..  /* move back to original workspace
&end  /* &do wkspace
&return
```

Notice the use of the &MESSAGES directive to control the printing of messages from ARC/INFO. By turning off system messages and creating your own messages, you control what the user sees. The program creates the desired workspaces and coverages, then returns the following information to the user:

```
Workspace:   venice1/melissa/tract1

Available coverages:
LAND1             ROAD1            VEG1

Workspace:   venice1/melissa/tract2

Available coverages:
LAND2             ROAD2            VEG2

Workspace:   venice1/melissa/tract3

Available coverages:
LAND3             ROAD3            VEG3
```

```
Workspace:    venice1/melissa/tract4

Available coverages:
LAND4           ROAD4           VEG4
```

When there are two loops nested, one is referred to as the *outside loop* and the other as the *inside loop*. The tricky part of creating nested loops is determining which should be placed inside the other.

The inside loop executes the greatest number of times. As an analogy, think of time: seconds nest within minutes. Minutes are the outer loop and seconds are the inner loop. In CREATEWORK.AML, three coverages are created every time one workspace is created, so the loop creating the coverages is placed inside the loop creating the workspaces. The following nested loop demonstrates this important point:

```
&type OUT   IN   /* column headings
&do out = 1 &to 5
    &do in - 1 &to 3
        &type %out%     %in%
    &end  /* &do in
&end   /* &do out
```

The result is as follows:

```
OUT       IN
 1         1
 1         2
 1         3
 2         1
 2         2
 2         3
 3         1
 3         2
 3         3
 4         1
 4         2
 4         3
 5         1
 5         2
 5         3
```

The inside loop repeats three times for each iteration of the outside loop. The outside loop repeats five times; therefore, the nested loop repeats a total of fifteen times. Because of the numerous iterations possible for the inside loop, keep lines of code that don't need to be repeated outside the loop. You can also nest any of the other loop types (e.g., &DO &WHILE, &DO &UNTIL, etc.). In these cases, the number of times the loop executes is based on a logical condition, not a counter.

You can nest more than two loops, and the same relationship between inside and outside loops holds true. The innermost loop always executes the greatest number of times, and the outermost, the fewest.

Question 5-7: Describe what the following code creates:

```
&do user &list tina melissa
     CREATEWORKSPACE %user%
     WORKSPACE %user%
        &do count = 1 &to 3
             CREATE LAND%count%
        &end
     WORKSPACE ..
&end
```

(Answer on page 5-39)

Creating indexed variables with nested loops

You can use loops to construct *indexed variables*. Indexed variables are grouped because they contain similar information, such as a group of coverages, images, or owner names. Give indexed variables a common name and unique reference number, or index number. For example, you might have three owner names and set the following variables:

```
&sv own1 = BOYD
&sv own2 = BRENNEMAN
&sv own3 = RIDLAND
```

Indexed variables simulate an *array* structure in AML. An array organizes similar data in groups while providing a means of accessing each element individually by its position in the group. Consider a post office with hundreds of mailboxes. The group of mailboxes is an array, and each mailbox is accessed by a specific address, name, or number.

In this way, you can group similar variables, assign them a common name, and append a unique index number to each variable in the group to differentiate it. The common name indicates the group to which the variable belongs, and the index number indicates the individual variable (e.g., COV5, COV6, IMAGE10, IMAGE24).

Counted loops can easily manage indexed variables; however, because you can't nest variables, you need the [VALUE] function to access and print them.

You might create a simple loop like this to print the owners' names:

```
&sv own1 = BOYD
&sv own2 = BRENNEMAN
&sv own3 = RIDLAND
&do number = 1 &to 3
   &type own%number%
&end
```

You'd expect the following output:

```
BOYD
BRENNEMAN
RIDLAND
```

but to your surprise, the list reads:

```
own1
own2
own3
```

You really want the *value* of OWN1, OWN2, and OWN3. To solve this problem, you might consider this ill-fated solution:

```
&type %own%number%%
```

Unfortunately, this format of a variable name is invalid. You must use the [VALUE] function to print the value of indexed variables. The [VALUE] function is the same as referencing a variable with percent signs. For example, the following lines of code produce the same result:

```
Arc: &type %cover%
WELL
Arc: &type [value cover]
WELL
```

In the example, the index number (NUMBER) needs to be evaluated before the index variable (OWN1, OWN2, etc.). INDEX.AML uses AML's order of interpretation to solve the problem; the index number is evaluated before the [VALUE] function.

```
/* indexvar.aml
&sv own1 = BOYD
&sv own2 = BRENNEMAN
&sv own3 = RIDLAND
&do number = 1 &to 3
  &type Value of own%number% = [value own%number%]
&end
&return
```

Notice the difference in the display between using OWN%NUMBER% and [VALUE OWN%NUMBER%].

```
Value of own1 = BOYD
Value of own2 = BRENNEMAN
Value of own3 = RIDLAND
```

Now look at a more complex example that uses indexed variables. Suppose you need to locate the node where two streams intersect. (This example could use streets, utility lines, or any linear data set.) The task involves setting up two simulated arrays of node numbers and comparing them. Each node is accessed using an indexed variable (i.e., NODE%INDEX%).

Don't let the length of the following code scare you away. The main components of this AML program are [SHOW SELECT], [EXTRACT], counted loops, and nested loops, which were all covered earlier in this chapter. This program, executed from ARCPLOT, is a good example of putting your knowledge together to solve a real geographic problem.

```
[1]   /* streamintersect.aml
[2]   CLEAR
[3]   ASELECT STREAM ARC
[4]   MAPEXTENT STREAM
[5]   &sv str1 = [getunique stream -arc strm_name 'Pick first stream']
[6]   &sv str2 = [getunique stream -arc strm_name 'Pick second
[7]   stream']
[8]   RESELECT STREAM LINE STRM_NAME = %str1%
[9]   &type Drawing %str1% in green...
[10]  ARCLINES STREAM 3
[11]  &sv numstr1 = [extract 1 [show select stream line] ]
[12]  /*  create first array
[13]  &do index = 1 &to %numstr1%
[14]    &sv str1node%index% = [show select stream line %index% item
[15]  tnode#]
[16]  &end
[17]  ASELECT STREAM ARC
[18]  RESELECT STREAM LINE STRM_NAME = %str2%
[19]  &type drawing %str2% in red...
[20]  ARCLINES STREAM 2
[21]  &sv numstr2 = [extract 1 [show select stream line] ]
[22]  /* create second array
[23]  &do index = 1 &to %numstr2%
[24]    &sv str2node%index% = [show select stream line %index% ~
[25]      item tnode#]
[26]  &end
[27]  /* search for matching nodes in arrays str1node and str2node
[28]  &sv intersection = 0
[29]  &sv qdone = .false.
[30]  &type /&Searching for intersection...
[31]  &do index1 = 1 &to %numstr1% &until %qdone%
[32]    &do index2 = 1 &to %numstr2% &until %qdone%
[33]      &if [value str1node%index1%] = [value str2node%index2%] ~
[34]        &then &do
[35]          &sv intersection = [value str2node%index2%]
[36]          &sv qdone = .true.
[37]      &end   /* if matched
[38]    &end     /* &do index2
[39]  &end       /* &do index1
[40]  &if %intersection% = 0 &then
[41]    &type /&No intersection between %str1% and %str2% was found.
      &else &type /&Intersection located at node %intersection%
      &return
```

Here's a walk-through of the previous code:

The selected set is cleared in line [3] to ensure that the user can select from all streams in the coverage.

In lines [5] and [6], the user is asked to pick two stream names from menus of unique stream names.

In line [7], all streams with the first name are selected.

Line [9] draws the selected stream on the screen.

Line [10] sets a variable containing the number of selected features returned from the [EXTRACT 1 [SHOW SELECT]]. (See chapter 1 for a more detailed discussion of nested functions.)

Lines [12], [13], and [14] create the first array. [SHOW SELECT] retrieves the TNODE# from each record and places it in the indexed variable STR1NODE%INDEX%.

Lines [15] through [24] repeat the procedure for the second stream chosen.

Variables are initialized in lines [26] and [27]. Lines [29] and [30] establish the counted loops to process each node in each simulated array.

In line [31], the two nodes are compared. If there's a match, the value is placed in the INTERSECTION variable [33], and QDONE is set to .TRUE. [34].

Once the matching node is found, both loops stop processing. This is accomplished by adding the &UNTIL QDONE to &DO. When QDONE is set to .TRUE., it signals the inside loop to stop executing and passes control to the outside loop, which also stops because QDONE is true.

If no match is found, INTERSECTION is still set to 0 [38], and the user is informed that there's no intersection [39]. Otherwise, the node where the two streams intersect is displayed [40] in the text window.

Question 5-8: In the last example, when matching nodes are *not* found, what causes the outside loop (INDEX1) to stop executing?

(Answer on page 5-39)

In summary

Compound clauses control the number of loop iterations and allow control to be based on information that can't be determined before the program executes.

Indexed variables simulate array processing in AML programming. Indexed variables help organize, store, and retrieve large groups of similar data by giving each variable a common name and distinguishing index or reference number. Counted loops efficiently handle indexed variables.

Answer 5-6: &UNTIL means *until the expression is true*, and QDONE is .TRUE.; therefore, the loop stops, even though the user asks to continue.

Answer 5-7: Two workspaces, one for TINA and one for MELISSA. Each contains three coverages: LAND1, LAND2, and LAND3.

Answer 5-8: The index variable (INDEX1) exceeds the value of NUMSTR1.

Exercises

Exercise 5.3.1

List the coverages that are drawn by the following AML program:

```
/* browsecov.aml
&sv feat
&sv qdone = .true.
&do tract = 1 &to 3
  &do cov &list soils streets zoning &while %qdone%
    /* test feature type
    &if [EXISTS %cov%%tract% -poly] &then
      &sv feat = polygons
    &else &do
      &if [EXISTS %cov%%tract% -arc] &then
        &sv feat = arcs
      &else &do
        &if [EXISTS %cov%%tract% -point] &then
          &sv feat = points
      &end
    &end
    /* draw and list feature attribute table
    &if not [null %feat%] &then &do
      CLEAR
      MAPEXTENT %cov%%tract%
      %feat% %cov%%tract%
      &s feat
    &end
    &sv qdone = [query 'Do you wish to continue {Yes|No}' .true.]
  &end    /* do cov
&end      /* do num
&return
```

Exercise 5.3.2

What's the advantage of appending a counted loop (&DO &TO &BY) with an &UNTIL QDONE?

Exercise 5.3.3

Modify INDEXVAR.AML (page 5-36) to set and print the variables in the loop. Change the counted loop to an &DO &LIST loop.

Exercise 5.3.4

The following triple-nested loop runs from ARCPLOT and creates several graphics files. List the workspace names and the coverage names. Also determine how many graphics files are created and how many of each size.

```
/* plot2sizes.aml
&messages &off
&type Creating graphics files...
shadeset colornames.shd
&do wksp = 1 &to 4
  &workspace tract%wksp%
  &do size = 1 &to 2
    &do cov &list zoning soils parcels
      display 1040
      %cov%%size%
      &select %size%
        &when 1
          PAGESIZE 8.5 11
        &when 2
          PAGESIZE 14 26
        &end
      MAPEXTENT %cov%
      POLYGONSHADES %cov% code %cov%.lut
      POLYGONS %cov%
    &end      /*  do cov
  &end        /* do size
&workspace ..
&end          /* do wksp
&messages &on
&return
```

Exercise 5.3.5

STREAMINTERSECT.AML (page 5-37) finds the intersection of two streams. Most of the procedure for the first stream repeats for the second stream. Insert the procedures in a counted loop.

Getting and using coordinates

Locational points called *coordinates* define ARC/INFO coverage features.
Coordinates also reference locations on a graphics window in a graphical
program. You can obtain coordinate data interactively using a mouse, screen
cursor, digitizer puck, or keyboard. Applications that run in a graphical
environment can use AML to store these coordinates and pass them to
ARC/INFO commands as needed.

Graphical operations rely on coordinate data. In ARCEDIT, for example, when
you select a coverage feature with the screen cursor, ARCEDIT identifies the
selection by matching its coordinates with the coordinates of features stored in an
ARC file. In ARCPLOT, the coordinates indicated with a two-point box can be
used to determine where in the graphics window to place a map legend.

This chapter covers the following topics:

Lesson 6.1—Getting coordinates and storing them in AML variables

- Introducing coordinates
- Capturing coordinate data in AML variables

Lesson 6.2—Capturing coordinates in the point buffer and watch files

- Loading and retrieving data from the point buffer
- Capturing coordinate data in an AML watch file

Lesson 6.3—Using mathematical and trigonometric functions

- Using mathematical functions with coordinate data
- Using trigonometric functions with coordinate data

Lesson 6.1—Getting coordinates and storing them in AML variables

In this lesson

This lesson defines coordinate data and describes how to capture it, store it in AML variables, and then use it in commands that require coordinate input.

Introducing coordinates

A coordinate is an x,y location in a two-dimensional coordinate system or an x,y,z location in a three-dimensional coordinate system. Cartesian coordinates describe an x,y location in terms of distance from a fixed reference. The fixed reference of a coordinate system, or *origin,* is the intersection of the x- and y-axes which is given an x,y value of 0,0. Three-dimensional x,y,z coordinates additionally represent elevation, specified by the z-value. The z-axis intersects the x- and y-axes at a right angle.

In ARC/INFO, coordinates represent the location of point features and define the location and shape of line and polygon features in a coverage. Most commonly, ARC/INFO coverages are stored in a Cartesian coordinate system using a *map projection,* which converts the spherical coordinates of the globe to two-dimensional coordinates. The x,y coordinates of a point in a coverage represent its location from the origin of the coordinate system. The x,y coordinate values usually represent units of measurement in feet or meters and are referred to as *map units.*

ARC/INFO software's graphics programs, such as ARCEDIT and ARCPLOT, use a two-dimensional coordinate system to reference locations in the graphics window, also referred to as the *graphics page.* The graphics page is the area on the screen of the display device where graphics can be drawn. The origin (i.e., an x,y value of 0,0) is the lower-left corner of the graphics page. Map elements can be selected from or placed on the graphics page by specifying the x,y distance relative to the lower-left corner. These coordinates are expressed in *page units,* which, by default, are inches.

Positioning a coverage on the graphics page is known as map-to-page transformation because it controls how coverage coordinates (in real-world units of feet or meters) are transformed to graphics page coordinates (in inches or centimeters) on your screen or other output device.

 FYI

To learn more about map coordinate systems, see *ARC/INFO Concepts >> Map Projections* in the online help. Also refer to *Cartography >> Map display and query using ARCPLOT* for more information about how to alter your displays using map and page parameters.

Capturing coordinate data in AML variables

Graphics programs such as ARCEDIT and ARCPLOT have many commands that require coordinate input. You can provide coordinate input to a command using a cursor device (mouse or digitizer) to specify a location on the graphics page, or you can enter coordinates from the keyboard. The device for coordinate input is specified with the COORDINATE command. The asterisk (*) argument in ARC/INFO commands indicates that the current input device will be used for coordinate input.

Coordinate input provided to commands isn't saved by the graphics program. If subsequent commands need coordinates, then you need a method of retrieving them. The example on the next page illustrates how the screen cursor can mark the position for a map legend.

```
Arcplot: KEYPOSITION *
Enter point
```

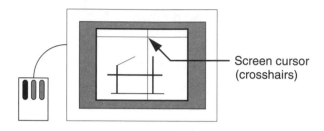

Screen cursor
(crosshairs)

```
Arcplot: KEYLINE STREET.KEY
```

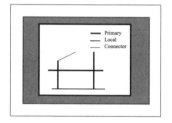

KEYPOSITION * uses the screen cursor to mark the location of the upper-left corner for a map legend that is then drawn with the KEYLINE command. This location, however, is only temporarily saved until the KEYLINE command is issued. Storing the location for later use allows you to add other graphic elements at locations relative to the legend.

AML provides an efficient method to store coordinates, like those generated with screen selections, so they can be passed to other commands. Three of AML's reserved variables store coordinate data. Their value corresponds to a cursor selection in a graphics window as follows:

Reserved variable	Value indicates
PNT$X	The x-location of the cursor
PNT$Y	The y-location of the cursor
PNT$KEY	The key pressed to enter the coordinates

Two AML directives set the PNT$ reserved coordinate variables: &GETPOINT and &GETLASTPOINT.

AML directive	Description
&GETPOINT	Gets a point from the user and stores the data in PNT$ variables
&GETLASTPOINT	Sets the PNT$ variables to values corresponding to the last use of the coordinate input device

The following example illustrates how &GETPOINT works:

```
Arcplot: MAPEXTENT STREETS
Arcplot: ARCS STREETS
Arcplot: &getpoint
Arcplot: &type %pnt$key% %pnt$x% %pnt$y%
1 2.263 5.225
```

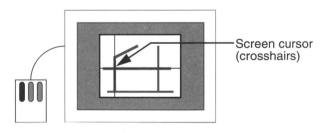

Screen cursor
(crosshairs)

When &GETPOINT is issued, the screen cursor appears in the graphic window. The PNT$ variables are set when you press a key on the current coordinate input device. The first value reported from the &TYPE directive (1) is the number of the key you pressed. The next values (2.263 and 5.225) are the x- and y-locations of the cursor when the key is pressed. Notice that the x- and y-values are in page units. In ARCPLOT, &GETPOINT retrieves coordinate values in page units by default.

Examine another example using &GETPOINT, this time from ARCEDIT:

```
Arcedit: MAPEXTENT STREETS
Arcedit: EDIT STREETS
Arcedit: DRAWENVIRONMENT ARCS
Arcedit: DRAW
Arcedit: &getpoint
Arcedit: &type &pnt$key% %pnt$x% %pnt$y%
1 6810113.792 1830065.516
```

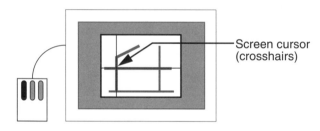

Screen cursor (crosshairs)

In ARCEDIT, &GETPOINT returns coordinates in map units by default. The &PAGE option to the &GETPOINT directive overrides the default and returns coordinates in page units. Use the &PAGE option when you want to pass coordinates to commands that manipulate graphic elements in the drawing environment (e.g., when you use the ARCEDIT command AP).

&GETPOINT also has an &MAP option, which returns PNT$X and PNT$Y values in map units. Use the &MAP option when you need to store coordinate data for passing to commands that require map coordinates (e.g., IDENTIFY in ARCPLOT, or MAPEXTENT in ARCEDIT or ARCPLOT).

To override the current coordinate input device, you can specify the method of coordinate entry when &GETPOINT is issued. Consider the following example:

```
Arcedit: COORDINATE KEYBOARD
Arcedit: &getpoint &map
AML ERROR - Cannot use &GETPOINT &CURRENT if the
     current coordinate source is KEYBOARD
AML ERROR - Unable to get point position
Arcedit: &getpoint &map &mouse
```

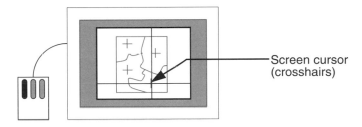

Screen cursor
(crosshairs)

An AML error message appears when &GETPOINT is first issued because &GETPOINT can't get coordinates from the keyboard. The &MOUSE option overrides the current input device (i.e., the keyboard), so the screen cursor becomes the method of coordinate input. All options available to the COORDINATE command are valid options for &GETPOINT as long as the device option name begins with an ampersand (e.g., &MOUSE, &DIGITIZER).

The &GETLASTPOINT directive has no arguments. It sets the PNT$ coordinate variables to the x- and y-values specified by the most recent use of the coordinate input device. Examine the examples on the following page.

```
Arcplot: IDENTIFY SOILS POLY
Arcplot: &getlastpoint
Arcplot: &type %pnt$key% %pnt$x% %pnt$y%
1 6510113.546 1298465.511
```

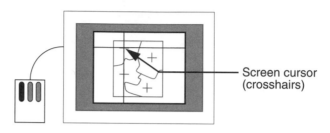

Screen cursor
(crosshairs)

```
Arcplot: MOVE 2, 8.5
Arcplot: TEXT 'Map of Soil Types'
Arcplot: &getlastpoint
Arcplot: &type %pnt$key% %pnt$x% %pnt$y%
1 2 8.5
```

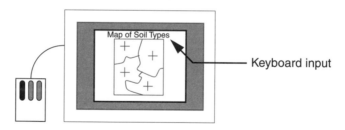

Keyboard input

When the current coordinate device is MOUSE, the IDENTIFY command activates the screen cursor for coordinate input. &GETLASTPOINT sets PNTKEY, PNTX, and PNT$Y to the coordinate values provided to the last command &TYPE. Notice that the IDENTIFY command uses map units.

Next, keyboard-entered coordinates in page units are provided to the MOVE command. &GETLASTPOINT retrieves those coordinates and sets the PNT$ variables.

The previous examples illustrate an important difference between &GETPOINT and &GETLASTPOINT. &GETPOINT can't use the keyboard as the source of coordinate input; however, &GETLASTPOINT can retrieve coordinates that were previously entered at the keyboard.

Question 6-1: Suppose you need to add a title to a map. You want to center the title and place it four page units above the reference point shown on the graphic below. Write the AML directive statement that activates the screen cursor for coordinate input and set the PNT$ variables.

```
Arcplot: _____
Arcplot: MOVE %pnt$x% [calc %pnt$y% + 4]
Arcplot: TEXT 'Street Network' LC
```

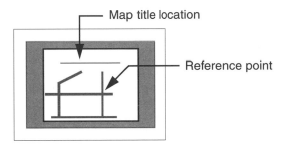

(Note: The MOVE command uses the PNT$ coordinates to mark a position for the map title. The [CALC] function adds 4 to the PNT$Y value. Lesson 6.3 discusses mathematical functions in detail.)

Question 6-2: Assume the current coordinate input device is KEYBOARD. Write an AML statement that activates the screen cursor, thus overriding KEYBOARD as the current device.

(Answers on page 6-13)

The following AML program demonstrates an efficient use of &GETPOINT to capture coordinates and pass them to the ARCPLOT command IDENTIFY:

```
/* identify.aml
/* Repeat the IDENTIFY command
CLEAR
&sv cov = [getcover]
MAPEXTENT %cov%
&sv feature = [getchoice polys arcs points]
%feature% %cov%
&type /&Select the feature -- Enter a 9 to quit
&getpoint &map
&do &while %pnt$key% ne 9
   IDENTIFY %cov% %feature% %pnt$x% %pnt$y%
   &getpoint &map
&end
&return
```

This example also shows the importance of the PNT$KEY variable. The value of PNT$KEY determines whether the loop should be entered and, if entered, whether or not the loop should continue. The screen cursor is activated when the program first encounters &GETPOINT &MAP (assuming the coordinate device is the mouse). The loop is entered unless the user presses the 9 key. The IDENTIFY command uses the coordinates in the PNT$X and PNT$Y variables to select the feature. &GETPOINT &MAP repeats inside the loop so the loop continues to execute until the user presses the 9 key.

The next example uses &GETLASTPOINT to set the PNT$ variables. This AML program also repeats the IDENTIFY command. Because &DO &UNTIL is used, the loop executes at least once. (If necessary, refer to chapter 5 for a review of looping.)

```
/* identify2.aml
/* Repeat the IDENTIFY command
CLEAR
&sv cov = [getcover]
MAPEXTENT %cov%
&sv feature = [getchoice polys arcs points]
%feature% %cov%
&type /&Select the feature -- Enter 9 to quit
&do &until %pnt$key% eq 9
   IDENTIFY %cov% %feature% *
   &getlastpoint
&end
&return
```

Coordinates are specified with the screen cursor when the IDENTIFY command line executes. &GETLASTPOINT sets the PNT$ variables to the key value and x- and y-values of the cursor selection. PNT$KEY determines whether the loop executes again.

The next example illustrates the use of the PNT$ variables in ARCEDIT. Suppose an engineering firm digitizes pipes from an engineering drawing but needs to add to the coverage a pipe that isn't on the plan. The new pipe connects to the end of the most recently digitized feature and is a specified length.

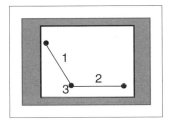

1 - Arc added from digitizer

2 - Arc added from keyboard

3 - To-node of arc #1

```
Arcedit: EDIT PIPES ARC
Arcedit: DRAWENVIRONMENT ARC
Arcedit: DRAW
Arcedit: EDITFEATURE ARCS
Arcedit: COORDINATE DIGITIZER
Arcedit: ADD
1 arc(s) added to /USER/PROJECT/PIPES
Arcedit: &getlastpoint
Arcedit: COORDINATE KEYBOARD
Arcedit: ADD
Enter Key,X,Y: 2 %pnt$x% %pnt$y%
Enter Key,X,Y: 2 [calc %pnt$x% + 1000] %pnt$y%
Enter Key,X,Y: 9 0 0
1 arc(s) added to /USER/PROJECT/PIPES
```

&GETLASTPOINT sets PNT$X and PNT$Y to the location of the ending node of the last arc digitized. You enter the beginning node of the new pipe feature with the keyboard using coordinates stored in the PNT$ variables. The [CALC] function performs a mathematical calculation to place the ending node of the new arc at 1,000 map units from the beginning node (PNT$X). The same y-value (PNT$Y) is used for the ending node.

Set the PNT$ reserved variables using the &GETLASTPOINT or
&GETPOINT directives only. Never set these variables using &SETVAR
or any other alternate method—doing so can produce unwanted results. If you
preset the PNT$ variables to some values, &GETPOINT and
&GETLASTPOINT won't override those values.

Question 6-3: Fill in the missing command lines to draw two concentric
circles, each with a different radius. The center of the first circle is specified
with the screen cursor.

```
Arcplot: USAGE CIRCLE
Usage: CIRCLE <xy radius | * {radius}>
Arcplot: CIRCLE * 1.5
Define the circle
Enter the center
Arcplot: _____        (Set the PNT$ variables to
coordinates provided to the CIRCLE command.)
Arcplot: CIRCLE _____  _____  2.0   (Draw the second circle
using the center coordinates of the first circle.)
```

(Answer on page 6-13)

In summary

Coordinates locate points on a map or graphics page. Coordinates are expressed in map units or graphics page units depending on which coordinate system (map or page) is referenced. Some ARC/INFO commands require input in map coordinates while others need page coordinates.

The &GETPOINT and &GETLASTPOINT directives set the PNT$ variables (i.e., PNTKEY, PNTX, and PNT$Y). These variables can supply coordinate data to commands that require graphic coordinate input. Remember that PNT$KEY can control the flow of operations (e.g., stopping a loop if the user presses the 9 key).

Answer 6-1:
```
Arcplot: &getpoint &page
Arcplot: MOVE %pnt$x% [calc %pnt$y% + 4]
Arcplot: TEXT 'Street Network' LC
```

Note: &GETPOINT can be used alone if the default units in ARCPLOT (PAGE) aren't changed.

Answer 6-2:
```
&getpoint &page &mouse
```

Answer 6-3:
```
Arcplot: USAGE CIRCLE
Usage: CIRCLE <xy radius | * {radius}>
Arcplot: CIRCLE * 1.5
Define the circle
Enter the center
Arcplot: &getlastpoint
Arcplot: CIRCLE %pnt$x% %pnt$y% 2.0
```

Exercises

Exercise 6.1.1

Suppose you have the following three coverages of the type indicated (i.e., arc, point, or poly):

```
STREETS   arc
SITES     point
ZONING    poly
```

Assume that the coverages are in map units of meters. Write an ARCPLOT AML program that allows the user to select a point from the coverages displayed on the screen. Using the selected point, have the program RESELECT all features from the three coverages that lie within 500 meters (1/2 km) of this point. If any part of a coverage feature is *within* the 1-km-diameter circle, it should be included in the selected set for that coverage. Reset the MAPEXTENT to the extent of the circle. Draw the selected features and the circle boundary in different colors.

Exercise 6.1.2

Write an ARCPLOT AML program like the one you wrote for the previous exercise with one change. In this case, use the selected point as the center of an area 1,000 meters square (1 square km) in which to RESELECT coverage features.

Exercise 6.1.3

COLORREF.AML (chapter 5, page 5-9) displays fifteen color selections at the bottom of the screen. Write an AML program that allows you to set a global variable called .COLOR$CHOICE with the color symbol selected with the cursor.

Lesson 6.2—Capturing coordinates in the point buffer and watch files

In this lesson

Some applications involve entering large amounts of coordinate data. Passing coordinate data manually, one command line at a time, is tedious and inefficient. This lesson shows how to store large sets of coordinate data in the point buffer and then pass them to commands when needed.

The previous lesson explained how AML variables can store coordinates for use during an ARC/INFO session. You lose the coordinates that you entered interactively when an ARC/INFO session ends. This lesson covers the technique of capturing coordinates in an AML-generated file (called a watch file) for use in subsequent ARC/INFO sessions.

Loading and retrieving data from the point buffer

AML allows you to store point data in a first-in, first-out *point buffer* that stores the device key used to enter the coordinates and their x,y coordinate values (i.e., the same data that's stored in the PNT$ variables). This point buffer is accessible from any of ARC/INFO's graphics programs and holds up to 500 point sets. A point set consists of the device key and x,y coordinate values.

The point buffer is empty each time you start an ARC/INFO session. Two AML directives, &PUSHPOINT and &GETPOINT with the &PUSH option, load the point buffer.

The point buffer takes priority over other sources of coordinate input. ARC/INFO looks first to the point buffer when it requires coordinate input; if the buffer is empty, ARC gets its input from the current coordinate device (e.g., mouse, keyboard, digitizer, etc.).

The &FLUSHPOINTS directive empties the point buffer. Before you add new point sets, always clear the buffer to make sure that AML removes any point sets remaining in the buffer from a previous operation.

&PUSHPOINT loads one point set in the point buffer. If you issue &PUSHPOINT without specifying the key and x- and y-values, the point buffer is loaded with the current values of the PNT$ variable; if the PNT$ variables aren't set, an error message appears in the dialog area of the text window.

&GETPOINT with the &PUSH option sets the PNT$ variables and loads these values into the point buffer. The following example demonstrates how points are loaded into the point buffer and used by a command that requires coordinate input:

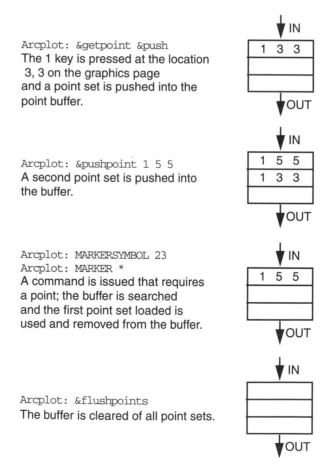

Arcplot: &getpoint &push
The 1 key is pressed at the location
3, 3 on the graphics page
and a point set is pushed into the
point buffer.

Arcplot: &pushpoint 1 5 5
A second point set is pushed into
the buffer.

Arcplot: MARKERSYMBOL 23
Arcplot: MARKER *
A command is issued that requires
a point; the buffer is searched
and the first point set loaded is
used and removed from the buffer.

Arcplot: &flushpoints
The buffer is cleared of all point sets.

The next example demonstrates using the point buffer to pass data to an ARCPLOT command. Consider an application that performs an analysis on habitat zones defined by a series of x,y coordinates. Assume that an ASCII file contains the following point data defining a habitat zone:

```
6804186.206 1838693.103
6805179.309 1831555.172
6802324.137 1830406.896
6800648.276 1835000.563
6798289.654 1831462.068
```

These coordinates are loaded into the point buffer, which is used as the source of input for the RESELECT command.

```
Arcplot: MAPEXTENT streams
Arcplot: ARCS streams
Arcplot: &flushpoints
Arcplot: &pushpoint 1 6804186.206 1838693.103
Arcplot: &pushpoint 1 6805179.309 1831555.172
Arcplot: &pushpoint 1 6802324.137 1830406.896
Arcplot: &pushpoint 1 6800648.276 1835000.563
Arcplot: &pushpoint 1 6798289.654 1831462.068
Arcplot: &pushpoint 9 0 0
Arcplot: RESELECT streams ARCS POLYGON * PASSTHRU
Selection polygon has an area 6614.860 and a perimeter 33663.725
/USR/GIS/ENVIRONMENT/STREAMS arcs: 7 of 13 selected
```

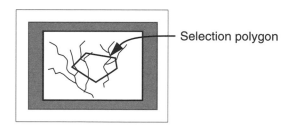 — Selection polygon

&FLUSHPOINTS ensures that the point buffer is clear. &PUSHPOINT loads the point sets into the point buffer. Each point set must include a key value and x- and y-coordinate values. The asterisk option (*) to the RESELECT command requires you to enter coordinate input using the current coordinate device; however, because the point buffer takes priority, it is used as the source for coordinates. The point set 9 0 0 is loaded into the buffer to end command input.

FYI

Loading coordinate data manually, as illustrated in the previous example, is a slow process. AML can read the ASCII file and load the point buffer automatically. See chapter 10 to learn how to open and read ASCII files with AML.

The point buffer only holds 500 point sets at a time. When you load more than 500 point sets into the point buffer, it's initialized (i.e., the point buffer is cleared of point sets and the loading process begins again). If this occurs, you can't rely on the coordinate data that the point buffer passes to commands.

The following AML program demonstrates how to use the point buffer with AML code to trap errors when graphically selecting features in ARCPLOT:

```
/* ensuresel.aml
/* Select features. Repeat the RESELECT command until
/* a valid selection is made.
&sv cov = [getcover]
MAPEXTENT %cov%
&sv feature = [getchoice poly arc point]
%feature% %cov%
&type /&Please select a feature with your cursor
&flushpoints
&getpoint &map &push
RESELECT %cov% %feature% ONE *
&do &while [extract 1 [show select %cov% %feature%]] = 0
  &type /&/&Please try another selection
  ASELECT %cov% %feature%
  &getpoint &map &push
  RESELECT %cov% %feature% ONE *
&end
&return
```

&GETPOINT activates the screen cursor and the user points at a feature. The &PUSH option pushes the point set into the point buffer for the RESELECT command to use. The &DO loop is entered if the user doesn't make a valid feature selection (e.g., the user doesn't point close enough to a feature to make a selection). The program prompts the user to try again and another &GETPOINT &MAP &PUSH directive is issued to load the point buffer. The loop repeats until the user selects a valid feature.

Question 6-4: Examine the command lines below and then answer the questions that follow:

```
Arcedit: &flushpoints
Arcedit: &pushpoint 2 1 10
Arcedit: &pushpoint 2 10 1
Arcedit: &pushpoint 9 0 0
Arcedit: &pushpoint 2 1 1
Arcedit: &pushpoint 2 10 10
Arcedit: &pushpoint 9 0 0
Arcedit: ADD
Arcedit: ADD
```

How many arcs are added after the first ADD command is issued? If the current coordinate device is MOUSE, what's the source of coordinate data for the second ADD command?

Question 6-5: Suppose the point buffer currently contains one point set: 1 3 3. Examine the command lines below and then answer the questions that follow:

```
Arcplot: MOVE 5 5   (The user issues MOVE to position text)
Arcplot: TEXT 'Solar Plant Site Suitability'
```

Where will the text be positioned? Which directive can be issued before the MOVE command to change the location where the text is drawn?

(Answers on page 6-23)

Capturing coordinate data in an AML watch file

An AML watch file can record the ARC/INFO program output that's displayed in the text area of the display terminal. Watch files are enabled using the &WATCH directive. &WATCH writes to a new file, writes over an existing file, or appends to an existing file. (Refer to chapter 1 if you need to review watch files.)

The &COORDINATES option used with the &WATCH directive records graphic input like coordinate data entered interactively with a cursor device (e.g., mouse or digitizer). Suppose that a user creates a map using the map composer in ARCPLOT. The &COORDINATES option is used with &WATCH to capture points entered with the screen cursor.

```
Arcplot: &watch coo.wat &coordinates
Arcplot: MAP DRAINAGE. MAP
Arcplot: MAPEXTENT STREATMS
Arcplot: ARCS STREAMS
```

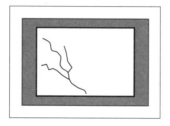

```
Arcplot: MFIT *
Define the box
(The user picks two points with the screen cursor.)
Arcplot: MAP END
Arcplot: &watch &off
Arcplot: QUIT
```

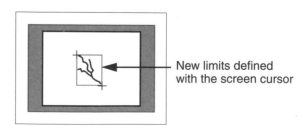

New limits defined with the screen cursor

Because the &COORDINATES option is specified, the coordinates generated by MFIT (defining the box) are included in the watch file, which looks like this:

```
Arcplot: |> MAP DRAINAGE.MAP <|
Arcplot: |> MAPEXTENT STREAMS <|
Arcplot: |> ARCS STREAMS <|
Arcplot: |> MFIT * <|
Define the box|>* 1.820866227149 7.135826587677 *<|
|>* 6.090058803558 1.021161437034 *<|
Arcplot: |> MAP END <|
Arcplot: |> &watch &off <|
```

Suppose the user wants to add coverages to the map composition later. The &CWTA directive can convert the watch file to an AML program. (Refer to chapter 1 if you need to review the &CWTA directive.) You can modify the AML program so that a coverage can be picked from a menu, added to the map composition, and then moved and scaled to register exactly with the streams coverage data.

```
Arc: &cwta coo.wat drainage_map.aml &coordinates
```

The resulting AML program, DRAINAGE_MAP.AML, looks like this:

```
MAP DRAINAGE.MAP
MAPEXTENT STREAMS
ARCS STREAMS
MFIT *
1.820866227149 7.135826587677
6.090058803558 1.021161437034
MAP END
&watch &off
```

As you can see, the coordinates transfer from the watch file to the AML program because the &COORDINATES option was used with the &CWTA directive. If you used the &COORDINATE option when creating the watch file, you must use the &COORDINATE option for &CWTA to ensure that the coordinates are in the final AML program. Notice that the asterisk is still present after the MFIT command. If the program runs as it currently exists, it prompts the user to select the box unless the coordinate environment is set to COORDINATE KEYBOARD. To avoid errors, edit the program to either remove the asterisk or ensure that the coordinate device is KEYBOARD when running the program.

Adding the [GETCOVER] and [GETCHOICE] functions as shown below makes the program more generic:

```
/* drainage_map.aml
/* scales and fits additional geographic data
/* to existing coverages
MAP drainage.map
&sv cov = [getcover]
&sv feature = [getchoice polys arcs points]
MAPEXTENT %cov%
%feature% %cov%
MFIT ~
1.820866227149 7.135826587677 ~
6.090058803558 1.021161437034
MAP END
&return
```

> Arcplot: &run drainage_map.aml
> *(The user picks a streams coverage from the menu.)*

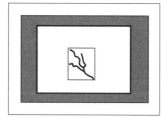

If the coordinate information from the MFIT operation in the first session isn't recorded in the watch file, the map image won't be placed in the same location when running the AML program.

FYI

Because watch files don't capture output from operating system commands or popup windows invoked through ARC, if popup windows are activated (i.e., &FULLSCREEN &POPUP), the output from ARC commands such as DESCRIBE isn't recorded. If you want to capture output from ARC commands in a watch file, be sure to disable popup windows by specifying &FULLSCREEN &OFF or &FULLSCREEN &ON. The LISTOUTPUT command can also store screen output in a text file.

In summary

The point buffer stores coordinate data and then passes it to programs as needed. Commands that require coordinate data use the coordinates stored in the point buffer before other sources. Using &PUSHPOINT or &GETPOINT with the &PUSH option are the two ways of loading the point buffer. Avoid incorrect results by using &FLUSHPOINTS to clear the point buffer before loading new point sets.

Using the &WATCH directive with the &COORDINATES option allows you to capture coordinate information in a watch file regardless of the input source.

Answer 6-4: One arc is added after the first ADD command. The point input buffer also provides coordinates to the second ADD command because it still contains data.

Answer 6-5: The text is positioned at 3,3 because the point buffer is used as the first source of coordinate input. If &FLUSHPOINTS is issued before the MOVE command, the coordinates specified with MOVE are used to position the text.

Exercises

Exercise 6.2.1

Suppose you have the following three coverages of the type indicated (i.e., arc, point, or poly):

```
STREETS   arc
SITES     point
ZONING    poly
```

Assume that the coverages are in the map units of meters. Fill in the missing parts of the following ARCPLOT AML program, which allows the user to reselect features from the three coverages using the POLY option. The program should select every feature from the three coverages that's at least partially contained within the specified polygon.

Note: You only specify the polygon boundaries once on the screen using the cursor. Reset the MAPEXTENT to contain the extent of the polygon and draw the selected features and the polygon boundary in different colors.

```
/* ex 6_2_1.aml
/* Polygon reselection from three different coverages
/* Polygon specified once by cursor
MAPEXTENT STREETS SITES ZONING
LINESET COLOR
UNITS MAP
/* Flush the point input buffer

&type Use 1 key to outline temporary area, 9 to close poly
/*   Use magenta to draw the outline while selecting vertices of
/*   the polygon.  Because you are using &GETPOINT, the polygon boundaries
/*   aren't drawn automatically as they are when you specify a polygon
/*   boundary with RESELECT.
LINESYMBOL 6
/* Choose a point in mapunits with the cursor

/* Initialize counter for array storage of points
&sv count 1
/* Place the selected point into the first element of x- and y-arrays
/* by assigning "array" variables to the key, x-, and y-coordinates
/* chosen.
&sv key%count% = _____
&sv x%count%   = _____
&sv y%count%   = _____
```

```
/* Make the last point equal to the first point for polygon closure
/* (XLAST and YLAST won't be loaded into the buffer).
&sv keylast  = _____
&sv xlast    = _____
&sv ylast    = _____
/* Set the variables CURRENT$X and CURRENT$Y equal to the first
/* point.
&sv current$x  = _____
&sv current$y  = _____
/* Keep accepting points until the user presses the 9 key.
&do &while _____ ne 9
/* Choose a point in map units with the cursor.

&sv count = %count% + 1
/* Place the selected point into the current element of the x- and
/* y-arrays by assigning "array" variables to the key, x-, and
/* y-coordinates chosen.
&sv key%count% = _____
&sv x%count%   = _____
&sv y%count%   = _____
/* If the user hasn't pressed a 9, draw a line from the last
/* x- y-coordinate (CURRENT$X,CURRENT$Y) to the coordinate just
/* chosen.  Then update the variables CURRENT$X and CURRENT$Y to
/* contain the coordinate just chosen.
&if _____ ne 9 &then &do
LINE _____
&sv current$x = _____
&sv current$y = _____
&end
&else &do
LINE %current$x% %current$y% %xlast% %ylast%
  &end
&end
&if %count% gt 3 &then &do
  /*  If the count is less than or equal to 3, not enough points were
  /*  chosen to make a polygon.  If the count is 3, the last one is
  /*  the 9-key, leaving only two points for the polygon.  Two
  /*  points doesn't define a polygon.  Clear the selected sets.
  ASELECT STREETS ARCS
  ASELECT SITES    POINTS
  ASELECT ZONING   POLYS
  /*  Push the selected points into the point buffer for
  /*  reselection of the first coverage.
  &do i = 1 &to %count%
      _____ [value key%i%]  [value x%i%]  [value y%i%]
  &end
  RESELECT STREETS ARC POLYGON * PASSTHRU
  /* The reselect emptied the buffer, but clear it just to be safe.
  _____
  /* Repeat the same procedure for the other two coverages.
```

```
&do i = 1 &to %count%
_____ [value key%i%]  [value x%i%]  [value y%i%]
&end
RESELECT SITES POINT POLYGON * PASSTHRU

_____
&do i = 1 &to %count%
_____ [value key%i%]  [value x%i%]  [value y%i%]
&end
RESELECT ZONING POLYS POLYGON * PASSTHRU
/* Now, using the same points in the array, find the MAPEXTENT
/* that completely contains the user-specified polygon.
/* NOTE:  The last point (i.e., the 9 key) doesn't count.
&sv mape_countlimit  %count% - 1
/* Set some ridiculously low and high numbers--make the initial
/* lows really high and the initial highs really low.
&sv xlow  = 9999999999.9
&sv ylow  = 9999999999.9
&sv xhigh = -9999999999.9
&sv yhigh = -9999999999.9
/* Loop through the points in the x- and y-arrays looking for
/* minimums and maximums.
&do i = 1 &to %mape_countlimit%
  &sv xlow  = [min %xlow%  [value x%i%] ]
  &sv ylow  = [min %ylow%  [value y%i%] ]
  &sv xhigh = [max %xhigh% [value x%i%] ]
  &sv yhigh = [max %yhigh% [value y%i%] ]
&end
MAPEXTENT %xlow% %ylow% %xhigh% %yhigh%
/* Redraw screen
CLEAR
ARCLINES STREETS 14
POINTMARKERS SITES 2
POLYGONLINES ZONING 3
/* Draw the boundary of the selected polygon, if you want to.
/* Use the line command with the point buffer.
/* Use the MAPE_COUNTLIMIT variable again to push vertices in
/* the buffer. Finally, push %keylast%, %xlast% and %ylast%
/* for the last vertex. End the line command by pushing these
/* values in the buffer: key=9,/* x=0, and y=0. The 9 key is
/* important in this final push, but the x and y coordinates
/* are ignored.

_____
&do i = 1 &to %mape_countlimit%
_____ [value key%i%]  [value x%i%]  [value y%i%]
&end
_____ %keylast%  %xlast%  %ylast%

_____
LINESYMBOL 6
LINE *
&end  /* &if %count% gt 3
```

```
&else &do
  &type Not enough points to form a polygon!!
  &type Exiting with no action performed!!
&end
/* CLEANUP - Clear the buffer

&return
```

Lesson 6.3—Using mathematical and trigonometric functions

In this lesson

Application programming often requires mathematical and trigonometric functions for tasks like determining distances and measuring angles. Many of these functions use coordinate input to perform a calculation. This lesson introduces the AML functions that perform these tasks and explains how to use the PNT$ variables and the point buffer to perform tasks more efficiently.

Using mathematical functions with coordinate data

Some of the AML mathematical functions used in applications that process numeric data are listed below:

AML function	Description
[ABS]	Returns the absolute value of a number
[CALC]	Returns the result of a numerical or logical expression
[CVTDISTANCE]	Converts metric units of distance to meters and imperial units of distance to feet
[INVDISTANCE]	Calculates the distance between two points
[MAX]	Returns the greater of two numbers
[MIN]	Returns the lesser of two numbers
[MOD]	Returns the remainder when one integer is divided by another
[OKDISTANCE]	Indicates whether a distance is in a valid ARC/INFO distance format
[ROUND]	Rounds a real number to an integer value (.5 or greater rounds up)
[SQRT]	Returns the square root of a given variable
[TRUNCATE]	Truncates a real number to an integer value

You must use the [CALC] function to evaluate all arithmetic expressions except those used with &SV, &IF, &DO &WHILE, or &DO &UNTIL. The next example uses mathematical expressions:

```
Arc: &sv a = 5
Arc: &sv b = 6
Arc: &type %a% + %b%
```
5 + 6 **(&TYPE does not force a mathematical calculation)**
```
Arc: &type [calc %a% + %b%]
```
11 **(CALC forces the mathematical calculation)**
```
Arc: &sv c = %a% + %b%  (&SV forces the calculation without [CALC])
Arc: &type %c%
```
11

The following example demonstrates using [CALC] with the PNT$ variables:

```
Arcplot: MOVE *
Arcplot: TEXT 'Stream Network'
```

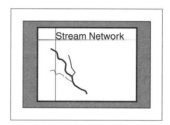

```
Arcplot: &getlastpoint
Arcplot: KEYPOSITION [calc %pnt$x% + 5] [calc %pnt$y% - 2]
Arcplot: KEYLINE stream.key
```

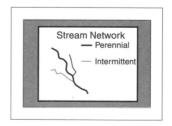

In this example, the screen cursor is used to designate the position of the map title. &GETLASTPOINT sets PNT$X and PNT$Y to the page unit coordinates of the cursor when a key is pressed. The [CALC] function specifies an explicit x- and y-offset for positioning the legend.

FYI

Some other math functions include [EXP], [LOG], [LOG10], and [RANDOM]. For a complete listing, see *Functional list of AML directives and functions* in the online help command references.

Q

Question 6-6: Assume that the user wants to draw a 3-inch-square box at a specific location. Examine the following command lines and then answer the questions:

```
Arcplot: USAGE BOX
Usage: BOX <* | xmin ymin xmax ymax>
Arcplot: &getpoint
Arcplot: BOX %pnt$x% %pnt$y% %pnt$x% + 3 %pnt$y% + 3
```

What's wrong with these command lines? Replace the incorrect line.

(Answers on page 6-35)

Suppose a user wants to query a coverage to determine the minimum straight-line distance between features. The following AML code determines the minimum distance between user-entered coordinate pairs. The coordinates are shown in the graphic below. The distances to calculate are between the coordinate pairs A,D, B,D, and C,D.

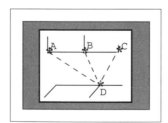

```
/* distances.aml
/* Determine the minimum distance between features in %cov%
&sv old_marker = [show markersymbol]
&sv old_line = [show linesymbol]
&sv old_markerset = [show markerset]
&sv old_lineset = [show lineset]
&sv old_units = [show units]
MARKERSET PLOTTER.MRK
MARKERSYMBOL 49
MARKERCOLOR YELLOW
LINESET CARTO.LIN
LINESYMBOL 111
LINECOLOR YELLOW
UNITS MAP
CLEAR
&sv cov [getcover]
MAPEX %cov%
&sv feature = [getchoice polys arcs points]
%feature% %cov%
&describe %cov%
/* Calculate the diagonal length of the map extent -- a
/* value larger than the distance between any two points
&sv mindistance = [invdistance %dsc$xmin% %dsc$ymin% ~
  %dsc$xmax% %dsc$ymax%]
&type /&Pick pairs of points with the cursor--Press 9 to quit
&getpoint &map
&do &while %pnt$key% ne 9
  &sv x_origin = %pnt$x%
  &sv y_origin = %pnt$y%

  MARKER %x_origin% %y_origin% /* Display point in yellow
  &getpoint &map
  MARKER %pnt$x% %pnt$y% /* Display second point in yellow
  /* Draw a yellow line between the point pair
  LINE %x_origin% %y_origin% %pnt$x% %pnt$y%
  /* Calculate the distance between the point pair
&sv distance = [invdistance %x_origin% %y_origin% ~
  %pnt$x% %pnt$y%]
  /* If the distance between the two points is smaller
  /* than any two points previously entered, then:
  &if %distance% < %mindistance% &then &do
    &sv mindistance = %distance%
    &sv x_point_origin = %x_origin%
    &sv y_point_origin = %y_origin%
    &sv x_point_end = %pnt$x%
    &sv y_point_end = %pnt$y%
  &end
&getpoint &map
```

```
&end
/* Display the point pair with the minimum distance in red
MARKERCOLOR RED
MARKER %x_point_origin% %y_point_origin%
MARKER %x_point_end% %y_point_end%
/* Display a line in red between the point pair
LINECOLOR RED
LINE %x_point_origin% %y_point_origin% %x_point_end% ~
  %y_point_end%
&type /&The minimum distance between point pairs~
is %mindistance%.~
/&These point pairs are displayed in red.
&type /&The x,y coordinates of these two points are:~
/&%x_point_origin% %y_point_origin% %x_point_end% %y_point_end%
&type
MARKERSET %old_markerset%
MARKERSYMBOL %old_marker%
LINESET %old_lineset%
LINESYMBOL %old_line%
UNITS %old_units%
&return
```

The key the user presses when the first &GETPOINT &MAP is encountered
determines whether or not the loop is entered. A key other than the 9 key enters
the loop and stores the coordinates specified from the first &GETPOINT
directive in the user-defined variables X_ORIGIN and Y_ORIGIN.

&GETPOINT is repeated, allowing the user to pick the second coordinate in the
pair. The x- and y-values of the second coordinate are stored in the reserved
variables PNT$X and PNT$Y.

The [INVDISTANCE] function determines the distance between the two
points. The distance between each coordinate pair is compared to the
previous distance. If the current distance is less than the last, the variables
MINDISTANCE, X_POINT_ORIGIN, Y_POINT_ORIGIN, X_POINT_END,
and Y_POINT_END update. When the loop terminates, the current values of the
variable generate the information in the &TYPE statements.

Using trigonometric functions with coordinate data

AML's trigonometric functions perform common operations, such as computing angles or returning an angle given its sine, cosine, or tangent. The following functions can be used to compute angles in a right triangle:

AML function	Name	Value
[SIN]	Sine	[SIN q] = a/c
[COS]	Cosine	[COS q] = b/c
[TAN]	Tangent	[TAN q] = a/b
[ASIN]	Arcsin	[ASIN a/c] = q, where -p/2 < q < p/2
[ACOS]	Arccos	[ACOS b/c] = q, where 0 < q < p
[ATAN]	Arctan	[ATAN a/b] = q, where -p/2 < q < p/2
[ATAN2]	Arctan	[ATAN2 a b] = q, where -p < q < p

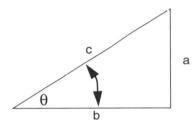

The following table lists other functions that deal with angles:

AML function	Description
[ANGRAD]	Converts an angle in a valid ARC/INFO format to radians
[INVANGLE]	Calculates the polar angle between two points
[OKANGLE]	Indicates whether an angle is in a valid ARC/INFO angle format
[RADANG]	Converts an angle measured in radians to an angle in a valid ARC/INFO format

ARC/INFO computes all angles in radians. If you want ARC/INFO to perform a computation on an angle expressed in degrees, you must use the function [ANGRAD] to convert from degrees to radians. To convert radians to degrees, use the [RADANG] function.

Use the [INVANGLE] function to compute the polar angle (in radians) between two points as shown in the following example:

```
Arc: &sv angle1 = [round [radang [invangle 1 1 5 5]]]
Arc: &sv angle2 = [round [radang [invangle 5 5 1 1]]]
Arc: &type angle1 is %angle1% and angle2 is %angle2%
angle1 is 45 and angle2 is 225
```

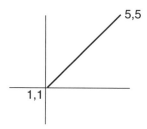

The next example demonstrates how the PNT$ variables can provide a trigonometric function with coordinate data to determine the angle between two points. You could use this AML program for a task such as determining the polar angle of a feature to align feature annotation.

```
/* angle.aml
/* Determine the angle between point pairs
&type /&Choose point pairs with your cursor--Press 9 to quit
&getpoint &map
&sv x_origin = %pnt$x%
&sv y_origin = %pnt$y%
&getpoint &map
&type /&The angle between your two points is [radang~
  [invangle %x_origin% %y_origin% %pnt$x% %pnt$y%]]
&return
```

Question 6-7: Fill in the missing code to set a variable to the sine of a 60-degree angle.

```
Arc: &sv angle = _____
```

(Answer on page 6-35)

In summary

AML provides functions for performing mathematical and trigonometric operations in AML programs. The PNT$ variables and point buffer are frequently used as sources of numeric data for these functions.

Answer 6-6: The mathematical expressions %PNT$X% + 3 and %PNT$Y% +3 can't be calculated without the [CALC] function. The correct line of code is as follows:

```
BOX %pnt$x% %pnt$y% [calc $pnt$x% + 3] [calc %pnt$y% + 3]
```

Answer 6-7: Arc: &sv angle = [sin [angrad 60]]

Exercises

Exercise 6.3.1

In ARCPLOT, use &GETPOINT to specify the center and then the edge of a circle. Draw the circle in red. Next, draw circles at one-quarter, one-half, and two times the diameter of the original circle. Draw each circle in a different color and display the chosen color and calculated distances for the user.

Exercise 6.3.2

Write an ARCPLOT AML that determines the x- and y-location of a point located 1,000 meters northeast (+45 degrees) from a chosen point (used as the origin on an x,y axis). Assume that the MAPUNITS are meters. Draw a point on the screen at the determined location.

Hint: There's more than one way to accomplish this:

- Using sin and cos
- Using [sqrt 2]
- Using polar coordinates

Exercise 6.3.3

Rewrite the AML program you wrote for the previous exercise to accept any angle and any distance. Again assume MAPUNITS and distances are measured in meters. Accept the angle in degrees where 0 degrees is east and +90 degrees is north.

Hint: Use trigonometry or polar coordinates.

Managing character strings

Character strings play an integral role in AML programming because a great deal of data is stored and returned to AML programs as character strings. AML variables store character strings. AML functions return character strings. Pathnames and error messages are also character strings. In this chapter, you'll learn how AML manipulates, queries, and formats character strings.

The AML character manipulation functions perform such tasks as extracting elements from and formatting characters in text strings. For example, you might use AML to extract the minimum and maximum x-coordinates from the string 4472.033,21001.020,88297.694,87095.460. (This is the format of the XMIN, YMIN, XMAX, YMAX returned by [SHOW MAPEXTENT].) AML can also modify character strings, determine the length of a string, and find the position of a substring within a string.

This chapter covers the following topics:

Lesson 7.1—Introducing character strings

- Interpreting character strings
- Concatenating character strings

Lesson 7.2—Manipulating character strings

- Modifying character strings
- Formatting character strings
- Evaluating length, position, or existence of characters in a string
- Parsing and returning parts of strings and lists

Lesson 7.3—Nesting string functions to perform complex tasks

- Determining the number of directories in a path
- Determining the length and data type of elements in a string
- Reordering elements in a string
- Determining the current platform

Lesson 7.1—Introducing character strings

In this lesson

In this lesson, you'll learn what AML recognizes as a character string. You'll also learn how AML concatenates strings to create new information.

Interpreting character strings

A *string* is a set of characters that the computer treats as a unit and interprets as text. A character string can contain any sequence of elements such as letters or numbers. Information returned by AML functions, values stored in AML variables, and environment settings are all character strings. Examine the following examples:

```
Arc: SHOW WORKSPACE
/home/judy/network
Arc: &sv coverages = [response 'Enter coverage names']
Enter coverage names: soils slope wells
Arc: &lv coverages
Local:   COVERAGES              'soils slope wells'

Arcplot: MAPEXTENT soels
Cannot read extent for geo_dataset SOELS
Arcplot: MAPEXTENT soils
Arcplot: SHOW MAPEXTENT
6794100,1825100,6807900,1844900

Arcedit: SHOW DRAWENVIRONMENT
ALL OFF, ARC ON, NODE ON, TIC IDS
```

All of the program responses shown above are character strings. Character strings can have embedded blanks (e.g., `Cannot read extent for geo_dataset SOELS`) and/or commas (e.g., `ALL OFF, ARC ON, NODE ON, TIC IDS`).

Blanks and commas affect how AML interprets the string. Any subset of characters separated from other characters in a string by a blank or comma is defined as one element of the string.

Strings can exist without quotes (e.g., /home/judy/network) or be quoted (e.g., 'soils slope wells'). Quotes around a string also affect how AML treats it. AML considers a quoted string as one element. Lesson 7.2 looks specifically at how AML deals with blanks, commas, and quotes.

 FYI

Values stored in AML variables are character strings; however, AML can report the different data types (i.e., character, integer, real, or logical) stored in the variable using the [TYPE] function.

Concatenating character strings

Data can be joined, or concatenated, by manipulating AML variables. You can concatenate character strings to include a maximum of 1,024 characters, the maximum length for any AML variable. Examine the following example:

```
Arc: &sv a = 123
Arc: &sv b = 456
Arc: &sv c = %a%%b%
Arc: &type %c%
123456
```

The concatenated string 123456 could be used as a number. You can embed blanks in a concatenated string by placing them between the variables as shown here:

```
Arc: &sv a = 123
Arc: &sv b = 456
Arc: &sv c = %a% %b%
Arc: &type %c%
123 456
```

123 456 is still one character string, but now AML recognizes two elements in this string. Because AML now recognizes this string as character data, it can't be used as a number. However, it could be used as a pair of numbers (for example, as two coordinates):

```
Arcplot: IDENTIFY SLOPE POLY %c%
```

The following code demonstrates how to concatenate a string of feature attribute items for labeling features in ARCPLOT by manipulating variables:

```
/* conclabel.aml
&sv cov = [getcover * -poly]
MAPEXTENT %cov%
POLYGONS %cov%
&sv first = .true.
/* Loop through the items in the PAT
&do item &list [listitem %cov% -polygon]
  /* If this is the first item in the PAT
  &if %first% &then &do
    &sv string = %item%
    &sv first = .false.
  &end
  /* If not the first item in the PAT
  &else &do
    &sv string = %string%\%item%
  &end
&end /* do item
POLYGONTEXT %cov% %string%
&return
```

Item names contained in the polygon attribute table (PAT) are first concatenated in the variable STRING. By separating the text with backslashes, a carriage return stacks the text when a labeling command is issued. This backslash is the same in function as the forward slash ampersand (/&) that was used with the &TYPE directive. The backslash is used with many ARCPLOT commands while /& is used only with the &TYPE directive. STRING is used for the {item} argument in the POLYGONTEXT command.

If the PAT contains the items AREA, PERIMETER, SOIL#, SOIL-ID, and SOIL-TYPE, the concatenated string equals AREA\PERIMETER\SOIL#\SOIL-ID\SOIL-TYPE and the attribute features stack inside the polygons as shown in the following diagram:

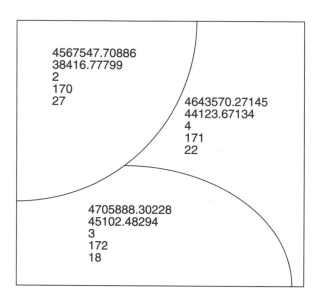

In summary

AML treats sets of characters (i.e., letters, numbers, and control characters) as character strings. Blanks, commas, and quotes affect how AML evaluates a character string. Multiple strings can be concatenated to create a new composite text string.

Exercises

Exercise 7.1.1

Write a program that accepts two strings from the user and returns the concatenation of the two as a single string output with the &TYPE directive.

Exercise 7.1.2

Change the program completed in exercise 7.1.1 to use the &LV directive to list out the new value of the variable. Is there a difference between the two programs if one of the strings contains spaces?

Exercise 7.1.3

Write a program that accepts a coverage name and a map scale from the user in ARCPLOT. The program should use the TYPE function to ensure that the input given for the map scale is of type integer. If not, keep prompting the user to enter a scale until it's of type integer. When the scale is satisfactory, set the MAPEXTENT to the chosen coverage and the MAPSCALE to the scale given.

Lesson 7.2 —Manipulating character strings

In this lesson

Character strings can be in the wrong format or contain too much data for programs and commands to handle. It's necessary, therefore, to have tools that can manipulate and query character strings. AML provides numerous functions for performing such operations as extracting a substring (i.e., a subset of a string) from a string or finding the position of a substring in a string. This lesson covers some of these in detail and provides you with functional lists of others. You'll also learn how AML treats strings that contain quotes, embedded blanks, or commas.

Modifying character strings

AML has tools that allow you to manipulate character strings in order to return data in a format acceptable to your programs.

Changing case

Some AML directives are case sensitive when evaluating character string data. You can reduce the amount of AML code necessary to perform certain actions if the information returned to the program is first changed to uppercase or lowercase.

The following functions change the case of character strings:

Function name	Purpose
[LOCASE]	Converts a string to lowercase
[UPCASE]	Converts a string to uppercase
[TRANSLATE]	Converts a string to uppercase, or translates a specified string into another in a target string

The [LOCASE] function converts a string to lowercase. Compare the following two code fragments:

```
/* example1
/* &SELECT without a function to change case
&sv choice = [response 'Enter feature type']
&select %choice%
&when poly,POLY,polys,POLYS,polygon,POLYGON,polygons,POLYGONS
  &do
     .
     .

/* example2
/* &SELECT with a function to change case
&sv choice = [response 'Enter feature type']
&select [locase %choice%]
&when poly,polys,polygon,polygons
  &do
     .
     .
```

The &SELECT directive is case sensitive. The first example was written to anticipate every possible response that a user might input. The second example, though, ensures that the response is in lowercase, thus reducing the number of options necessary to perform the conditional testing.

If you don't put any arguments other than the string to be modified into the [TRANSLATE] function, it acts exactly like [UPCASE] and converts the string to uppercase. The [TRANSLATE] function can also change selected characters to a different character set.

The examples below demonstrate how to use [UPCASE] and [TRANSLATE]:

Change owner name to uppercase:

```
Arc: &sv owner = [response [quote Enter owner's name]]
Enter owner's name: Hill
Arc: &type %owner%
Hill
Arc: &type [upcase %owner%]
HILL
Arc: &type [translate %owner%]
HILL
```

Translate lowercase "a" and "i" into uppercase "A" and "I":

```
Arc: &type [translate arc/info AI ai]
Arc/Info
```

Translate "-" to "$" and "." to "_".

```
Arc: &type [translate first-check-cov.status $_ -.]
first$check$cov_status
```

Manipulating pathnames

Pathnames are character strings. The AML functions listed in the following table can return different elements from pathnames to provide programs with the information they require:

Function name	Purpose
[PATHNAME]	Returns a fully expanded file specification
[DIR]	Returns the directory part of the file specification
[ENTRYNAME]	Returns the file name part of the given file specification
[JOINFILE]	System-independent method for establishing pathnames

Suppose the following AML program is run from another AML program to return information to the calling program:

```
/* path.aml
&sv .cov = [getcover /usr/network/data]
/*GETCOVER returns the full pathname for a coverage.
&type /&The coverage you have chosen is [entryname %.cov%]
&type The workspace in which [entryname %.cov%] resides ~
  is [dir %.cov%]
&return
```

If the user picks a coverage named STREETS, the following &TYPE statements are returned:

```
The coverage you have chosen is streets
The workspace in which streets resides is /usr/network/data
```

Without [ENTRYNAME], the program would return the complete pathname to the user:

```
The coverage you have chosen is /usr/network/data/streets
The workspace in which /usr/network/data/streets resides is
/usr/network/data
```

The calling program can use the global variable .COV as needed. For example, querying an INFO file in an external workspace such as ITEMS %.cov%.aat requires a complete pathname.

The only problem with PATH.AML is that the pathname to the data is hard coded in the AML program. If this AML program is executed from Windows NT, the correct data may not be found because the path lacks a drive letter. The best remedy is to set up a system variable that contains the path to the data.

```
UNIX
tavarua:tom[22]% setenv DATA_PATH /usr/network/data
Windows NT
Click on Main -> Control Panel -> System
Variable: data_path
Value:    d:\usr\network\data
```

This system variable must be set up before the ARC/INFO session is launched. Once the variable is set, it can be used anywhere in an ARC/INFO session just like a global variable. The only difference between the system variable and the global variable is the dot in front of the global variable name and that the system variable cannot be deleted using &DELVAR.

Outside of the drive letter, Windows NT and UNIX pathnames can look exactly the same. UNIX only accepts forward slashes, but Windows NT accepts both forward and backslashes within an ARC/INFO session. Windows NT accepts all of the following formats:

```
d:/usr/network/data
d:\usr\network\data
d:/usr\network/data
```

PATH2.AML incorporates the modifications to check for the existence of a system variable and inform the user if such a variable doesn't exist:

```
/* path2.aml
&term 9999
&if [null [value DATA_PATH]] &then ; &do
  &type /&You do not have the correct system variable set.
  &if [upcase [extract 1 [show &os]]] cn 'NT' &then ; &do
    &type Quit from ARC, then use the NT Program Manager.
    &type Main -> Control_panel -> System
    &type Set variable DATA_PATH to <path_to_database>.
    &return
  &end
  &else ; &do
    &type Quit from ARC, then type:
    &type "setenv DATA_PATH <path_to_database>"
    &type from unix prompt, restart ARC and re-run PATH2.AML.
    &return
  &end
&end
&sv .cov = [getcover %DATA_PATH%]
/*GETCOVER returns a full pathname
&type /&The coverage you have chosen is [entryname %.cov%]
&type The workspace in which [entryname %.cov%] resides ~
  is [dir %.cov%]
&return
```

In the previous AML program,[NULL [VALUE DATA_PATH]] is used to find out if the system variable exists. The [VALUE] function returns a null value if the variable doesn't exist. If [VALUE] wasn't used, %DATA_PATH% would return an error because the variable wouldn't yet be defined. Lesson 3 of this chapter discusses the [EXTRACT] function and how it can be used here to find the operating system.

You may have noticed that the variable name DATA_PATH is in uppercase. This is mostly just a standard convention for identifying system variables from other variables. UNIX is case sensitive, so it recognizes DATA_PATH as a variable, but not data_path. Windows NT, on the other hand, isn't case sensitive and recognizes the variable as long as it is spelled the same.

Prior to ARC/INFO Version 7.1, UNIX system variables had to be referenced with a $. To reference DATA_PATH you would have to type $DATA_PATH. At Version 7.1, system variables can be referenced with percent signs on both the UNIX and Windows NT platforms.

FYI

The [JOINFILE] function was established at ARC/INFO Version 7.0 to take care of path problems between platforms. [JOINFILE] inserts the correct path delimiter between two directory names. [JOINFILE] is most useful in dealing with the VMS operating system, which uses brackets and periods as path delimiters. ARC/INFO no longer supports VMS as of Version 7.0.4. [JOINFILE] doesn't help compatibility at Version 7.1 because both Windows NT and UNIX accept front slashes as pathname delimiters. This doesn't mean that you should never use [JOINFILE]. By using [JOINFILE] in your AML programs, you can make your application compatible with earlier versions of the software, as well as with upcoming versions that might support other platforms.

Handling quotes and embedded blanks or commas

ARC/INFO treats a quoted character string as a single element or command argument even if it contains embedded blanks. For example, suppose you want to enter a map title using the TEXT command in ARCPLOT. This command expects only a single element as an argument, so you must quote the map title (character string) if it contains more than one word or embedded blanks.

Some ARC/INFO commands require multiple elements as input such as MAPEXTENT <XMIN YMIN XMAX YMAX>. In this case, the input character string is an unquoted string consisting of four elements separated, or delimited, by a comma or a space.

Examine the character strings and compare the number of elements in each listed in the following table:

Character string	Number of elements
67,18,68,19	Contains four elements—commas separate the elements
67 18 68 19	Contains four elements—spaces separate the elements
67, 18, 68, 19	Contains four elements—commas and spaces separate the elements
'67 18 68 19'	Contains one element—treats characters inside quotes as one element

A consecutive set of characters with no embedded blanks or commas is a character string with a single element. Each part of a string that is delimited with a blank or comma is considered a separate element within that string. AML treats character strings within single quotes as single elements regardless of the presence of embedded blanks or commas.

Quoting and unquoting strings controls how AML evaluates character strings to provide required input to programs. Use the following functions to manipulate quotes in strings:

Function name	Purpose
[QUOTE]	Places quotation marks around quoted or unquoted strings
[UNQUOTE]	Removes quotation marks from each end of a quoted string
[QUOTEEXISTS]	Determines if the argument is quoted, has quoted strings, or has quoted characters

Examine the following code fragment that uses the [RESPONSE] function to allow the user to provide needed input to the program. Below, [RESPONSE] defines the variable MAP_EXTENT with data received from the user. This data is the coordinate information required by the ARC/INFO MAPEXTENT command.

```
&sv map_extent = [response 'Enter map extent coordinates']
MAPEXTENT %map_extent%
```

The minimum and maximum x,y coordinates needed by the MAPEXTENT command are entered interactively using the [RESPONSE] function. As you've seen, the coordinate data must be delimited by spaces or commas before the MAPEXTENT command can recognize the input as four separate elements.

```
Enter map extent coordinates: 6794100 1825100 6807900 1844900
```

In this case, MAPEXTENT fails as follows:

```
Cannot read extent for geo_dataset 6794100 1825100 6807900 1844900
```

MAPEXTENT fails because [RESPONSE] places quotes around input containing spaces. The coordinate data is separated by spaces so the entire string is quoted and treated as one element:

```
'6794100 1825100 6807900 1844900'
```

The MAPEXTENT command requires the four separate elements for its coordinate data argument, so the string must be unquoted before it's passed. This is accomplished using the [UNQUOTE] function as follows:

```
&sv map_extent = [response 'Enter map extent coordinates']
MAPEXTENT [unquote %map_extent%]
```

FYI

> If you entered the coordinates separated by commas instead of spaces, as follows:
>
> ```
> 6794100,1825100,6807900,1844900
> ```
>
> then AML doesn't quote the string (it contains no spaces) and the MAPEXTENT command accepts it with no modification.

The next AML code fragment illustrates a use of the [QUOTE] function:

```
/* Create the map title
MOVE 4.5 7.0
TEXT [quote Map of [entryname %cov%]]
```

The TEXT command needs the character string Map of %cov% to be quoted because it contains blanks. However, if you place quotes around this string (i.e., 'Map of %cov%'), the quotes suppress AML's interpretation of the variable COV. Using the [QUOTE] function to quote the string enables COV to be evaluated and expanded to the coverage name before the necessary quotes are placed around the entire string. AML's order of interpretation is as follows: variables, functions, directives, and commands.

FYI

If you want to return a character string that includes an apostrophe, use two single quotes instead of an apostrophe. Consider this example:

```
Arcplot: MOVE 4 4
Arcplot: TEXT 'Map of the Planning Commission's ~
Arcplot: preferences for new school sites'
Usage: TEXT <text_string> {LL | LC | LR | CL | CC | CR | UL | UC | UR}
```

The usage for TEXT is returned because the command can't handle the triple occurrence of single quotes in the string (i.e., before and after the string plus the apostrophe). This problem is corrected below using two single quotes to indicate the apostrophe to AML.

```
Arcplot: TEXT 'Map of the Planning Commission''s ~
Arcplot: preferences for new school sites'
```

Now AML correctly includes the apostrophe in the returned string as follows:

```
Map of the Planning Commission's preferences for new school sites
```

AML automatically places single quotes around user-provided character strings that contain embedded blanks. Use the [UNQUOTE] function to remove the quotes for commands that require unquoted strings.

Consider another AML program code fragment that asks the user for input for the DRAWENVIRONMENT command:

```
/* edit.aml
ARCEDIT
EDIT [getcover]
&sv display = [response 'Enter the features you wish to ~
 display']
DRAWENVIRONMENT [unquote %display%]
DRAW
&return
```

If the user enters the features delimited by commas, AML doesn't quote the string and DRAWENVIRONMENT executes normally. If a user enters the features delimited by spaces, AML places quotes around the string. In this case, [UNQUOTE] is needed to strip the quotes so that the command receives the correct number of arguments.

In many situations, the user must quote character string data, so you'll need to use the [QUOTE] and [UNQUOTE] functions frequently in your AML programs.

You can also use [QUOTEEXISTS] to check for the existence of a variety of different situations for quotes within a string. [QUOTEEXISTS] can check for bounding quotes, quotes within a string, or single quotes. The returned value is always a logical statement, stating whether the condition exists or not. The following are some examples of how [QUOTEEXISTS] functions:

```
Arc: &type [quoteexists -bound 'abc def ghi']
.TRUE.
Arc: &type [quoteexists -string abc 'def' ghi]
.TRUE.
Arc: &type [quoteexists -char abc de'f ghi]
.TRUE.
Arc: &type [quoteexists -bound abc 'def' ghi]
.FALSE.
```

See the online help command references for more information on the [QUOTEEXISTS] function.

FYI

Checking for the presence or absence of quotes can help diagnose an AML program that executes unsuccessfully. The &LISTVAR (&LV) directive returns a list of variable values; quoted variable values are displayed with quotes. To display the value of a single variable, specify the variable name along with the &LV directive. Look at the following example:

```
Arc: &sv covers = [response 'Enter coverage names']
Enter coverage names: soils slope streets
Arc: &lv covers
Local:   covers                        'soils slope streets'
```

Because the user input contains embedded blanks, AML quotes the string. &LV displays the quotes when the value of the variable is listed.

Be careful; &TYPE does *not* display the quotes on a quoted string. Examine the following variation of the example in the previous FYI:

```
Arc: &sv covers = [response 'Enter coverage names']
Enter coverage names: soils slope streets
Arc: &type %covers%
soils slope streets
```

Question 7-1: Fill in the missing code to provide a properly formatted character string to the program.

```
/* drawcov.aml
&sv covers = [response 'Enter coverages you wish to display']
/* Set the map extent to the coverages
MAPEXTENT _____
&sv old_linesymbol = [show linesymbol]
&sv linecolor = 1
/* Loop through the list of coverages
&do cov &list _____
  LINECOLOR %linecolor%
  ARCS %cov%
  &sv linecolor = %linecolor% + 1
&end
LINESYMBOL %old_linesymbol%
&return
```

(Answer on page 7-31)

Other AML functions that modify character strings

The following AML functions can also be used to modify character string data:

Function name	Purpose
[SUBST]	Substitutes one specified string for another in a target string
[TRIM]	Removes any occurrences of a specified character from the ends of a target string

The [SUBST] function replaces all occurrences of a specified string with another string. Suppose, for example, you need to create a program that accepts space-delimited input and replaces the spaces with commas. You'll often need this when you output data to other systems that require comma separators between data values. The following code uses [SUBST] to accomplish this:

```
&sv old_string = [response 'Enter string to modify']
&sv new_string = [subst %old_string% ' ' ,]
```

If the following string is entered at the prompt:

```
4562189c 2 RES1 410
```

the data in the variable NEW_STRING is as follows:

```
&lv new_string
4562189c,2,RES1,410
```

This kind of operation is most often accomplished by reading data from an ASCII file, modifying it as shown above, and then writing the result to a new ASCII file for use by the other system. Chapter 10, "Reading and writing files," discusses this functionality.

The [TRIM] function strips specified characters from either or both ends of a character string. Account numbers, for example, often contain leading zeros. The following AML code fragment demonstrates how to strip those leading zeros:

```
/* Strip leading zeros from account numbers
&sv record = [trim %acctnum% -left 0]
```

If the account number is 00005-3461-562968, then

```
&lv record
5-3461-562968
```

Formatting character strings

You may want to format the display of data for some of your applications. For example, you might want to display numerical data with a specified number of decimal places or incorporate explicit text with data returned by AML.

The &FORMAT directive and the [FORMAT] function format the display of data.

Name	Purpose
&FORMAT	Sets the number of decimal places for display of real numbers
[FORMAT]	Formats the display of a character string

&FORMAT sets the number of decimal places for displaying real numbers returned by ARC/INFO commands, such as DESCRIBE and TOLERANCE in ARC, MEASURE in ARCPLOT, and WHERE in ARCEDIT. The default for &FORMAT is three decimal places. Examine the following example:

```
Arcedit: DISTANCE
345.678
Arcedit: &format 2
Arcedit: DISTANCE
345.68
```

The &FORMAT setting doesn't affect the display of real numbers in the following situations:

- Display returned by the family of SHOW commands and functions, for example:

```
Arcplot: MAPEXTENT PARCELS
Arcplot: &format 2
Arcplot: SHOW MAPLIMITS
0,0,10.92519,7.849409
```

- Display returned from the evaluation of the variable, for example:

```
Arcplot: &format 2
Arcplot: &getpoint &map
Arcplot: &type %pnt$x% %pnt$y%
6805148.276122 1835775.862212
```

- Display returned from listings of INFO data files, for example:

```
Arcplot: &format 2
Arcplot: LIST parcels.pat info
                1
AREA                       =    -50635867.54752
PERIMETER                  =       277374.23715
PARCELS#                   =    1
PARCELS-ID                 =    0
.
.
.
```

In each example, even though &FORMAT specifies that the data should be displayed to only two decimal places, the display isn't affected. This is because &FORMAT doesn't affect AML displays. You also cannot format INFO file listings with &FORMAT. INFO has its own FORMAT command.

To control the display of real numbers returned by AML variables or functions, use the [FORMAT] function in conjunction with the &FORMAT directive. Examine the usage for the [FORMAT] function:

```
[FORMAT <format_string> {argument...argument}]
```

[FORMAT] substitutes arguments in a format string. The <format_string> is a quoted string that can contain tokens that function as placeholders for variables contained in the argument list. These tokens, or format variables, are written as %1%, %2%, etc., up to %10%; they indicate where the arguments contained in {argument . . . argument} should be substituted. The argument list is a space-delimited list of variables that are surrounded on either side by percent signs. Consider the following example:

```
&sv string = [FORMAT '   %1% %2%' %cover% %item%]
```

[FORMAT] places the value of %COVER% at the location of %1% and the value of %ITEM% at the location of %2% in the quoted format string. AML quotes the string returned by the [FORMAT] if it contains blanks. If %COVER% = FLOOD and %ITEM% = ZONE, the contents of the STRING variable is as follows:

```
%string% = 'FLOOD ZONE'
```

The next example shows how [FORMAT] formats the display of data returned from the evaluation of numbers when &FORMAT doesn't:

```
Arcplot: &format 2
Arcplot: &getpoint &map
Arcplot: &type x = %pnt$x% y = %pnt$y%
x = 6805148.276122 y = 1835775.862212
Arcplot: &type [format 'x = %1% y = %2%' %pnt$x% %pnt$y%]
x = 6803876.27 y = 1835775.86
```

[FORMAT] places the value of %PNT$X% at the location of %1% and the value of %PNT$Y% at the location of %2% in the quoted format string.

FYI See lesson 10.3 for more about the [FORMAT] function.

[FORMAT] also controls field width and justifies output. To accomplish this task, specify a second integer value in the format variable, separated from it by a comma. The following AML program formats the display of the items and their definitions from an INFO file:

```
/* display_item.aml
&sv cover = [getcover]
&sv feature = [getchoice poly line point]
&type
&type        Item Definition for [entryname %cover%]
&type        ---------------------------
&type        Item Name    Item Definition
/* Loop through the items in the attribute table
&do item &list [listitem %cover% -%feature%]
  /* Format the item names and their definitions
&type [format '%1,-14%    %2%' %item% [iteminfo %cover%~
   -%feature% %item% -definition]]
&end
&return
```

The –14 that follows the first format variable means that the argument substituted for this variable displays right-justified in a field that's fourteen characters wide. A positive integer specifies left justification. The spaces in the format string `'%1,-14%    %2%'` align the arguments with the column titles Item Name and Item Definition.

Suppose the AML program runs and the user chooses a coverage called SOILS and the feature class POLY. The result might look like this:

```
Item Definition for soils
---------------------------
Item Name    Item Definition
     AREA    8,18,F,5
PERIMETER    8,18,F,5
   SOILS#    4,5,B,0
 SOILS-ID    4,5,B,0
SOIL_TYPE    2,2,I,0
```

FYI

There is also the [FORMATDATE] function, which allows you to format a date string independent of its present format. This function is primarily used with ArcStorm's historical database capabilities. See the online help command references for more information on [FORMATDATE].

Question 7-2: Fill in the missing code to provide a formatted string similar to the following: The width of the coverage is 10891.0 and the height is 5779.3.

```
&sv cov = [getcover]
&describe %cov%
&sv width = %dsc$xmax% - %dsc$xmin%
&sv height = %dsc$ymax% - %dsc$ymin%
/* Set the decimal digit display to 1
_____
/* Format the string
&type _____
     _____
&return
```

(Answer on page 7-31)

Evaluating length, position, or existence of characters in a string

Several AML functions can determine the characteristics of a string. This string query is necessary for such tasks as conditional processing or finding a substring in a string. The existence of a certain character string, or an element therein, may act as a condition for other actions.

The following functions perform these query tasks:

Function name	Purpose
[INDEX]	Returns the position of the left-most occurrence of a specified string in a target string
[KEYWORD]	Returns the position of a keyword in a list of keywords
[LENGTH]	Returns the number of characters in a string
[NULL]	Indicates whether a string is all blanks (i.e., null) or if it contains characters
[SORT]	Sorts a list of characters
[SEARCH]	Returns the position of the first character of a search string in a target string
[VERIFY]	Returns the position of the first character in a target string that doesn't occur in a search string

Use the [NULL] function for conditional processing by testing for the existence of characters in a string. The following AML code fragment enters a loop only if the variable FILE contains data:

```
/* Pick a file to process or none
&sv file = [getfile * -file -none]
&do &while not [null %file%]
    .
    .
    .
  &sv file = [getfile * -file -none]
```

The -NONE option to the [GETFILE] function offers _NONE_ as one of the choices on the file selection menu. If the user picks _NONE_, the variable FILE is set to null (no value). The [NULL] function tests for the existence of a value in FILE. [NULL] returns .TRUE. when FILE is null and .FALSE. when FILE contains characters. In the example, the loop is entered only if a file is picked from the menu. The line &sv file = [getfile * -file -none] inside the loop allows the user to continue processing files until _NONE_ choice is selected.

The [INDEX] function is useful for finding the position of specified characters in a string. [INDEX] returns an integer indicating the starting position of the left-most occurrence of the specified characters in the string.

The [NULL] and [VALUE] functions can be used together to verify whether a variable has been set or not. If [NULL] tests a variable by itself, it returns an error whenever that variable hasn't been defined:

```
&type [null %some_var%]
AML ERROR - Undefined variable: some_var

&type [null [value some_var]]
.TRUE.
```

In the second example, the function [VALUE] returned a null value that [NULL] interpreted to .TRUE. This combination of functions was used on page 7-12 when the program checked for the existence of a global variable. This functionality is also useful for checking to see if an argument was passed to an AML program. Passing arguments is covered in chapter 8.

Suppose you have a list of owners' names listed with the last name first in a file. Examine the following AML code fragment, which uses [INDEX] to read the names that begin with the character the user specifies:

```
/* Read owner names beginning with a specified letter
&sv first_letter = [response 'Enter the first letter of ~
  the last names to be processed']
/* Get owner name from a file and store the data in the
/* variable, NAME. The syntax for this is shown in
/* chapter 10.
    .
    .
    .
/* If the letter specified as %first_letter% is the
/* first character in the last name
 &if [index %name% %first_letter%] = 1 &then &do
    .
    .
```

If you want to search a string for a unique match, use the [KEYWORD] function. [KEYWORD] returns the element number for a word within a string. [KEYWORD] is most useful in conjunction with &SELECT blocks because it isn't case-sensitive and it also accepts unique abbreviations. See page 4-26 for an example of [KEYWORD] used with a &SELECT block.

Question 7-3: Fill in the missing code to locate the position of a specified item in an INFO file.

Note: [LISTITEM] returns a space-delimited string of the items in an INFO file.

```
/* item_pos.aml
&sv cover = [getcover]
&sv feature = [getchoice LINE POINT POLYGON]
&sv item = [getitem %cover% -%feature%]
&sv count - [_____ %item% [listitem %cover% -%fea-
ture%]]
&lv count
&return
```

(Answer on page 7-31)

Parsing and returning parts of strings and lists

So far in this lesson, you've learned to modify character strings for performing such tasks as changing the case of strings, placing or removing quotes around strings, changing the format of strings, and finding the position of an element in a string. Many applications need to return specified substrings or elements to a program once their position is determined. *Parsing,* or the ability to break data into smaller pieces so that a program can use the data, is a process that AML accomplishes with the following functions:

Function name	Purpose
[AFTER]	Returns the string after the first occurrence of the search string in the target string
[BEFORE]	Returns the string before the first occurrence of the search string in the target string
[EXTRACT]	Extracts a specified element from a list of elements
[SUBSTR]	Extracts a substring from a string starting at a specified character position
[TOKEN]	Allows elements in a list to be manipulated

The [AFTER] and [BEFORE] functions are useful when the information you need is located entirely before or after a known substring. On the Windows NT platform, you may want to find only the drive letter on which your current workspace is located.

```
Arc: &type [show &workspace]
c:\users\default
```

This gives you the entire path of your current workspace, but you only want the drive letter. If you extract the characters before the first backslash, you're left with only the drive letter.

```
Arc: &type [before [show &workspace] \]
c:
```

Likewise, in UNIX, if you wanted to return the first directory in your path, you could do so with the [BEFORE] and [AFTER] functions.

```
Arc: &type [show &workspace]
/tavarua2/tom/workbook/aml_71/amlwb/ch07/ls2
```

```
Arc: &sv path = [after [show &workspace] /]
Arc: &type %path%
tavarua2/tom/workbook/aml_71/amlwb/ch07/ls2
Arc: &type [before %path% /]
tavarua2
```

or:

```
&type [before [after [show &workspace] /] /]
```

Use the [EXTRACT] function when you know the position of the element you want to extract from a string and return to a program. Consider an application that allows the user to change the map extent using the screen cursor. Suppose the program needs the width and height of the new map extent. The following AML code fragment returns this information to the program by extracting the needed coordinates from the output produced by the ARC/INFO command, SHOW MAPEXTENT:

```
/* Determine the width/height of the new map extent
&s mapex  = [SHOW MAPEXTENT]
&s width  = [EXTRACT 3 %mapex%] - [EXTRACT 1 %mapex%]
&s height = [EXTRACT 4 %mapex%] - [EXTRACT 2 %mapex%]
```

The [SHOW MAPEXTENT] function returns a string consisting of these four elements: XMIN, YMIN, XMAX, and YMAX. The width is calculated by extracting the values of XMAX and XMIN and performing the required subtraction. The process is repeated using the YMAX and YMIN values for the height.

Another function useful for extracting information from character strings is the [SUBSTR] function. [SUBSTR] extracts a substring whose starting position in a string is known. The following AML code fragment performs conditional processing based on the exchange value in a telephone number. In the example, the variable PHONE_NUMBER (set earlier in the program) stores values like 909-793-2853. The substring to be returned is the string 793. This action is completed by having the [SUBSTR] extract a string starting at position 5 and extracting three characters from the data in PHONE_NUMBER.

```
/* determine the exchange value in %phone_number%
&sv exchange = [substr %phone_number% 5 3]
&select %exchange%
  &when 793; &do
    .
    .

    .
  &when 792; &do
    .
    .
    .
```

One of the most useful functions for manipulating the elements in a character string is the [TOKEN] function. [TOKEN] can perform a wide variety of operations which are determined by a number of arguments that work within the function. The following table is a list of those arguments:

[TOKEN] Arguments	Purpose
-COUNT	Returns the number of elements in a list
-FIND	Returns the position of the element within the list
-MOVE	Moves an element from one position to another
-INSERT	Inserts a string at a given position
-DELETE	Deletes an element at a given position
-REPLACE	Replaces an element at a given position with a new string
-SWITCH	Switches the position of two elements

Suppose you wanted to assign all the line coverages in your current workspace a different color. The next example uses the [TOKEN] function to find out how many line coverages exist in the current workspace:

```
/* draw.aml
&s covs = [unquote [listfile * -cover -line]]
&s number = [token %covs% -count]
&if %number% ne 0 &then
  &do
    &do count = 1 &to %number%
      &s cover = [extract %count% %covs%]
      clear
      mape %cover%
      linecolor %count%
      arcs %cover%
      &pause
```

```
        &end
      &end
  &return
```

The function [LISTFILE] returns a comma-delimited list of files fitting a certain description. In DRAW.AML, line coverages are the files that are returned. [TOKEN] counts up the number of elements in the string that [LISTFILE] created. [EXTRACT] is then used within the loop to pull out one coverage at a time for drawing.

Here are some more examples of [TOKEN] at work:

```
Arc: &s text = ESRI produces the software ARC/INFO

Arc: &type [TOKEN %text% -SWITCH 4 5]
ESRI produces the ARC/INFO software

Arc: &type [TOKEN %text% -REPLACE 5 ArcView]
ESRI produces the software ArcView

Arc: &type [TOKEN %text% -INSERT 4 excellent]
ESRI produces the excellent software ARC/INFO
```

Question 7-4: Fill in the missing code that returns the symbol number assigned to a background coverage in ARCEDIT.

[show backcoverage 1] returns the back coverage name and its symbol number. These values are returned as a single string delimited by a comma (e.g., /usr/home/network/street,3).

```
  /* Determine the symbol number for the third backcoverage
  &sv backcov_symbol = [_____[show backcoverage 3]]
```

(Answer on page 7-31)

In summary

AML can manipulate character strings to return needed data to your programs. Common string manipulation tasks include changing the case of a character string, substituting strings, finding the position of substrings, and extracting substrings from strings.

A character string containing embedded blanks must be quoted for AML to treat it as one element, and strings with quotes must be unquoted to satisfy the data input requirements of some ARC/INFO commands. Remember that AML treats sets of characters separated by blanks or commas as separate elements in a string.

In the next lesson, you'll see how nesting functions provides more powerful character string manipulations.

Answer 7-1:

```
/* drawcov.aml
&sv covers = [response 'Enter coverages you wish to display']
/* Set the map extent to the coverages
MAPEXTENT [unquote %covers%]
&sv old_linesymbol = [show linesymbol]
&sv linecolor = 1
/* Loop through the list of coverages
&do cov &list [unquote %covers%]
  LINECOLOR %linecolor%
  ARCS %cov%
  &sv linecolor = %linecolor% + 1
&end
LINESYMBOL %old_linesymbol%
&return
```

Answer 7-2:

```
/* description.aml
&sv cov = [getcover]
&describe %cov%
&sv width = %dsc$xmax% - %dsc$xmin%
&sv height = %dsc$ymax% - %dsc$ymin%
/* Set the decimal digit display to 1
&format 1
/* Format the string
&type [format 'The coverage width is %1% and ~
  the height is %2%' %width% %height%]_
&return
```

Answer 7-3:

```
/* item_pos.aml
&sv cover = [getcover]
&sv feature = [getchoice LINE POINT POLYGON]
&sv item = [getitem %cover% -%feature%]
&sv count = [keyword %item% [listitem %cover% -%feature%]]
&lv count
&return
```

Answer 7-4:

```
&sv backcov_symbol = [extract 2 [show backcoverage 3]]
```

Exercises

Exercise 7.2.1

Write the code to concatenate two quoted strings. The resulting string should contain a space between the two original strings and no embedded quotes. For example, if the two strings are `'ARC ON'` and `'NODE ERRORS'`, the resulting string should be: `'ARC ON NODE ERRORS'`

Exercise 7.2.2

Write a program that counts the number of AML programs and MENU files within the current directory. If there are no AML programs or MENU files within the current directory, then the program should return a different message informing the user that there are none of the specified file. The output should look something like this:

```
Arc: &r ex7_2_2
There are 4 AML(s) in this directory
There are no menus in this directory
```

Hint: The function [LISTFILE *.aml -FILE] returns a comma-delimited list of files that have the .AML extension in the current directory.

Exercise 7.2.3

Write the code to extract three values from an ASCII file formatted with a fixed length. You don't yet know the ins and outs of file input/output (covered in chapter 10), so treat the problem as one line of text in the file. The goal is to extract the SECTION, BLOCK, and LOT numbers from the string. They are located together in the string from position 7 through 22, with the following lengths:

```
SECTION: 6
BLOCK:   4
LOT:     6
```

Write the code assuming that the string is contained in a variable named STRING. Assume that the string always contains data in positions 7 through 22 that corresponds to the data you're searching for.

```
i.e., STRING = 0117mb0616tb757311463r

Section = 0616tb
BLOCK = 7573
LOT = 11463r
```

Exercise 7.2.4

Write a program that displays the minimum and maximum x- and y-coordinates for a coverage chosen from a menu. Format the coordinates to print three decimal places and right-justified. The output should look like this example:

```
The bounding coordinates from the coverage shoreline are:

Xmin =      -322856.750  Ymin =        10656824.243

Xmax =      1775052.875  Ymax =        12151713.714
```

Exercise 7.2.5

Suppose you have the following three coverages of the type indicated (i.e., arc, point, or poly):

```
STREETS    arc
SITES      point
ZONING     poly
```

Sets of features have been reselected from each of the three coverages in ARCPLOT. The task is to set the MAPEXTENT to the area that is the maximum extent of all of the features in the selected set across all three coverages. Write an AML procedure that sets the MAPEXTENT you want.

Hint: The line MAPEXTENT <feature> cover1 cover2 cover3 won't work because it considers only one feature type, and you have three feature types here. The line MAPEXTENT cover1 cover2 cover3 also won't give you what you want because it sets the map extent to the maximum bounding rectangle (BND) of the combined coverages, not to the set of selected features across all three coverages.

Exercise 7.2.6

You're given the following &SETVAR lines at the beginning of an AML program:

```
&sv  x1   1.234
&sv  y1   10.567
&sv  x2   88.7654
&sv  y2   100.09876
&sv  x3   9876.65
&sv  y3   10012.111
&sv  x4   1234567.8901
&sv  y4   2345678.9012
```

Write the rest of the AML using &FORMAT, [FORMAT], and &TYPE to produce the following output on the screen:

```
x =         1.23   y =         10.57

x =        88.77   y =        100.10

x =      9876.65   y =      10012.11

x =   1234567.89   y =   2345678.90
```

Note: Numeric fields are ten columns wide and right-justified.

Hint: See the [FORMAT] function in the *AML User's Guide*.

Lesson 7.3—Nesting string functions to perform complex tasks

In this lesson

This lesson shows how nesting the string functions discussed in the previous lessons performs complex tasks and reduces the amount of code needed to perform those tasks. If you need to review nested functions, refer to chapter 1. Remember that AML evaluates nested functions from the inside out.

Determining the number of directories in a path

Storing data too many levels removed from the root directory can make database management difficult. Suppose you want to limit the number of directory levels under the root level to which data can be written. The following code fragment determines the number of directories in a pathname. This code could be used to ensure that the data isn't written at a level more than a specified number of directories deep.

```
&sv dir_num = [token [subst [dir [pathname *]] / ,] -count]
```

If the application runs with the user located at /usr/home/network/data, the nested functions are evaluated and the variable dir_num is set as follows:

- [pathname *] returns: /usr/home/network/data/*

  [PATHNAME] returns a fully expanded file specification.

- [dir [pathname *]] returns: /usr/home/network/data

  [DIR] returns the directory portion of the file specification.

- [subst [dir [pathname *]] / ,] returns: ,usr,home,network,data

  [SUBST] replaces the directory delimiters with commas. Remember: Comma delimiters allow directories to be treated as separate elements.

- [token [subst [dir [path *]] / ,] -count] returns: 4

  [TOKEN -COUNT] gives a count of the total amount of elements in the string.

Determining the length and data type of elements in a string

The following example demonstrates how character string functions can determine how many elements are in a character string and the length and data type of each element. The program evaluates the string stored in the variable, RECORD, set previously in the program. This character string could be provided by the user or read from an ASCII file.

```
[1]    &sv num_elements = [token [unquote %record%] -count]
[2]    &type /&The record has %num_elements% elements which are ~
       of the following length and data type:
[3]    &do i = 1 &to %num_elements%
[4]      &sv length = [length [extract %i% [unquote %record%]]]
[5]      &type /&Element %i% is %length% characters in length
[6]        &select [type [extract %i% [unquote %record%]]]
[7]          &when 1
[8]             &type Data is character
[9]          &when 2
[10]            &type Data is boolean
[11]         &when -1
[12]            &type Data is integer
[13]         &when -2
[14]            &type Data is numeric
[15]      &end /* select
[16]   &end
```

AML evaluates the character string functions as follows:

In line [1], [UNQUOTE] removes quotes that AML placed on the string if it contained embedded blanks. [TOKEN -COUNT] counts the number of elements in the string %record%.

In line [4], [UNQUOTE] removes any quotes that are present. [EXTRACT] retrieves each element, one at a time, beginning with the first element. [LENGTH] determines the length of each element after it's retrieved.

In line [6], [UNQUOTE] and [EXTRACT] perform the same function as in previous steps. [TYPE] determines the data type of the element. Finally, the &SELECT block of code uses the value returned by [TYPE].

Reordering elements in a string

Sometimes it's necessary to exchange the positions of elements in a character string. For example, assume a user has an ASCII file containing data about property owners delimited with spaces. The owners' last and first names occupy the second and third columns of data in the file, respectively. An application needs the names returned with the first name first. The following AML code fragment changes the position of the first and last names for each record of owner data stored in the variable RECORD. The first letter of the first and last names is also capitalized.

This variable RECORD is set by reading the file of owner data one record at a time. Reading the file isn't covered in this example, but you'll examine reading and writing files in chapter 10.

```
&s record = '33j07 BRENNEMAN TOM'
&s temp_string = [unquote %record%]
&s apn = [extract 1 %temp string%]
&s uplname = [extract 2 %temp_string%]
&s upfname = [extract 3 %temp_string%]
&s last_name = [upcase [substr %uplname% 1 1]]~
[locase [substr %uplname% 2]]
&s first_name = [upcase [substr %upfname% 1 1]]~
[locase [substr %upfname% 2]]
&s new_string = %apn% %first_name% %last_name%
```

- [UNQUOTE] is used to drop the quotes off of record.

  ```
  &s temp_string = 33j07 BRENNEMAN TOM
  ```

- Each element is then extracted and placed in separate variables.

  ```
  &s apn = 33j07
  &s uplname = BRENNEMAN
  &s upfname = TOM
  ```

- [SUBSTR] is then used to disjoin the first letter of the last name from the rest of the last name. The first letter is capitalized with [UPCASE] while the rest of the name is reduced to lowercase. When extracting the first letter, notice that 1 is specified as the number of characters to extract. To pull out the rest of the name, you don't specify how many characters to extract so [SUBSTR] extracts the rest of the characters in the string.

  ```
  &s last_name = [upcase B][locase RENNEMAN]
  &s last_name = Brenneman
  ```

- The same operation is repeated on the first name.

- The variable new_string is then set to the new string.

```
&s new_string = %apn% %first_name% %last_name%
&s new_string = '33j07 Tom Brenneman'
```

Determining the current platform

You may have to create an AML application that is frequently used on both the UNIX and Windows NT platforms. In order to optimize efficiency, you may create different modules to execute depending on the platform. At this point, it may be necessary to make a decision in your application based on which platform you're using. The following block of code uses the [SHOW] function to establish the current platform:

```
/* platform.aml
&if [extract 1 [show &os]] lk *nt* &then
   &run NT
&else
   &run UNIX
&return
```

- First the [SHOW &OS] function is evaluated.

```
&if [extract 1 Windows_NT,1057,3.51,Intel,CAMBRIA] lk *nt*
&then
```

- [EXTRACT 1] pulls out the platform name, which is the first element in the string.

```
&if Windows_NT lk *nt* &then
```

- The string "Windows_NT" does contain the letters "nt." The logical expression evaluates to .TRUE. and the Windows NT version of the application is executed. For more information on logical operators, see page 4-4 of this workbook or the online help index under *operators, AML.*

Question 7-5: Fill in one line of code using the character string functions [UNQUOTE], [EXTRACT], and [SUBSTR] to set a variable to the first letter in a client's last name. Clients' names are stored in an ASCII file formatted with first name, middle initial, and last name respectively. Each element is delimited by a space. The variable RECORD stores the client's name.

```
&sv first_let = [_____[_____[_____%record%]]_____
```

(Answer on page 7-39)

In summary

Nesting character string functions allows you to accomplish complex tasks, such as determining the number of directories in a path, reordering elements in a string, and determining the length and data type of elements in a string. AML always evaluates nested functions from the inside out.

Answer 7-5: [substr [extract 3 [unquote %record%]] 1 1]

Exercises

Exercise 7.3.1

Write a single line of code that returns the major revision number for ARC/INFO. For example, if you're running Version 7.1.1, then the line should return the number 7.

Hint: Check out [SHOW VERSION].

Exercise 7.3.2

Write a program that accepts a person's first and last name in a [RESPONSE] function and transforms it to uppercase using first initial and last name. For example, if the user types this:

```
judy boyd
```

the following is output:

```
J. BOYD
```

Exercise 7.3.3

Write a program that allows the user to select a RELATE by name from a menu. Return the type of relate access to the user.

Hint: Refer to the RELATE subsection under the [SHOW] function (ARC, ARCEDIT, or ARCPLOT) in the *AML User's Guide*.

Exercise 7.3.4

Write a program that prompts the user to enter a coverage and map scale with the [RESPONSE] function. Make sure that the data entered is actually a coverage name and an integer scale. If not, tell the user and end the program. If the data is correct, use it for the ARCPLOT commands MAPEXTENT and MAPSCALE.

Exercise 7.3.5

Write a program that allows the user to select an INFO file from a menu and then an item from the chosen file. The program should report the ITEM TYPE to the user.

8 ▽ Divide and conquer

This chapter introduces two programming techniques: modular programs and routines. These techniques represent different approaches to programming, but aren't mutually exclusive. Modular programs perform specific tasks as needed in a larger application. Routines also accomplish specific tasks as needed, but are blocks of code in a larger program. Lesson 8.3 explains the advantages of each method and shows how to use them together in one application.

Applications usually execute many tasks. Some tasks need to be performed by different parts of the application or more than one application. You can store blocks of code that perform discrete tasks in separate programs accessible to one or more applications. This style of writing programs is known as *modular programming*. Modular programming allows you to break a large application down into smaller, more manageable parts. You can use these parts, called modules, repeatedly throughout your application. If you find yourself copying blocks of code from one program to use in others, you'll appreciate this technique. This chapter introduces modular programs and explains techniques for passing data between them.

This chapter also introduces another programming technique whereby blocks of code that perform discrete tasks are placed at the end of the main body of a program and then executed from the program as needed. These blocks of code are called subprograms or *routines*.

This chapter covers the following topics:

Lesson 8.1—Dividing complex programs into discrete tasks

- Writing modular programs
- Calling other AML programs and menus

Lesson 8.2—Passing data between programs and menus

- Using global variables to pass data between programs
- Passing data between a program and a menu
- Passing arguments: Please pass the &ARGS

Lesson 8.3—Using routines in programs

- Getting into an &ROUTINE
- Combining modularity with routines

Lesson 8.1—Dividing complex programs into discrete tasks

In this lesson

This lesson introduces modular programming. Modular programming is a technique that divides a complex program or application into independent programs called modules. This lesson shows you how to use modular programming techniques for building applications. You'll also learn how to invoke AML programs and menus from other programs and menus.

Writing modular programs

A *module* is a collection of instructions that perform a discrete task. You should design and write modules in such a way that several applications can use them. A modular program accepts data from other programs, generates its own data, and provides data to the application. Some advantages of writing modules instead of one large program that does everything include the following:

- Modules can be reused in any number of applications.
- Modules can perform system- or site-dependent functions so that the rest of an application remains generic.
- Modules are less complex and easier to read and understand than one long program.
- Modules are easier to maintain and errors easier to isolate.

The following diagram illustrates a modular approach. Individual programs perform the tasks of getting, drawing, and querying a coverage:

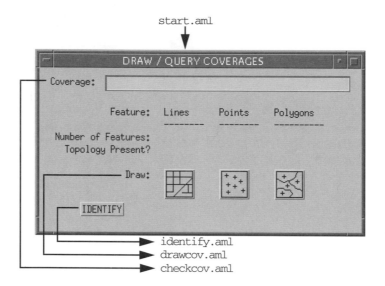

START.AML invokes the DRAW/QUERY COVERAGES menu. From the menu, you can retrieve a coverage and information about it (e.g., the type of topology or the number of features), identify specific features in the coverage, and draw lines, points, and polygons. The programs CHECKCOV.AML, DRAWCOV.AML, and IDENTIFY.AML are modules that perform discrete tasks. These programs could be called from other parts of the application or other applications to perform the task for which they were written.

You could write the code needed to invoke the menu and complete these tasks in one program, but modularity offers several advantages. For example, assume DRAWCOV.AML brings up a menu of coverages and allows the user to pick one and draw it using default symbols. With a nonmodular approach, the AML code needed to perform this task must be repeated in every application that includes this functionality. If you want to expand the program so that the user can draw coverages using symbols based on feature attributes, you need to incorporate the change in every application. With a modular approach, however, only DRAWCOV.AML needs to be modified to incorporate the new functionality in every application that uses it.

Calling other AML programs and menus

Building a complex application using modular programming requires that you design and write many separate programs. Programs and menus are invoked in the application as needed. The &RUN and &MENU directives (introduced in chapters 1 and 2) invoke programs and menus, respectively. They can be issued from the command line or from AML programs and menus. Study the following example:

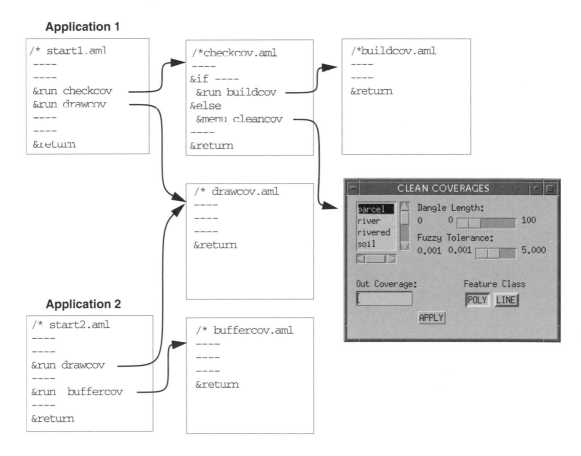

The MAKE TOPOLOGY application (application 1) accesses several programs that perform discrete tasks. Notice that applications 1 and 2 both access DRAWCOV.AML.

Question 8-1: Which of the two techniques shown below uses the modular programming approach? _____

Technique 1:

```
/* Start1.aml
-----
-----
-----
/* Draw coverages
ARCPLOT
&sv cov = [getcover * -poly -none]
&do &while not [null %cov%]
  MAPEXTENT %cov%
  POLYGONS %cov%
  &sv cov = [getcover * -poly -none]
  CLEAR
&end
QUIT
-----
-----
&return
```

Technique 2:

```
/* Start2.aml
-----
-----
-----
/* Draw coverages
&run drawcov
-----
-----
&return
```

```
/* Drawcov.aml
ARCPLOT
&sv cov = [getcover * -poly -none]
&do &while not [null %cov%]
  MAPEXTENT %cov%
  POLYGONS %cov%
  &sv cov = [getcover * -poly -none]
  CLEAR
&end
QUIT
&return
```

List four advantages of modular programming:

_____

_____

_____

_____

(Answers on page 8-7)

In summary

Modular programs perform discrete tasks or a group of related tasks. Modular programming reduces the amount of redundant code needed to run your applications and makes maintaining your applications easier.

Because modular programming creates many smaller programs instead of one large one, these programs need to call other programs and menus. Use &RUN and &MENU directives in your AML programs to call other programs and invoke menus.

Answer 8-1: Technique 2
(1) Modules are usable more than once in an application and in other applications.
(2) Modules can isolate system- or site-dependent functions so that the rest of an application remains generic.
(3) Modules are less complex and easier to read and understand.
(4) Modules are easier to maintain and errors easier to isolate.

Exercises

Exercise 8.1.1

Divide the following AML program into modules that perform discrete tasks. The application sets environments, edits coverages, builds or updates topology, and performs polygon overlay between coverages. One program should call the programs you create as they're needed. Use your own naming convention for the programs. Refer to the ARC/INFO online command references for help with commands you aren't familiar with.

```
/* do_everything.aml
&terminal 9999
&fullscreen &popup
DIGITIZER 9100 ttya
COORDINATE DIGITIZER
ARCEDIT
&sv cov = [getcover * -all 'Choose a cover to edit']
EDIT %cov%
MAPEXTENT %cov%
DRAWENVIRONMENT [unquote [response 'Specify features to display']]
DRAW
/* Use Editcov.menu to perform editing tasks
&menu editcov &position &ur &screen &ur
SAVE %cov% %cov%ed
&type /&Your edited coverage is [entryname %cov%ed]
QUIT
&sv cov = [getcover * -all 'Choose the cover you wish to build']
COPY %cov% %cov%bd
&sv topology = [getchoice POLY LINE POINT -prompt 'What type of
  topology to build?']
BUILD %cov%bd %topology%
&type /&Your built coverage is [entryname %cov%bd]
&sv incov = [getcover * -all 'Choose an in cover']
&sv cov_type = [getchoice poly line point -prompt 'What type of ~
  cover is your in cover?']
&sv identity_cover = [getcover * -poly 'Pick the cover to~
  combine']
IDENTITY %incov% %identity_cover% compositecov %cov_type%
&return
```

Exercise 8.1.2

Write the AML directive statement that invokes a program called 911.AML for emergency service analysis when the user chooses 911 from the pulldown menu shown here:

Lesson 8.2—Passing data between programs and menus

In this lesson

Modular programming requires data to pass between programs. For example, a program might need data from the calling program to perform its tasks, or the called program might generate data needed by another program or menu. In this lesson, you'll learn how to pass data between AML programs and between programs and menus.

Using global variables to pass data between programs

Global variables are one vehicle for passing data between AML programs. Global data is available to any AML program or menu and is accessible from every ARC/INFO subsystem within the same session. *Scope* is a term used to describe the extent to which a given variable can be referenced from inside and outside of programs. Global variables are within the scope of all ARC/INFO and AML programs in a given ARC/INFO session. Refer to chapter 1 for more discussion about global variables.

Examine the following example:

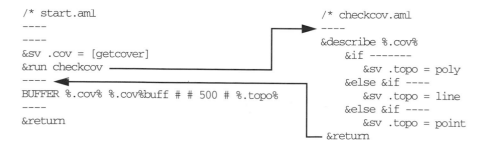

```
/* start.aml                                      /* checkcov.aml
----                                              ----
----                                              &describe %.cov%
&sv .cov = [getcover]                                 &if -------
&run checkcov                                             &sv .topo = poly
----                                                  &else &if ----
BUFFER %.cov% %.cov%buff # # 500 # %.topo%               &sv .topo = line
----                                                  &else &if ----
&return                                                   &sv .topo = point
                                                      &return
```

CHECKCOV.AML accesses the value of the global variable .COV, which is defined in START.AML. In turn, START.AML performs an operation based on the value of the global variable .TOPO, which is defined in CHECKCOV.AML.

Use caution when you define global variables. Global data can cause errors that are hard to trace because any program can access and manipulate it. Examine the following AML programs:

```
/* edit.aml              /* drawcov.aml            /* describe.aml
ARCEDIT                  ARCPLOT                   &sv .cov = [getcover]
&sv .cov = [getcover]    MAPEXTENT %.cov%          &describe %.cov%
                         POLYGONS %.cov%
      .                                                  .
      .                        .                         .
&return                        .                   &return
                         &return
```

EDIT.AML and DESCRIBE.AML both define a variable named .COV. DRAWCOV.AML accesses the data stored in .COV. The data stored in .COV, however, depends on which of the two programs was last executed. With this in mind, consider the following scenario which demonstrates, on a small scale, how global data can corrupt other programs.

Suppose that an AML startup program sets the global variable .COV to a coverage named PARCELS, which is accessed later by an application that uses DRAWCOV.AML. Imagine that the user first runs EDIT.AML to correct some spatial errors in a coverage named SOILS. Next, the user runs the application that uses DRAWCOV.AML, expecting it to draw features for the PARCELS coverage. Instead of the PARCELS coverage, the SOILS coverage displays because the EDIT.AML program reset .COV. If the user isn't familiar with the programs that define .COV, tracking the problem to its source is difficult.

Manage global data carefully. One advantage of the modular approach is that you can protect data from becoming corrupted by controlling access to it. Overusing global variables defeats the purpose of modular programming by making data accessible to every module.

Following these guidelines helps prevent data corruption:

- Define global variables only when they're necessary
- Delete global variables when they're no longer needed

Consider another example:

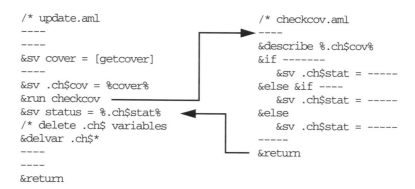

```
/* update.aml                    /* checkcov.aml
----                             ----
----                             &describe %.ch$cov%
&sv cover = [getcover]           &if -------
----                                 &sv .ch$stat = -----
&sv .ch$cov = %cover%            &else &if ----
&run checkcov                        &sv .ch$stat = -----
&sv status = %.ch$stat%          &else
/* delete .ch$ variables             &sv .ch$stat = -----
&delvar .ch$*                    -----
----                             &return
----
&return
```

The global variables in this example are all named using the same prefix (.CH$) and deleted immediately after serving their purpose. Using a naming convention for variables that pass related data (e.g., CH$ for coverage-checking information) helps manage global data. There's no significance to the dollar sign ($) other than that it's a convenient separator and likely to keep the name unique. This type of naming convention reduces the chance of using the same variable name for different purposes and ensures that global variables can be traced and deleted as a group.

The ArcTools naming convention for global variables is .TOOLNAME$VARIABLE_NAME. For example, the APPEND tool has a global variable .append$in_subclass. With this method, it's very easy to identify out-of-place global variables and delete all global variables for a specific tool. If you use this naming convention, be careful not to exceed the thirty-two-character limit for variable names.

Question 8-2: Fill in the missing code in the blanks below that facilitates passing data between the two AML programs and then deletes the global data.

```
/* plot.aml
  .
  .
&sv .plot$cov = [getcover]
&run checkcov
/* If the coverage width and height are greater than 3000 feet
&if _____ > 3000 and _____ > 3000 &then
   PAGESIZE 24 20
&else
   PAGESIZE 17 11
/* Transfer .plot$cov to a local variable
&sv cover = %.plot$cov%
/* Delete the .plot$ variables
_____
  .
  .
&return


============================================================
/* checkcov.aml
  .
  .
/* Set the DSC$ variables for the coverage
&describe _____
/* Set global variables to the width and height of the
/* coverage
&sv .plot$width = %dsc$xmax% - %dsc$xmin%
&sv .plot$height = %dsc$ymax% - %dsc$ymin%
&return
```

(Answer on page 8-22)

Passing data between a program and a menu

Sharing data between an AML program and a menu is handled differently than sharing data between AML programs. The following rules apply:

- When a program calls a menu, data stored in the local variables is shared.
- When a menu calls a program, data stored in local variables is not shared.

The following diagram illustrates the scope of local variables. Menus can call AML programs, but local variables remain within their own scope.

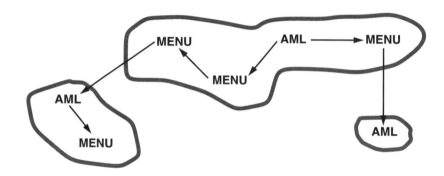

The following example demonstrates how you can pass data between a program and a menu:

```
/* drawcov.aml invokes Draw_query.menu
ARCPLOT
&sv cov = [getcover * -all 'Choose a coverage']
&sv feature_resp = [getchoice -pairs polygons polygon~
 arcs arc points point -var feature~
 -prompt 'Choose a feature type']
&menu draw_query
&delvar .st$*
RESELECT %cov% %feature% MANY *
  .
  .
  .
&return
```

DRAWCOV.AML starts the application and sets two local variables, COV and FEATURE. The program then invokes DRAW_QUERY.MENU, which allows the user to draw features for a coverage, retrieve attributes, and calculate statistics for an item. When DRAW_QUERY.MENU is dismissed, the program continues to execute, issuing the RESELECT command. DRAW_QUERY.MENU, shown below, is a form menu. (Chapter 12 discusses form menus in detail.)

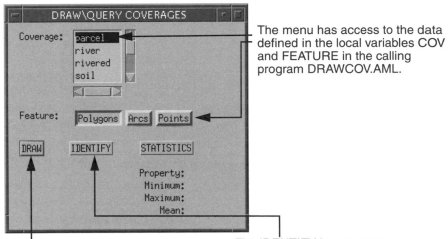

The menu has access to the data defined in the local variables COV and FEATURE in the calling program DRAWCOV.AML.

If the user presses the DRAW button, two commands execute that use the local variables:
MAPEXTENT %COV% and
%FEATURE% %COV%.

The IDENTIFY button uses local variables defined in the calling program:
IDENTIFY %COV% %FEATURE% *.

An AML menu invoked from an AML program shares data with the program through local variables. In this case, if the user changes the value of the variables COV and FEATURE by interacting with the menu, the program can access this new data. For example, refer back to the code for DRAWCOV.AML. Notice that the RESELECT command line includes the variables COV and FEATURE. Changes to the value of these variables made in the menu are used when RESELECT executes.

When the user presses the STATISTICS button, the DRAW/QUERY COVERAGES menu invokes an AML program called STATISTICS.AML. A program can't access local variables from the menu that invoked it, so it needs another method to provide the data it requires. In this case, global variables pass data to STATISTICS.AML. Examine the following diagram:

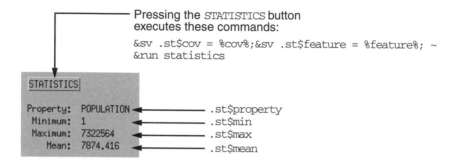

Pressing the STATISTICS button executes these commands:

```
&sv .st$cov = %cov%;&sv .st$feature = %feature%; ~
&run statistics
```

STATISTICS

Property:	POPULATION
Minimum:	1
Maximum:	7322564
Mean:	7874.416

.st$property
.st$min
.st$max
.st$mean

STATISTICS.AML accesses data from the calling menu, so global variables are used as follows:

```
/* statistics.aml
&sv .st$property = [getitem %.st$cov% -%.st$feature% ~
 -noncharacter 'Choose a Property']
STATISTICS %.st$cov% %.st$feature%
MIN %.st$property%
MAX %.st$property%
MEAN %.st$property%
END
&sv .st$min =  [show statistic 1 1]
&sv .st$max =  [show statistic 2 1]
&sv .st$mean = [show statistic 3 1]
&return
```

The global variables .ST$COV and .ST$FEATURE pass the coverage name and feature type to STATISTICS.AML. The menu needs data generated in STATISTICS.AML and stored in the global variables .ST$PROPERTY, .ST$MIN, .ST$MAX, and .ST$MEAN. The menu accesses these global variables and displays their values as shown on the graphic.

The names of these global variables use the .ST$ prefix, so you can delete them as a group when they're no longer needed. Inserting the &DELVAR directive on the line that's executed when control returns to DRAWCOV.AML accomplishes this. Notice how this is coded on page 8-14. (Chapter 12 includes a more thorough discussion on deleting variables.)

Passing arguments: Please pass the &ARGS

AML provides an alternative to using global variables for passing data to a program. The {argument...argument} option for the &RUN directive allows one or more arguments to pass to an AML program in the &RUN statement. These arguments are AML variables or strings that the program needs. Passing arguments with &RUN is an efficient way to set multiple variables. See the following usage:

```
&run <aml file> {argument...argument}
```

To receive the arguments from the &RUN statement, the program must include the &ARGS directive and specify variable names to store the data passed as arguments.

Consider the task of writing an AML program to create a plot file of a given coverage at a given scale. Two pieces of information are needed to complete the task: the name of the coverage and the scale at which to display it. In the next example, an AML program named PLOTMAP.AML accepts a coverage name and scale entered by the user as arguments in the &RUN statement as follows:

```
Arcplot: &run plotmap 24000 soils
```

To receive the arguments, the first executable line of PLOTMAP.AML reads as follows:

```
/* plotmap.aml
&args scale covername
DISPLAY 1040
%covername%.gra
PAGESIZE 24 17
MAPE %covername%
MAPSCALE %scale%
ARCS %covername%
&return
```

The &ARGS directive acts like two &SETVAR directives. SCALE and COVERNAME are the variables that store the data provided as arguments. &ARGS sets the value of SCALE to 24000 and the value of COVERNAME to soils. The order of the arguments specified with &RUN corresponds to the order in which &ARGS receives them.

With few exceptions, the &ARGS directive should be the first line in your AML program. AML allows only comments and the &SEVERITY (see chapter 9) and &ECHO directives to appear before &ARGS.

PLOTMAP.AML assumes that the user entered all the required arguments in the &RUN statement. If the user forgets, the command fails, and it may not be obvious which argument is missing. In the next example, PLOTMAP.AML is modified using AML directives and functions that you learned in previous chapters to check whether all the necessary arguments were entered.

PLOTMAP.AML expects data from two arguments when &RUN initiates the program. The &ARGS directive sets SCALE and COVERNAME to the values provided as arguments. If either or both of the arguments are missing, the program fails because one of the variables contains a null value. To keep the user from receiving a cryptic error message, check to see if data exists in the variables. The example below shows how you would verify this.

```
/* plotmap2.aml modified to check for arguments
&args scale covername
&if [null %scale%] &then
  &return &error \You did not enter a scale--\~
    USAGE: &RUN PLOTMAP <scale> <coverage>
&if [null %covername%] &then
  &return &error \You did not enter a coverage--\~
```

```
       USAGE: &RUN PLOTMAP <scale> <coverage>
DISPLAY 1040
%covername%plot
PAGESIZE 24 17
MAPE %covername%
MAPSCALE %scale%
ARCS %covername%
&return
```

After the arguments are received, PLOTMAP2.AML tests the variable SCALE.
If SCALE contains a null value, then the user didn't provide an argument. In this
case, the &IF statement is true, so the &RETURN statement terminates the
program with an error message explaining the problem and providing the usage
needed to run the program. If scale is provided, but a coverage name isn't, the
user gets a similar result.

The commands that create the plot only execute if both arguments are valid.
PLOTMAP2.AML doesn't test for valid data, so if the user enters the
arguments in the wrong order, or enters one or more invalid values, the
program fails. Checking to see if the user entered valid data requires error-
handling techniques beyond the scope of this chapter. Chapter 9 describes
how to handle this kind of error.

Question 8-3: In PLOTMAP2.AML, what happens if the user enters the
&RUN statement shown below?

```
&run plotmap2 soils
```

(Answer on page 8-22)

PLOTMAP2.AML checks for the existence of data in the variables SCALE and COVERNAME. Because both arguments are required, data in the second argument guarantees that there's also data in the first. This is useful when programs need to accept many arguments; instead of checking each one, check only the last one. If the last variable contains data, all the variables that precede it in the list are also defined. PLOTMAP3.AML shows how this is done:

```
/* plotmap3.aml modified to check last argument
&args scale covername
&if [null %covername%] &then
  &return &error USAGE: &RUN PLOTMAP <scale> <coverage>
DISPLAY 1040
%covername%plot
PAGESIZE 24 17
MAPE %covername%
MAPSCALE %scale%
ARCS %covername%
&return
```

When the number of arguments needed to run a program is constant, it's easy to code the &ARGS statement in the program. For instance, in PLOTMAP.AML, the program always needs two arguments in the &RUN statement to execute successfully.

Sometimes, however, a program needs a varying number of arguments specified by a user to complete a task. For example, suppose a program is to receive the arguments for the ARCEDIT DRAWENVIRONMENT command. Specifying a fixed number of variables in the &ARGS statement isn't practical because the user may specify one or many arguments for the command, depending on which coverage features display.

To accommodate this situation, AML can assign multiple arguments to a single variable in the &ARGS statement. The :REST option assigns any argument not yet assigned to the last variable in the &ARGS statement. The following program demonstrates how to use the :REST option:

```
/* edit.aml establishes the edit environment
/* USAGE: &RUN EDIT <coverage> <feature_class>
/*        <drawenvironment>
&args cover feature display:rest
ARCEDIT
EDIT %cover%
EDITFEATURE %feature%
```

```
DRAWENVIRONMENT [unquote %display%]
DRAW
.
.
&return
```

Invoking EDIT.AML sets three variables. The variable COVER is set to the value of the first argument, FEATURE to the second, and all remaining arguments are assigned to the last variable, DISPLAY. For example, the following &RUN directive statement might be issued to invoke the program EDIT.AML:

```
&run edit streets arc arc ids node errors anno
```

In this case, COVER is set to streets, FEATURE is set to arc, and DISPLAY accepts the remaining arguments (i.e., arc ids node errors anno). When multiple arguments are assigned to a single variable with &ARGS, AML places single quotes around the element (e.g., 'arc ids node errors anno'). Use the [UNQUOTE] function to strip the quotes before passing the data to the DRAWENVIRONMENT command. See chapter 7 if you need to review how to manage quoted strings with AML.

Question 8-4: The following program expects arguments from &RUN. The variable COVERS must accept multiple coverages as arguments. Write the &ARGS statement so that the program accepts arguments from &RUN.

```
/* plotcov.aml draws multiple coverages
/* USAGE: &RUN PLOTCOV <scale> <covers>
&args _____
ARCPLOT
MAPEXTENT [unquote %covers%]
MAPUNITS FEET
MAPSCALE %scale%
&sv line_color = 1
&do cov &list [unquote %covers%]
  LINECOLOR %line_color%
  ARCS %cov%
  &sv line_color = %line_color% + 1
&end
&return
```

(Answer on page 8-22)

In summary

Global variables can pass data between AML programs and menus. An AML program and any menu that it invokes can share data stored in local variables. A program that uses the &ARGS directive can receive local data passed in the form of arguments with the &RUN statement.

Answer 8-2:

```
      .
      .
&sv .plot$cov = [getcover]
&run checkcov
/* If the coverage width and height are greater than 3000 feet
&if %.plot$width% > 3000 and %.plot$height% > 3000 &then
  PAGESIZE 24 20
&else
  PAGESIZE 17 11
/* Transfer .plot$cov to a local variable
&sv cover = %.plot$cov%
/* Delete the .plot$ variables
 &dv .plot$*
      .
      .
&return

/* checkcov.aml
      .
      .
/* Set the DSC$ variables for the coverage
&describe %.plot$cov%
/* Set global variables to width and height of the coverage
&sv .plot$width = %dsc$xmax% - %dsc$xmin%
&sv .plot$height = %dsc$ymax% - %dsc$ymin%
&return
```

Answer 8-3: The invalid value, SOILS, is assigned to the variable, SCALE. If this is the only error in the &RUN statement, AML accepts it and the program fails when the variable is used with the MAPSCALE command. In this case, however, AML first detects the null value for the COVERNAME argument and returns the coded error message You did not enter a coverage.

Answer 8-4: &args scale covers:rest

Exercises

Use the following programs to complete exercises 8.2.1 and 8.2.2.

```
/* drawcov2.aml-- symbolizing features using attribute values
/* Set a variable to a coverage
_____ = [getcover]
&run checkcov
/* Check the type of toplogy
&select _____
  &when polys
    .
    .
  &when lines
    .
    .
  &when points
    .
    .
  &otherwise
    .
    .
&end
&return



/* checkcov.aml
/* Describe the coverage
&describe _____
/* Set a variable to the status of topology and features
/* Polygons and topology present
&if %dsc$qtopology% and %dsc$polygons% > 1 &then
  _____ = polys
/* Arcs and topology present
&else &if %dsc$aat_bytes% > 0 and %dsc$arcs% > 0 &then
  _____ = lines
/* Points and topology present
&else &if %dsc$xat_bytes% > 0 and %dsc$points% > 0 &then
  _____ = points
&else
    .
    .
&return
```

Exercise 8.2.1

DRAWCOV2.AML must pass a value for a coverage to CHECKCOV.AML, which must pass the type of topology back to DRAWCOV2.AML. Using global variables, fill in the code that accomplishes these tasks.

Exercise 8.2.2

Write the code to pass the name of the coverage from DRAWCOV2.AML to CHECKCOV2.AML as an argument in the &RUN statement.

Exercise 8.2.3

Examine the following code:

```
/* edit2.aml
ARCEDIT
&sv .cov = [getcover]
EDIT %.cov%
EDITFEATURE [getchoice arc label node anno -prompt 'Choose~
  a feature to edit']
DRAWENVIRONMENT [unquote [response 'Enter the features to display']]
DRAW
MAPEXTENT *
&sv .extent = [show mapextent]
MAPLIMITS *
&sv .limits = [show maplimits]
MAPPOSITION LL LL
&sv .position = [show mapposition]
/* Perform edits

      .
      .
save %.cov% %.cov%ed
QUIT
&run plotmap
&return

/* plotmap.aml--creates a graphics file of edited coverages
ARCPLOT
MAPEXTENT %.extent%
MAPLIMITS %.limits%
```

```
MAPPOSITION %.position%
DISPLAY 1040
ARCS %.cov%ed
   .
   .
&return
```

(a) Which change would make the global variables easier to manage?

(b) Add the code the program needs to delete all the global variables after the graphics file is created.

Use the following diagram for exercises 8.2.4 and 8.2.5.

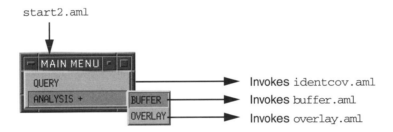

START2.AML invokes MAIN.MENU. From the main menu, buttons invoke AML programs as shown above.

Exercise 8.2.4

START2.AML must get a coverage name and pass it to MAIN.MENU, which must pass the coverage name to IDENTCOV.AML, BUFFER.AML, or OVERLAY.AML when a button is pressed. (a) Write the lines of code for START2.AML, (b) the menu directive lines that invoke the programs, and (c) the lines needed by the three programs invoked by the menu to pass the coverage name. Don't use global variables unless you have to.

Exercise 8.2.5

OVERLAY.AML creates an output coverage and must pass the name to START2.AML. Assume that OVERLAY.AML uses the UNION command to create the output coverage. The usage for UNION is as follows:

```
Usage: UNION <in_cover> <union_cover> <out_cover>
             {fuzzy_tolerance}{JOIN | NOJOIN}
```

Write the code for OVERLAY.AML to accomplish the following tasks: OVERLAY.AML should get the <in_cover> from the &RUN statement that invokes it, obtain the <union_cover>, create the <out_cover>, and pass the name of the <out_cover> to START2.AML.

Lesson 8.3—Using routines in programs

In this lesson

A section of code written in a program in such a way that it can be invoked whenever necessary is called a *routine* or *subprogram*. A routine is a program inside another program. Routines usually contain reusable code that a program needs to execute more than once to accomplish its task. This lesson describes how to include routines in your AML programs, a programming technique that broadens the scope of code stored in a modular program.

Getting into an &ROUTINE

Routines perform tasks like:

- Displaying the usage for commands and directives
- Setting environments in ARC/INFO or one of its subsystems
- Restoring environments after a program finishes executing
- Cleaning up after a program fails
- Initializing application variables and menu interfaces

Routines must be located at the end of your AML program after the &RETURN that terminates the program upon completion. In AML, a routine has a name, often called a label, and is specified using the &ROUTINE directive and the routine label (e.g., &ROUTINE EXIT).

The two directives shown in the table below are used to call routines.

AML directive	Description
&CALL	Transfers control to the block of code following the &ROUTINE directive that has the same label referenced with &CALL
&SEVERITY	Specifies what to do when a warning or error condition occurs. The &ROUTINE option can pass control to a routine.

Routines can execute actions, transfer control to other routines (&CALL), and invoke other programs (&RUN) or menus (&MENU). After the actions in the routine execute, control transfers back to the main body of the program. Execution continues with the statement following the one that called the routine.

Examine the following examples:

```
/* edit2.aml
&sv old_coordinate = [show coordinate]
&sv old_display = [show display]
&sv edit_cov = [getcover]
EDIT %edit_cov%
    .
    .
&call exit
&return
/*
/* Restore environments, save if necessary,
/* then exit ARCEDIT
&routine exit
COORDINATE %old_coordinate%
DISPLAY %old_display%
&if not [exists %edit_cov%ed -cover] &then
  /* If coverage hasn't been saved
  &do
    SAVE %edit_cov% %edit_cov%ed
  &end
&if [show program] = ARCEDIT &then
  &do
    QUIT /* From ARCEDIT
  &end
&return
```

In EDIT2.AML, &CALL EXIT transfers control to the block of code following the label &ROUTINE EXIT. The exit routine executes until &RETURN transfers control to the line of code in the main body of the program following the line that called the routine (i.e., &CALL EXIT). In this case, control passes to the &RETURN line that terminates the program.

The next example illustrates how a routine can clean up and gracefully exit if an AML program fails:

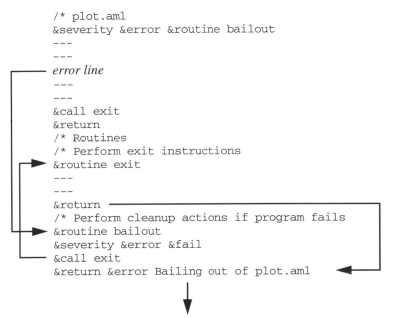

```
/* plot.aml
&severity &error &routine bailout
---
---
error line
---
---
&call exit
&return
/* Routines
/* Perform exit instructions
&routine exit
---
---
&return
/* Perform cleanup actions if program fails
&routine bailout
&severity &error &fail
&call exit
&return &error Bailing out of plot.aml
```

Transfers control to the program that invoked PLOT.AML

Using any of the &RETURN directive's options (i.e., &INFORM, &WARNING, or &ERROR) in a routine terminates the program after the routine executes instead of returning control to the main body of the AML program.

PLOT.AML includes severity-handling instructions. Chapter 9 covers severity handling in detail —for now, it's enough to understand that an error occurring on any line after the &SEVERITY statement causes control to transfer to the routine BAILOUT. The bailout routine calls another routine, EXIT, using the &CALL directive. After the exit instructions execute, control transfers back to the bailout routine. The &ERROR option to &RETURN in the bailout routine terminates PLOT.AML. Control transfers to the program that invoked PLOT.AML.

> Always place your routines at the end of a program; the &ROUTINE directive isn't allowed in the main body of an AML program.

Whether you write task-specific code as routines or as modules depends on your application. The code for a routine resides as a subprogram instead of as a separate program. When deciding between the two methods, consider these points:

- &CALL in one program can't invoke a routine found in another program.
- A block of code used only in one program can be called as a routine to save the overhead of maintaining a separate program.
- AML indexes routines, facilitating quick access to the code.
- Modular programming is a good technique to use when the modules need to be used by more than one program.

Question 8-5: The program shown below needs a coverage name passed as an argument in the &RUN statement. Fill in the missing code so a routine named USAGE is called if the argument isn't provided.

```
/* Drawcov.aml
&args cover
/* If no coverage argument, call the usage routine
&if [null %cover%] &then
  &do
    _____

  &end
  .
  .
&return
/* Routines
/* Provide usage to run drawcov.aml
_____
&return USAGE: &RUN DRAWCOV <COVERAGE>
```

(Answer on page 8-34)

Combining modularity with routines

There's a way to combine modular program style and AML routines to create modularity in a single AML program. Consider a program that invokes a menu. A strictly modular approach requires that a separate program exists for each choice on the menu. Depending on which menu choice the user makes, the appropriate program runs from beginning to end and then returns to the menu. When the user exits, the menu returns control to the calling program.

As an alternative, consider what happens if you place the module in the calling program as routines. In this case, a menu choice accesses the appropriate routine in the calling program. This type of program uses modular components, but doesn't need to execute from start to finish as programs usually do.

The following example demonstrates how you might integrate modularity and routines using a task common to most applications, changing workspaces.

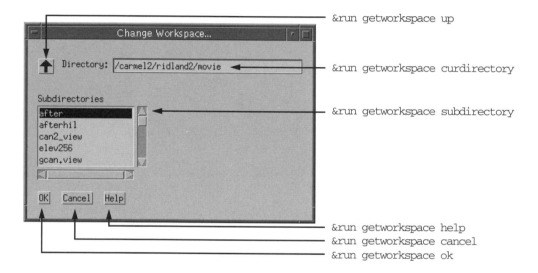

Using the CHANGE WORKSPACE menu, the user can navigate the file system to a desired location. Each choice on the menu runs the AML program GETWORKSPACE.AML with a different argument (e.g., UP, OK, HELP, etc.). Examine the example below as a skeleton of this program:

```
/* getworkspace.aml
&args routine
&call %routine%
&return
/* Routines
/*
&routine up
    .

    .
&return
/*
&routine curdirectory
```

```
      .
      .
&return

/*
&routine subdirectory
      .
      .
&return
/*
&routine ok
      .
      .
&return
/*
&routine cancel
      .
      .
&return
/*
&routine help
      .
      .
&return
```

Blocks of code that perform specific tasks are written like modules but stored in GETWORKSPACE.AML as routines. The program never executes from top to bottom, but instead accepts a routine name as an argument and executes the appropriate routine with the &CALL directive.

Other programs could access the routines in this program by passing the routine name as an argument in the &RUN statement. This technique reduces the number of programs you have to maintain. You'll examine this programming style closely in chapter 15, where ArcTools are introduced.

In summary

Routines are blocks of code written as subprograms that perform specific tasks. The &ROUTINE directive and routine name, or label, reference a routine in a program. &CALL or &SEVERITY, along with the routine name, are used to call routines. Depending on the application, modules are sometimes better written as routines in a program. Using routines reduces the number of programs you have to maintain.

Answer 8-5:
```
&call usage
&routine usage
```

Exercises

Exercise 8.3.1

What's wrong with the following code? Make the necessary corrections.

```
/* zoompan.aml
&args changeview
&select %changeview%
  &when zoom_in
    &routine zoom_in
  &when zoom_out
    &routine zoom_out
  &when pan
    &routine pan
&end
&return
/* Routines
zoom_in
    .
    .
&return
/*
zoom_out
    .
    .
&return
/*
pan
    .
    .
&return
```

Exercise 8.3.2

Rewrite the following AML program to check for the existence of the COVERAGE argument. If the argument is null, the code fragment following the program should be called as a routine.

```
/* edit.aml
&args coverage
ARCEDIT
EDIT %coverage%
DRAWENVIRONMENT [unquote [response 'Specify features to draw']]
EDITFEATURE [unquote [response 'Specify feature class to edit']]
    .
    .
&return
=========================
/* Get a coverage
&sv coverage = [getcover]
```

Expecting the unexpected

Most new programmers don't anticipate errors as they write programs. They write a powerful AML program that does it all, but find themselves surprised when an error message appears at run time. Well, nobody's perfect. This chapter shows you some time-saving techniques for testing your program's logic, known as *debugging,* without having to execute all the ARC/INFO commands it contains. You'll learn ways to isolate and correct problems in your programs until they run error-free.

And even if your application executes flawlessly each time *you* run it, that doesn't necessarily mean that it will run properly for your user. Because many ARC/INFO commands contain several arguments, it's important to realize that users may enter incorrect or incompatible parameters. This chapter also shows you how to anticipate these kinds of errors and design your program to take the appropriate actions. Finally, you'll learn how to handle the errors that occur between AML programs—what happens, and what you should do when an AML program runs another program that fails.

This chapter covers the following topics:

Lesson 9.1—Debugging AML programs

- Common debugging tools
- The &TTY directive

Lesson 9.2—Expecting and handling errors

- Premeditated error handling
- The &SEVERITY directive
- Saving and restoring environments and using routines with &SEVERITY
- The AML$MESSAGE reserved variable

Lesson 9.3—Handling errors between AML programs and menus

- Handling errors between AML programs
- Handling errors between menus and AML programs

Lesson 9.1—Debugging AML programs

In this lesson

Why does your program do what it does instead of what it's supposed to do? Every programmer occasionally writes a program that doesn't work. This lesson examines some debugging techniques you can use on your AML programs. You'll learn how to locate errors by testing a program before actually running it, and how to watch a program run to learn how the code is interpreted.

Common debugging tools

The &TEST directive checks your programming logic by running your AML program without executing the ARC/INFO commands it contains. &TEST keeps you from having to wait for time-consuming command executions (e.g., BUFFER). Use this method to make sure you assigned variables properly and used functions correctly.

The following code starts ARCEDIT and prompts the user to pick a coverage containing a route system. After the user chooses a coverage, the program provides the user with a list of route systems from which to choose, then specifies the edit feature and sets the drawing environment accordingly.

```
/* example.aml
ARCEDIT
&sv cov = [getcover * -route]
EDITCOVERAGE %cov%
&describe %cov%
&sv routfeat = [getchoice %dsc$routenames% -prompt ~
  'Please select a route-system to edit']
DRAWENVIRONMENT ROUTE.%routefeat%
EDITFEATURE ROUTE.%routefeat%
&return
```

Using &TEST &ON with &ECHO &BRIEF reveals how this program executes.

```
Arc: &test &on
(&TEST) Arc: &echo &brief

(thread0001, tty)

  24: &echo &brief
(&TEST) Arc: &r example
  25: &r example

(thread0001, example.aml)

  2: ARCEDIT
  3: &sv cov = /amlwb/ch09/ls1/streets
  4: EDITCOVERAGE /amlwb/ch09/ls1/streets
  5: &describe /amlwb/ch09/ls1/streets
  6: &sv routfeat = BUS
AML ERROR   - Undefined variable: routefeat
               line 8 of file
/amlwb/ch09/ls1/example.aml
Line causing error : DRAWENVIRONMENT route.%routefeat%
AML MESSAGE - Stopping execution of AML file due to ERROR condition
(&TEST) Arc:
```

&TEST reports an undefined variable on line 8. A typographical error on line 6, where the variable was named incorrectly (ROUTFEAT was the variable set, not ROUTEFEAT), caused the error. Of course, running this program without &TEST would reveal the same error, but &TEST avoids executing the ARC/INFO commands. ARCEDIT, therefore, never actually starts and you aren't left there without knowing which environments were set before the program failed.

FYI

Using &TEST &ON with &ECHO &BRIEF or &ON won't produce the results you want if the program you're testing includes the &MESSAGES directive with the &OFF option. In the example below, notice that &MESSAGES is set to &OFF in line 2. Although &ECHO is set to &ON, messages past this line in the program are suppressed.

```
Arc: &test &on
(&TEST) Arc: &echo &on
(&TEST) Arc: (thread0004, tty/39) &echo &on
(&TEST) Arc: &r example
(&TEST) Arc: (thread0004, tty/41) &r example
(&TEST) Arc: (thread0004, example.aml/1) ARCEDIT
(&TEST) Arc: (thread0004, example.aml/2) &messages &off
AML ERROR    - Undefined variable: routefeat line 7 of file
   /pathname/example.aml
```

It's difficult to debug a program if you can't see it, so comment out (/*) the &MESSAGES line until the program runs without errors.

Question 9-1: In the previous example, the second AML program error isn't reported and there's no Arc: prompt. This isn't a printing error; there's no prompt after this program fails. Why?

(Answer on page 9-11)

Here's another helpful debugging technique: combine &LISTVAR and &PAUSE to temporarily stop a program. Place &LISTVAR followed by &PAUSE before the section of the program where you think there's a problem. &LISTVAR lists local, global, and program variables, while &PAUSE temporarily halts the program flow until you enter a carriage return. This method is especially useful if you used &ECHO; without &PAUSE, your program continues to run and scrolls by—variables and all.

```
/* example2.aml
ARCEDIT
&sv cov = [getcover * -route]
EDITCOVERAGE %cov%
&describe %cov%
&sv routfeat = [getchoice %dsc$routenames% -prompt ~
  'Please select a route-system to edit']
&listvar
&pause
DRAWENVIRONMENT ROUTE.%routefeat%
EDITFEATURE ROUTE.%routefeat%
&return
```

Now when you run the program, it lists its variables and waits for a carriage return before proceeding.

```
Arc: &test &on
(&TEST) Arc: &echo &brief
  64: &echo &brief
(&TEST) Arc: &r example2
  65: &r example2

(thread0001, example2.aml)

  2: ARCEDIT
  3: &sv cov = /amlwb/ch09/ls1/streets
  4: EDITCOVERAGE /amlwb/ch09/ls1/streets
  5: &describe /amlwb/ch09/ls1/streets
  6: &sv routfeat = BUS
  8: &listvar
Local:
  ROUTFEAT                        BUS
  COV
    /amlwb/ch09/ls1/streets
Global:
  No global variables defined
Program:
  :PROGRAM                        ARC
  9: &pause
Hit <return> to continue:
```

Using &LISTVAR and &PAUSE can be awkward if you have many variables but only want to see a few. In this case, you have two options: specify the variables you want to see in a space- or comma-delimited list as arguments to &LISTVAR, or use &TYPE as shown in the following example:

```
/* example3.aml
ARCEDIT
&sv cov = [getcover * -route]
EDITCOVERAGE %cov%
&describe %cov%
&sv routfeat = [getchoice %dsc$routenames% -prompt ~
  'Please select a route-system to edit']
&type the variable COV = %cov% /&
&pause
DRAWENVIRONMENT ROUTE.%routefeat%
EDITFEATURE ROUTE.%routefeat%
&return
```

When this program executes, instead of listing all the variables, &TYPE lists only the value of the specified variable, COV.

```
  72: &r example3

(thread0001, example3.aml)

  2: ARCEDIT
  3: &sv cov = /amlwb/ch09/ls1/streets
  4: EDITCOVERAGE /amlwb/ch09/ls1/streets
  5: &describe /amlwb/ch09/ls1/streets
  6: &sv routfeat = BUS
  8: &type the variable COV = /amlwb/ch09/ls1/streets

the variable COV = /amlwb/ch09/ls1/streets

  9: &pause
Hit <return> to continue:
```

The line feed character (/&) is often useful for separating important output text from the surrounding output from &ECHO.

Question 9-2: What kind of typographical errors does &TEST miss?

(Answer on page 9-11)

FYI

You can use &GOTO to skip incomplete sections of code by jumping to &LABEL and continuing from there. For example:

```
&goto skip
(Buggy code fragment)
&label skip
```

This method is useful for general debugging. Suppose you have an AML program that performs some complex process that takes an hour to run, and suppose that somewhere in the last few lines it fails. Instead of running the entire program to test it, just add &GOTO near the beginning of the program and insert &LABEL somewhere before the place it fails.

Now when you run the program it skips most of the time-consuming process, allowing you to troubleshoot the problem areas. Ensure that portions of the program that you run don't depend on anything that you skip and that any coverages or data created on the previous run are available if needed.

The &TTY directive

Like &LISTVAR or &TYPE, the &TTY directive also halts your program so you can examine the variable data. Instead of simply observing when the program stops, &TTY provides a prompt to give you access to the program. At this prompt, you can type any appropriate ARC/INFO command (i.e., don't type ARCEDIT commands from the Arc: prompt), including AML directives. Using &TTY in conjunction with &LISTVAR halts the program, lists the variables, and allows you to change the values of these variables or assign new ones. To return to the program, type &RETURN. In the next example, &TTY replaces &TYPE and &PAUSE.

```
/* example4.aml
ARCEDIT
&sv cov = [getcover * -route]
EDITCOVERAGE %cov%
&describe %cov%
&sv routfeat = [getchoice %dsc$routenames% -prompt ~
  'Please select a route-system to edit']
&tty
DRAWENVIRONMENT route.%routefeat%
EDITFEATURE route.%routefeat%
&return
```

In the next two examples, notice the use of &ECHO &ON. Both &ECHO &ON and &BRIEF give the same type of output, just in different formats. When the &TTY line executes, an Arc: prompt appears.

```
Arc: &test &on
(&TEST) Arc: &echo &on
(&TEST) Arc: (thread0001, tty/6) &echo &on
(&TEST) Arc: &r example4
(&TEST) Arc: (thread0001, tty/7) &r example4
(&TEST) Arc: (thread0001, example4.aml/2) ARCEDIT
(&TEST) Arc: (thread0001, example4.aml/3) &sv cov =
/amlwb/ch09/ls1/streets
(&TEST) Arc: (thread0001, example4.aml/4) EDITCOVERAGE
/amlwb/ch09/ls1/streets
(&TEST) Arc: (thread0001, example4.aml/5) &describe
/amlwb/ch09/ls1/streets
(&TEST) Arc: (thread0001, example4.aml/6) &sv routfeat = BUS
(&TEST) Arc: (thread0001, example4.aml/8) &tty
(&TEST) Arc:
```

At this point you can examine the variables and set the variable ROUTEFEAT equal to the value in the variable ROUTFEAT. This corrects the variable references on lines 7 and 8. Typing &RETURN allows the program to finish executing (successfully, in this case).

```
(&TEST) Arc: &lv
(&TEST) Arc: (thread0001, tty/1) &lv
Local:
   ROUTFEAT                           BUS
   COV
      /amlwb/ch09/ls1/streets
Global:
   No global variables defined
Program:
   :PROGRAM                           ARC
(&TEST) Arc: &sv routefeat = %routfeat%
(&TEST) Arc: (thread0001, tty/2) &sv routefeat = BUS
(&TEST) Arc: &return
(&TEST) Arc: (thread0001, tty/3) &return
(&TEST) Arc: (thread0001, example4.aml/9) DRAWENVIRONMENT
route.BUS
(&TEST) Arc: (thread0001, example4.aml/10) EDITFEATURE route.BUS
(&TEST) Arc: (thread0001, example4.aml/11) &return
(&TEST) Arc:
```

Use &TTY for debugging, as shown, or to test unfinished programs. You can manually set variables that will be set later by portions of the program that aren't yet written.

Question 9-3: &TTY is a useful debugging tool, but it might confuse new users (or anyone who is unfamiliar with &TTY) if you use it in your application program. Why?

(Answer on page 9-11)

In summary

The &TEST directive is useful for checking program logic and testing programs without entering the ARC/INFO subsystems. &ECHO assists in the debugging process by allowing you to see your program as it runs. You can also stop a program and view variable data using &LISTVAR, &TYPE, and &PAUSE. &TTY allows you to access and define data for your AML program.

Answer 9-1: There's no visible Arc: prompt because the prompt itself is considered a message. At first glance, you might think that your program is hung up. In reality, your program has ended, but you can't see the prompt. To remedy this situation, type &MESSAGES &ON and your prompt reappears. In lesson 9.2, you'll learn to reset &MESSAGES and other environments when your AML program fails.

Answer 9-2: Because &TEST doesn't execute ARC/INFO commands, it misses any typographical or syntax errors in their usage. Thus, even after you have thoroughly debugged your program in &TEST mode, you may still encounter errors.

Answer 9-3: New users aren't likely to know which ARC/INFO command to issue at the Arc: prompt, nor are they likely to know to type &RETURN to get back to the program. If you use &TTY in your application as a way to accept input, be sure to include &TYPE statements that give users the information they'll need to execute commands and return to the program.

Exercises

Exercise 9.1.1

Write an AML program that accepts arguments for the BUFFER command. The program should accept either the {buffer_item} {buffer_table} arguments or the {buffer_distance} argument as input.

```
BUFFER <in_cover> <out_cover> {buffer_item} {buffer_table}
{buffer_distance} {fuzzy_tolerance} {LINE | POLY | POINT | NODE} {ROUND |
FLAT} {FULL | LEFT | RIGHT}
```

Use any method to get the input (i.e., RESPONSE, GETCOVER, etc.).

Use &LISTVAR and &PAUSE to examine the values of variables before executing the BUFFER command. Use &TEST and &ECHO when you run the program to assist you with debugging.

Exercise 9.1.2

Write a program that starts ARCPLOT and reports the minimum, maximum, and mean depth for a WELL coverage. Use &TTY to execute STATISTICS from the command line. Enter the following on the command line:

```
Arcplot: STATISTICS WELL POINT
Statistics: MIN DEPTH
Statistics: MAX DEPTH
Statistics: MEAN DEPTH
Statistics: END
```

After you &RETURN to the program, use [SHOW STATISTIC] to assign variables to the values you generated.

(Hint: Remember how &TEST &ON works. If you use &TEST, the STATISTICS command won't execute.)

Exercise 9.1.3

Write the same program as in exercise 9.1.2 but, instead of using &TTY, code the STATISTICS syntax in your AML program. Also use &MESSAGES so that the user doesn't see all of the output from the STATISTICS command.

Lesson 9.2—Expecting and handling errors

In this lesson

Good programmers expect errors. In this lesson, you'll learn two methods for expecting and handling errors. You use the first when you know exactly what errors to expect and when they will occur. Unfortunately, not all errors are like this, so you have a second option, which enables you to make default actions for all errors.

Premeditated error handling

You can expect certain errors. ARC/INFO error messages like these stop perfectly good programs dead in their tracks:

```
Cannot open PAL file
... does not have point feature class.
```

Consider the following program, which produces these error messages whenever the user chooses a coverage and feature class that don't match:

```
/* drawcov.aml
&sv cov = [getcover * -all 'Select a coverage to draw, or~
 "none" to quit' -none]
&do &while not [null %cov%]
  MAPEXTENT %cov%
  /* Find out what kind of display the user wants.
  &sv choice = [getchoice Point Line Poly -prompt ~
   'What kind of display']
  &select [locase %choice%]
    &when point; &do
      MARKERSYMBOL 71   /* green star
      POINTS %cov%
    &end /* when
    &when line; &do
      LINECOLOR red
      ARCS %cov%
    &end /* when
    &when poly; &do
```

```
        LINECOLOR green
        POLYGONS %cov%
        POLYGONTEXT %cov% [entryname %cov%]-ID
     &end /* when
        &end /* select
  /* Get another coverage
  &sv cov = [getcover * -all 'Select a coverage to draw, or~
    "none" to quit' -none]
  CLEAR
&end /* do while
&return
```

The program prompts the user to choose a coverage and a feature class. The
program fails if the coverage doesn't contain the feature class chosen. For
example, if the user chooses CITIES (a POINT coverage) and then chooses the
POLY option, the program fails and issues the error message Coverage
/MAZATLAN3/MANNION/WELL does not have polygon feature class. This is a
common place for users to make mistakes.

Use the [EXISTS] function in conjunction with the [GETCHOICE] function to
trap this kind of error. In the example below, [EXISTS] determines which types
of topology are present. The program then builds the appropriate list of feature
types to present to the user with the [GETCHOICE] function. Using
[GETCHOICE] in this manner presents the user with only valid feature types for
the chosen coverage. The [GET...] functions play a helpful role in avoiding user
errors by presenting specific data to choose from.

```
/* drawcov2.aml
&sv cov = [getcover * -ALL 'Select a coverage to draw,~
 or "none" to quit' -none]
/* Create the list of valid feature types
/*in the variable: LIST
&do &while NOT [null %cov%]
  &sv list =
  &if [exists %cov% -line] &then &sv list = %list% LINE
  &if [exists %cov% -poly] &then &sv list = %list% POLY
  &if [exists %cov% -point] &then &sv list = %list% POINT
  MAPEXTENT %cov%
  /* Find out what kind of display the user wants.
  &sv choice = [getchoice %list% -prompt 'What kind of display']
  &select [locase %choice%]
    &when point; &do
      MARKERSYMBOL 71  /* green star
      POINTS %cov%
    &end /* when
```

```
     &when line; &do
       LINECOLOR red
       ARCS %cov%
     &end /* when
     &when poly; &do
       LINECOLOR green
       POLYGONS %cov%
       POLYGONTEXT %cov% [entryname %cov%]-ID
     &end /* when
   &end /* select
/*
/* Get another coverage
/*
   &sv cov = [getcover * -all 'Select a coverage to draw, or~
  "none" to quit' -none]
   CLEAR
&end /* do while
&return
```

This method is obviously useful, but you can't anticipate every error. What happens if you forget to account for a potential user error, or there's an error in your code? In this case, you'll want to take more formal measures to capture the error.

Question 9-4: Which function would you use to provide a lookup table to an AML program? Write an example.

(Answer on page 9-22)

The &SEVERITY directive

The &SEVERITY directive allows you to specify the action to take when your program encounters a warning or error condition. The directive has the following default settings when you start ARC/INFO:

```
&severity &warning &ignore
&severity &error &fail
```

When the program encounters an error condition, &SEVERITY &ERROR &FAIL causes execution to terminate, or *bail out*. &SEVERITY &WARNING &IGNORE causes execution to continue when the program encounters a warning. In most cases, you won't want to alter the &SEVERITY &WARNING &IGNORE setting. Just be aware that it's possible to make your program fail over minor details.

As the programmer, you have the power to change these parameters at any time. Novice programmers may think that setting &SEVERITY &ERROR &IGNORE solves all of their problems. This isn't the case. You can, however, use &SEVERITY &ERROR to your advantage. For example, if you have a program or code fragment that you want to run without checking for errors, you can set &SEVERITY as follows:

```
&severity &error &ignore
(Buggy code fragment)
&severity &error &fail
```

Saving and restoring environments and using routines with &SEVERITY

As you write AML programs, consider the environments set before your program executes. A user-friendly AML program doesn't leave the user in a system other than the one where the program initiated. It won't give the user 300 global variables or change environments (i.e., &MESSAGES, &FULLSCREEN, DISPLAY, etc.). The following example shows a number of the environments that a program can save using the [SHOW] function. AEENV.AML preserves the user's ARCEDIT environment, then runs another AML program to launch a specific application. When the application finishes, control returns to AEENV.AML, which resets the environments.

```
/* aeenv.aml
&if [show program] ne ARCEDIT &then
     &return This program must be run from ARCEDIT
&type /&/&You are about to enter one of the city of Bomphaq's~
 editing applications
&sv oldfull = [show &fullscreen]
&sv oldwksp = [show &workspace]
&sv oldamlpath = [show &amlpath]
&sv oldcoor = [show coordinate]
&sv oldec = [show editcoverage]
```

```
&sv oldde = [show drawenvironment]
&sv oldintersect = [show intersectarcs]
&sv oldarcsnap = [show arcsnap]
&sv oldduparcs = [show duplicatearcs]
&sv oldnodesnap = [show nodesnap]
&sv appchoice = [getchoice Parcel Sewer Roads -prompt ~
 'Please select an editing application:']
&select %appchoice%
  &when Parcel
    &r parcellaunch.aml [username]
  &when Sewer
    &r sewerlaunch.aml [username]
  &when Roads
    &r roadslaunch.aml [username]
&end  /* select
&if [show program] ne ARCEDIT &then &do
  &if [show program] ne ARC &then QUIT
  ARCEDIT
&end
&fullscreen %oldfull%
&workspace %oldwksp%
&amlpath %oldamlpath%
COORDINATE %oldcoor%
EDITCOVERAGE %oldec%
DRAWENVIRONMENT %oldde%
INTERSECTARCS %oldintersect%
ARCSNAP %oldarcsnap%
DUPLICATEARCS %oldduparcs%
NODESNAP %oldnodesnap%
&return
```

FYI

> AEENV.AML can run into some errors when it attempts to reestablish the old environments. If ARCSNAP is set to OFF, or there's no edit coverage when AEENV.AML is executed, an error will occur. The other &SEVERITY options, covered in the rest of this lesson, handle these errors.

By saving these environments, your program can restore them before exiting, leaving users with the original environments.

Saving and restoring environments, as shown in the previous example, doesn't work if the program fails before reaching the section that restores environments. In these cases, the user is left wherever things go wrong. This is frustrating to experienced users and both frustrating and problematic to novice users.

You can include routines in your program that avoid this problem. Two such routines, often named BAILOUT and EXIT, are discussed next.

The &SEVERITY directive has another option: &ROUTINE <routine_name>. The &ROUTINE option is the most versatile option to use with &SEVERITY. Examine the following line of code:

```
&severity &error &routine bailout
```

In this case, when the program encounters an error, it executes a named routine instead of failing. For consistency, call this routine BAILOUT throughout this book; however, this name isn't required.

The BAILOUT routine &CALLs a routine named EXIT. The exit routine includes the code to restore the environment settings.

```
/* aeenv2.aml
&severity &error &routine bailout
&if [show program] ne ARCEDIT &then
  &return This program must be run from ARCEDIT
&type /&/&You are about to enter one of the city of Bomphaq's~
 editing applications.
&sv oldfull = [show &fullscreen]
&sv oldwksp = [show &workspace]
&sv oldamlpath = [show &amlpath]
&sv oldcoor = [show coordinate]
&sv oldec = [show editcoverage]
&sv oldde = [show drawenvironment]
&sv oldintersect = [show intersectarcs]
&sv oldarcsnap = [show arcsnap]
&sv oldduparcs = [show duplicatearcs]
&sv oldnodesnap = [show nodesnap]
&sv appchoice = [getchoice Parcel Sewer Roads -prompt ~
 'Please select an editing application:']
&select %appchoice%
     &when Parcel
       &r parcellaunch.aml [username]
     &when Sewer
       &r sewerlaunch.aml [username]
     &when Roads
       &r roadslaunch.aml [username]
&end  /* select
&call exit
&return
/*
/*-----------------Routine Exit-----------------
```

```
&routine exit
&if [show program] ne ARC and [show program] ne ARCEDIT~
  &then QUIT; ARCEDIT
&if [show program] eq ARC &then ARCEDIT
&fullscreen %oldfull%
&workspace %oldwksp%
&amlpath %oldamlpath%
COORDINATE %oldcoor%
EDITCOVERAGE %oldec%
DRAWENVIRONMENT %oldde%
INTERSECTARCS %oldintersect%
ARCSNAP %oldarcsnap%
DUPLICATEARCS %oldduparcs%
NODESNAP %oldnodesnap%
&return
/*
/*------------------Routine Bailout-------------
&routine bailout
&severity &error &ignore
&call exit
&return &error Bailing out of AEENV2.AML
```

The exit routine is called in two cases. First, if the program terminates successfully, the exit routine is called by &CALL EXIT (placed just before the &RETURN statement). Second, if the program fails, the exit routine is called by the routine called BAILOUT. This ensures the restoration, under any circumstance (short of a system crash), of the environment settings in effect before the program executed.

It's a good habit to write your AML programs to clean up after themselves. Use routines like EXIT to do any needed cleanup before programs terminate, including restoring environments and deleting global variables or temporary data created by the program.

In the example, the line `&severity &error &routine bailout` calls the bailout routine when an error is encountered. The program goes to BAILOUT and executes the statements found there. BAILOUT calls the exit routine to reset environments. Finally, BAILOUT executes `&return &error Bailing out of AEENV2.AML`, causing the program to terminate with a message to the user.

Recall that &RETURN at the end of a routine returns the user to the line in the program following the one in which &CALL executed. In this case, you don't want to return to the program (e.g., the line after which the error was encountered), but terminate the program. The &ERROR option on &RETURN terminates the program.

FYI

Using any option after the &RETURN directive causes the program to terminate. Therefore, if you want to send a message after a routine finishes, use an &TYPE statement and issue the &RETURN without specifying an option:

```
. . .
&type Operation successful.
&return
```

Always include `&severity &error &routine bailout` as one of the first lines in your programs. (Note: &SEVERITY and &ECHO are the only directives that can precede the &ARGS directive in a program.) Also, set `&severity &error &ignore` in the first line of your bailout routine so that it ignores any errors encountered within itself. This eliminates the possibility of an endless loop if there's an error in your bailout or exit routine (e.g., the program encounters an error and calls the bailout routine, the bailout routine has an error in it, calls itself, and encounters an error).

Question 9-5: Suppose many of the AML programs in your application include the following code in their exit routines:

```
/*----------------
&routine EXIT
/*----------------
&if [show program] ne ARC and [show program] ne ~
   ARCEDIT &then QUIT; ARCEDIT
&if [show program] eq ARC &then ARCEDIT
&fullscreen %oldfull%
&workspace %oldwksp%
&return
```

How could you change the exit routines to reduce the redundancy in the programs?

(Answer on page 9-22)

The AML$MESSAGE reserved variable

There may be times when you don't want a program to stop even though there's an error. Suppose you don't want to exit the program just because you receive the `Cannot open PAL file` message. AML$MESSAGE, a reserved variable, stores the last error message issued by any command. The next example shows how to use the contents of this variable to choose the appropriate action to take.

In the following bailout routine, the contents of AML$MESSAGE alerts the user that the coverage doesn't have polygon topology. This returns the user to the place in the program where the error occurred without bailing out.

```
&routine bailout
/* Reset severity
&severity &error &ignore
/* Test for the "Cannot open PAL file" error message
/* Don't bail out in this case
&if [QUOTE %aml$message%] cn 'open PAL file' &then &do
     &type Coverage does not have polygon topology.
     &severity &error &routine bailout
&return
&end
&call exit
&return &error Bailing out of DRAW.AML
```

Testing messages isn't the best way to take care of error handling. This is because messages can sometimes be updated between versions of the software. If you have an older AML program with problems testing messages, you may want to check out the new syntax on the error messages. The following reserved AML$ variables may be useful for error handling:

Variable Name	Description
AML$ERRORLINE	The line number of the last error
AML$ERRORFILE	The file that the last error occurred in
AML$ERRORTHREAD	The thread name where the last error occurred

Consult the online help index for a more complete list of AML$ variables.

The next lesson introduces another useful reserved variable, AML$SEV.

In summary

You need to anticipate potential errors in your programs and handle them *before* they cause your programs to fail. Programs can expect and react to errors intelligently. Using &SEVERITY with routines enables the program to handle unexpected errors. A user-friendly program saves and restores the environment settings that are in place when it's executed.

Answer 9-4: To avoid errors, the [GETFILE] function would be the best function to use. Here's an example:

```
&sv lut = [getfile *.lut -info 'Select a lookup table']
```

Answer 9-5: If you create an AML program that contains the code contained in the exit routine:

```
/* cleanup.aml
&args oldfull oldwksp
&if [show program] ne ARC and [show program] ne ~
   ARCEDIT &then QUIT; ARCEDIT
&if [show program] eq ARC &then ARCEDIT
&fullscreen %oldfull%
&workspace %oldwksp%
&return
```

you could change the code in the exit routines in all the programs to be:

```
/*---------------
&routine EXIT
/*---------------
&r cleanup %oldfull% %oldwksp%
&return
```

Exercises

Exercise 9.2.1

Rewrite DRAWCOV2.AML on page 9-14 to include &DESCRIBE for the situation where there are no features present. The program should first accept the coverage and then check for the number of features present. If features are present, prompt the user for a feature class and draw the features. If there's only one feature type in the coverage, just draw it.

(Hint: Don't specify the MAPEXTENT for a coverage unless you know that it's going to be drawn.)

Exercise 9.2.2

In addition to restoring program environments and environment settings, name some other tasks that your exit and/or bailout routines should handle.

Exercise 9.2.3

Create a nested AML program that runs CLEAN or BUILD on a polygon coverage if the coverage has been edited. If BUILD fails, ask if the user wants to clean the coverage or cancel the process.

(Hint: Search for the string Use CLEAN, which is returned as a warning.)

Lesson 9.3—Handling errors between AML programs and menus

In this lesson

Modular programming requires programs to handle errors between, as well as within, programs. Failures make managing a system of programs and menus complex. In this lesson, you'll examine how severity handling works with interdependent programs. The techniques you learn here enable you to produce more stable applications.

In the last lesson, you saw how AML$MESSAGE stores the last error message issued by any command. In this lesson, you'll use another AML$-reserved variable, AML$SEV, to resolve problems encountered between programs without bailing out. AML$SEV stores the severity code indicating the status of the last line executed by the AML processor as follows:

Returned value	Description
0	No error or warning occurred
1	A warning occurred (includes usages)
2	An error occurred (includes fatal errors)

Handling errors between AML programs

The &SEVERITY directive sets the severity handling for a single AML program. Though it can be used in conjunction with &SEVERITY directives in other programs, it's important to remember that, just as programs can't share routines, they can't share severity.

The following diagram shows the normal interaction between modular programs:

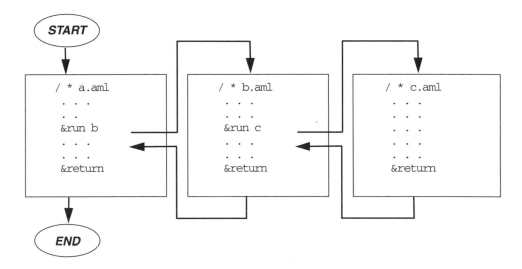

After an AML program executes, it returns control to the source that called it—another program, a menu, or a TTY. That is, if all goes well. Unfortunately, sometimes programs fail. When they do, control is returned before the program finishes executing.

The next diagram shows what happens when a program fails with no provision for handling errors.

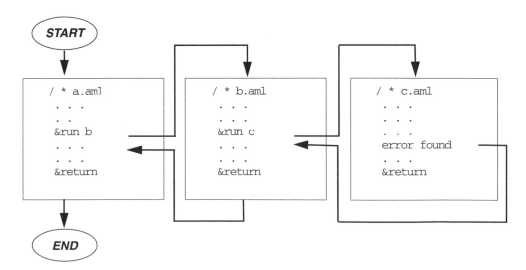

In this case, program C.AML encounters an error and returns prematurely to B.AML. There's no error checking, so C.AML doesn't clean up environments or global variables. C.AML passes the AML$SEV variable back to B.AML with a value of 2, signifying that an error occurred. Program B.AML doesn't bail out, but finishes before returning to the program that called it, A.AML.

B.AML doesn't register the error in C.AML. Because the &SEVERITY directive is local, even if program B.AML included error checking (e.g., &SEVERITY &ROUTINE &BAIL), it still wouldn't bail out. An error condition in one AML program can't trigger an error condition in another, error-free, AML program without returning control to the calling AML through the use of &RETURN &ERROR.

As the programmer, you can check the status of AML$SEV and act accordingly. You may choose to ignore the error, send a message to the user, or bail out. The following code fragment shows one such solution applied to the last problem:

```
...
&r c.aml
&if %aml$sev% = 2 &then &do
  &type PROGRAM C.AML failed
  &if [query 'Would you like to bail out of the application'] ~
    &then &do
      &return &error Bailing out of B.AML
  &end
&end
...
```

The next diagram shows what happens when all programs in an application have error-handling routines. &RETURN &ERROR triggers the error-handling mechanism in the calling program.

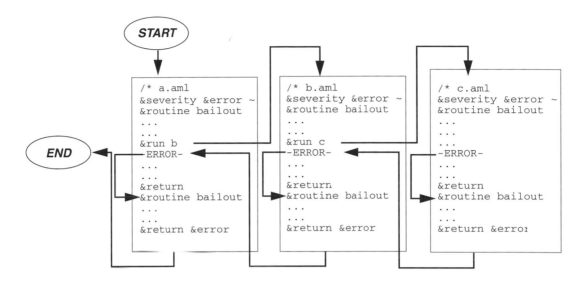

Whether or not you want the entire application to shut down when a program fails depends on the role of the failed program. Programs that retrieve and display data may not pose a threat to the rest of the application if they fail; however, if the failed program generates data that other programs use, make sure that the application bails out.

The following code adds a choice to the previous example by allowing you to ignore the error:

```
&severity &error &routine &bailout
...
&severity &error &ignore
&r c.aml
&if %aml$sev% = 2 &then &do
  &type PROGRAM C.AML failed
  &if [query 'Would you like to bail out of the application'] ~
    &then &do
       &call bail
  &end
&end
&severity &error &routine bailout
...
```

Handling errors between menus and AML programs

A menu adds another level of complexity to error handling in an application. Menus ignore errors as shown in the next example.

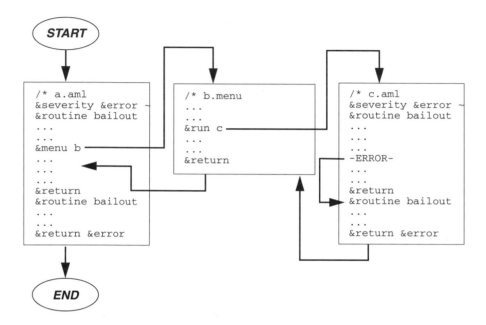

B.MENU executes C.AML. When C.AML fails, control is returned to the menu. The menu stays functional until it's dismissed, then returns control to A.AML, which doesn't detect an error. A program that fails when executed from a menu won't cause the menu itself to fail, even if it returns with &RETURN &ERROR.

FYI

Syntax errors prevent menus from executing. This results in an error in the calling program because the menu can't be created (e.g., the line &MENU B fails). In this case you may see an error message like this:

```
AML ERROR - Cannot initialize alpha menu
```

Because menus ignore errors, there's no way to capture exactly why or when an error occurred in the called AML program. You can only create error-handling routines in AML programs. These routines should be able to shut down the entire application—not just return to the calling menu—when programs encounter crucial errors. You need to know your application thoroughly to know when to halt processing.

In summary

Because an application is made up of many parts, you must handle errors that occur between AML programs and between AML programs and menus. The &SEVERITY directive sets the error handling for a single AML program. Ultimately, it's your knowledge of your application that enables you to anticipate errors and handle them with your programs.

Exercises

Exercise 9.3.1

The following AML program (GETCOV.AML) launches DRAWCOV.AML and IDENTIFY.AML.
GETCOV.AML passes the variable %cov% to DRAWCOV.AML. Write DRAWCOV.AML to allow the
user to pick the feature class to draw and the color to draw it in, and then to draw the coverage. Set
the feature class to the global variable .FEAT. If an error exists in DRAWCOV.AML, you don't want
IDENTIFY.AML to execute because it will also error, so have the application stop before
IDENTIFY.AML is executed.

```
/* getcov.aml
&severity &error &routine bailout
&if [show program] nc ARCPLOT &then
   &return This program must run from Arcplot
&s old_mape = [show mapextent]
&s old_shade = [show shadeset]
&s old_line = [show lineset]
&s old_mark = [show markerset]

SHADESET CONTRAST
LINESET COLOR
MARKERSET OILGAS
&s cov = [getcover *]
&r drawcov %cov%
&r identify %cov%
&call exit
&return

&routine exit
MAPEXTENT %old_mape%
SHADESET %old_shade%
LINESET %old_line%
MARKERSET %old_mark%
&dv .feat
&return

&routine bailout
&severity &error &ignore
&call exit
&return &error
```

10 / Reading and writing files

AML can read, write, create, and delete ASCII, or system, files to transfer data between programs and access data produced by other software. Additionally, AML can dynamically create menus from AML programs for immediate use.

This chapter shows how to use the file-handling tools provided by AML. The discussion covers the following functions as they pertain to file input and output (i.e, reading and writing files):

[ACCESS]	[LISTUNIQUE]
[CLOSE]	[OPEN]
[COPY]	[WRITE]
[DELETE]	[READ]
[LISTFILE]	[RENAME]
[LISTITEM]	[SCRATCHNAME]

This chapter covers the following topics:

Lesson 10.1—Basic file handling

- Creating a system file containing a list of objects
- Opening and reading system files

Lesson 10.2—Writing system files

- Writing to a new file
- Writing to an existing file
- Managing system files

Lesson 10.3—Using AML to write menu files

- Creating menus dynamically from AML programs

Lesson 10.1—Basic file handling

In this lesson

AML can create system files in several ways. This lesson shows how the AML functions [LISTFILE], [LISTITEM], and [LISTUNIQUE] can automate the creation of these files. You'll also use the [OPEN] and [READ] functions to examine the contents of the system files you create.

Creating a system file containing a list of objects

There are times during an ARC/INFO session when you need a list of available objects (e.g., files, directories, coverages, grids, INFO items, or item values). You might want to access all the objects in a list to make global changes to the set of objects, one at a time. Or, you might want to locate only certain objects in a list (e.g., find all coverages with a TMP suffix so you can KILL them).

The [LISTFILE], [LISTITEM], and [LISTUNIQUE] functions allow you to write to the screen or create a system file containing a list of specified objects. [LISTFILE] creates a list of files, directories, coverages, or grids that match a wildcard character. The following is the syntax for [LISTFILE]:

```
[LISTFILE <specifier> <-type> {-FULL} {output_file}]
```

The following code sets the variable FILES to contain a comma-delimited list of file names. This list includes all AML files, menu files, graphics files, the LOG file, key files, projection files, and any other object that's not a subdirectory found in the current directory. (The * is a *wildcard character* that represents all filenames.)

```
&sv files = [listfile * -file]
```

You can display the contents of this variable with a simple &TYPE statement:

```
Arc: &type %files%
atout,ch103ex1.aml,colorpick.aml,colorpick.menu,
coverlist.aml,covinfo.lis,ex4-2,ex4-2.c,getpoints.aml,
identify.aml,log,modal.aml,msinform.aml,msinform.menu,
nwcoverlist.aml,origrunlist.aml,outline,stwell.view,
template.aml,template.menu,ts.aml
```

You could also use the optional OUTPUT_FILE option of [LISTFILE] to create a file containing all the file names:

```
&sv files = [listfile * -file file.list]
```

In this example, FILE.LIST contains the following information. Notice that each entity is on its own line:

```
atout
ch103ex1.aml
colorpick.aml
colorpick.menu
coverlist.aml
covinfo.lis
ex4-2
ex4-2.c
getpoints.aml
identify.aml
log
modal.aml
msinform.aml
msinform.menu
nwcoverlist.aml
origrunlist.aml
outline
stwell.view
template.aml
tcmplate.menu
ts.aml
```

You can control the search by using a wildcard to list only certain files. If, for example, you want the output file to contain only AML program files, use [LISTFILE] like this:

```
&sv numobs = [listfile *.aml -file aml.list]
```

The resulting file, AML.LIST, contains only those files whose names end with the .AML filename extension.

```
ch103ex1.aml
colorpick.aml
coverlist.aml
getpoints.aml
identify.aml
modal.aml
msinform.aml
nwcoverlist.aml
origrunlist.aml
template.aml
ts.aml
```

Use the <-type> option to specify the object type (e.g., INFO files, grids, workspaces, libraries, and coverages) or the coverage type (e.g., line, point, poly, etc.) you want to list. For example, if you want a list of all the polygon coverages in a workspace, use [LISTFILE] as follows:

```
&sv numobs = [listfile * -cover -poly poly.list]
&lv numobs
Local:    NUMOBS              4
```

The file POLY.LIST contains a list of polygon coverages like this example:

```
parcel
polystreet
soil
sub174
```

Question 10-1: How could you use the [LISTFILE] function to create a file containing all of the grids in a workspace named /cannes/jack/models?

(Answer on page 10-14)

When a file name is specified, [LISTFILE] returns an integer to the variable NUMOBS that indicates whether or not AML successfully completed the intended operation (i.e., creating a file with a list of objects). If the operation is successful, NUMOBS is set to the number of objects written to the file. In the last example, NUMOBS = 4 because four coverages were written to the POLYLIST file. The following table shows all other possible return codes:

Returned value	Indicates	Reason
−1	Couldn't create file	No write access to the output file directory No file created
0	File created, but no objects written	No objects in the directory or workspace that match the request
> 0	The number of objects written	File created with number of objects listed as the returned value

[LISTITEM] and [LISTUNIQUE] work almost identically to [LISTFILE]. They can create a comma-delimited list of elements or a file that contains the elements. When these functions create an output file, the returned values are the same as [LISTFILE].

FYI

[FILELIST] also creates a list of files, but it always creates an output file. With [FILELIST], the output file name is required instead of optional as with [LISTFILE]. There aren't any benefits to using [FILELIST] over [LISTFILE]. [FILELIST] remains as a function for backward compatibility.

Opening and reading system files

You can use AML to read system files like those created by [LISTFILE], [LISTITEM], or [LISTUNIQUE]. This four-step process, outlined below, applies to reading any type of file and is discussed in more detail on the following pages.

1. *Verify that the file exists* with the correct permissions before you try to read it (optional).

2. *Open the file for reading.* This guards against adding data to a file by mistake.

3. After the file opens, you can *read the file* one record (line) at a time.

4. *Close the file* when you finish reading it.

Step 1. Verify that the file exists with the correct permissions.

Use the [EXISTS] and [ACCESS] functions to verify that a file exists with the correct permissions. (If you need to review [EXISTS], refer to chapter 4.) [ACCESS] allows you to check for read permission before you attempt to open the file for reading.

```
/* Verify that the file exists with the correct permissions
&if not [exists %file%] &then &do
  &call exit
  &return File %file% doesn't exist...
&end
&if not [access %file% -read] &then &do
  &call exit
  &return You do not have read access to %file%...
&end
 /* Open and Read the file
```

Step 2. Open the file for reading.

Use the [OPEN] function to open the file for processing. This function opens a file in one of three modes: -READ, -WRITE, or -APPEND. (The -READ mode is discussed here; lesson 10.2 covers the -WRITE and -APPEND modes.)

The mode in which you open the file depends on what kind of file-processing functions you want to use. To read the file using the [READ] function, you must open the file in the -READ mode. Consider the following example:

```
&sv amlunit = [open %file% openstat -read]
```

Like all functions, [OPEN] returns a string value. In addition, [OPEN] sets a status variable whose name is specified *inside* the function brackets (e.g., OPENSTAT). This is also true of the other file-processing functions, [READ] and [WRITE]. The status variable indicates whether or not the file opened successfully. The following table lists all possible values:

Returned value of status_variable	Indicates
0	File opened successfully
100	No free AML units
101	File not found (-READ)
102	Couldn't open file (-READ, -APPEND)
103	Couldn't create file (-WRITE, -APPEND)

This status variable is called OPENSTAT in this workbook, but this name isn't required. The status variable doesn't appear within percent signs because it's set, not referenced, by the function.

It's a good idea to check the value of the status variable before you attempt to process a file. You can't access a file for processing if it's not open.

An easy way to find out if the file opened successfully is to check to see if the status variable equals zero:

```
&sv amlunit = [open %file% openstat -READ]
&if %openstat% ne 0 &then &do
  &call exit
  &return Error opening file, error code: %openstat%
&end
/* Read the file
```

Alternatively, you could use an &SELECT block with OPENSTAT to check for the value of every possible completion status and act accordingly.

[OPEN] returns an AML unit number (1–20) used by functions such as [READ], [WRITE], and [CLOSE] for subsequent references to the file. In the previous example, the variable named AMLUNIT stores this unit number. AML allows only twenty files open at one time; if you try to open more, the value of the status variable equals 100 (no free AML units) and the value of the [OPEN] function equals zero.

Returned value of [OPEN]	Indicates
0	AML unable to open the file
1 through 20	AML file unit number (twenty maximum open)

Keep track of the variable that's assigned the value returned by [OPEN] because you'll need it to use other file input/output functions. For instance, you'll see that the [READ] function requires the <aml_file_unit> value.

Step 3. Read the file.

Now that the file is open, use the [READ] function to read the data from the file. You must use the AML unit number (the variable AMLUNIT) to access the file:

```
&sv record = [read %amlunit% readstat]
```

Like the [OPEN] function, [READ] sets a status variable (e.g., READSTAT) in addition to the return value. The table below lists the six-integer values that the status variable can return.

Returned value of the status variable	Indicates
0	Record read successfully
100	File not open for reading
101	Error during read
102	End of file
103	Unable to reopen file (files are closed between ARC programs)
104	String truncated to 1,024 characters

[READ] returns a string equal to the record (line) read from the file. This return string can contain a maximum of 1,024 characters. If the return string contains blanks, it's quoted so its maximum length becomes 1,022 (with the quotes, 1,024).

Each time the [READ] function executes, it accesses another record in the file, usually accomplished using an &DO loop as shown in the following example:

```
&if not [exists %file%] &then &do
  &call exit
  &return File %file% doesn't exist...
&end
&if not [access %file% -read] &then &do
  &call exit
  &return You do not have read access to %file%...
&end
&sv amlunit = [open %file% openstat -READ]
&if %openstat% ne 0 &then &do
  &call exit
  &return Error opening file, error code: %openstat%
```

```
&end

&sv record = [read %amlunit% readstat]
&do &while %readstat% = 0
  /* Do something with each record, one at a time
  &sv record = [read %amlunit% readstat]
&end  /* &do &while
```

Use the READSTAT variable, as shown in the next example, to perform error checking and determine whether the entire file is read. When READSTAT isn't equal to zero, it should equal 102, signaling that the file was read successfully. If its value isn't equal to either zero or 102, there was a problem reading the file.

```
&sv amlunit = [open %file% openstat -READ]

&if %openstat% NE 0 &then
  &call exit
  &return  Error opening file, error code: %openstat%
&sv record = [read %amlunit% readstat]

&do &while %readstat% = 0
  /* Do something with the record
  &sv record = [read %amlunit% readstat]
&end /* &do &while
 &select %readstat%
  &when 100,101,103
    &do
      &type Complete file not read. Error code ~
       %readstat% encountered
      /* Do what is necessary: Continue, call your bailout
      /* routine, or execute some contingency plan
    &end
  &when 104
    &do
      &type Complete record not read. Error code~
       %readstat% encountered
      /* Do what is necessary: Continue, call your bailout
      /* routine, or execute some contingency plan
    &end
&end /* &select
```

Question 10-2: What is the loop control structure for reading a file until the end or until a line longer than 1,024 characters is encountered?

(Answer on page 10-14)

Step 4. Close the file.

Finally, close the file when [READ] finishes processing, successfully or not. Use the [CLOSE] function with the AML file unit number (assigned by the [OPEN] function) as follows:

```
&sv closestat = [close %amlunit%]
```

[CLOSE] returns the completion status of the close request as follows:

Returned value of [CLOSE]	Indicates
0	File closed
101	Bad unit number (number given is > 20)
102	Unit not open (number given is <= 20 but the unit named isn't open)

Always close files that you aren't using so others can access them. Multiple users can open a file with -READ access, but when a file is opened using the -WRITE or -APPEND options, no other user can access it and the file can't be read until it's closed.

 FYI

Always keep track of whether a file opened successfully before attempting to access or close it. When an application uses [OPEN] and [CLOSE], test the section of code that uses the [OPEN] function before you execute the code that uses [CLOSE]. This helps you avoid errors and saves debugging time.

During debugging, you'll find [CLOSE] with the -ALL option useful because it closes all open files. During normal operation, however, it's better to keep track of the AML file unit numbers and close each file individually, when it's not in use, so that others can access it. Bailout routines can also use the -ALL option to close all files if a program fails.

With the addition of the [CLOSE] function, the example is now complete. Notice that after the file successfully closes, the variable that held the unit number is deleted.

```
&if not [exists %file%] &then &do
  &call exit
  &return File %file% doesn't exist...
&end
&if not [access %file% -read] &then &do
  &call exit
  &return You do not have read access to %file%...
&end
&sv amlunit = [open %file% openstat -READ]
&if %openstat% ne 0 &then &do
  &call exit
  &return Error opening file, error code. %openstat%
&end

&sv record = [read %amlunit% readstat]
&do &while %readstat% = 0
  /* Do something with the record
  &sv record = [read %amlunit% readstat]
&end
&select %readstat%
  &when 100,101,103
    &do
      &type Complete file not read. Error code %readstat% ~
        encountered
      /* Do what is necessary: Continue, call your
      /* bailout routine, or execute some contingency
      /* plan
    &end
  &when 104
    &do
      &type Complete record not read. Error code %readstat% ~
        encountered
      /* Do what is necessary: Continue, call your
      /* bailout routine, or execute some contingency
      /* plan
    &end
&end /* &select
&sv closestat = [close %amlunit%]
&if %closestat% = 0 &then
  &dv amlunit
&else
  &type Error closing unit %amlunit% Error code: %closestat%
```

Many AML programmers use [CLOSE] in their exit routines. This serves a twofold purpose. First, the exit routine closes files when the program terminates naturally. Second, files are closed even if the program encounters an error and calls the bailout routine. Here's an example of an exit routine with this functionality:

```
&routine exit
/* If file is open then close it
&if [variable amlunit] &then &do
  &sv closestat = [close %amlunit%]
  &if %closestat% = 0 &then
    &dv amlunit
  &else
    &type Error closing unit %amlunit% Error code: %closestat
&end /* &if [variable amlunit]
&return
----------------------------
&routine bailout
&severity &error &ignore
&call exit
&return &error Bailing out of SOME.AML
```

FYI

The exit routine uses an AML file input/output coding standard: programs should delete the unit variables whenever a file closes successfully or opens unsuccessfully. This ensures that if a file was closed in the main body of the program, or if it wouldn't open, the exit routine won't attempt to close it because [VARIABLE AMLUNIT] is .FALSE..

FYI

Use the &LISTFILES directive to determine which files are open:

```
Arc: &listfiles
Unit 1   Record 0   Access R   File   /PATHNAME/FILE.LIST
```

The output shows that the file named FILE.LIST is assigned AML file unit number 1, has zero records read, and is open with read access. The record number increments by one as each record is read. Use &LISTFILES during debugging to determine whether the appropriate files closed when your program finished or failed.

ARC Macro Language

To read all the files in a particular directory, you can use the [LISTFILE] function in conjunction with &DO &LIST. If [LISTFILE] isn't given an output file name, it produces a comma-delimited list of file names that &DO &LIST can step through. If you wanted to use the previous file-reading example to read all the files in the current directory, you could simply insert that code within an &DO &LIST loop:

```
&do file &list [listfile * -file]

  /* Check to see if the file exists
  /* Test whether we have read access to the file
  /* Open the file
  /* Test whether the file was opened correctly

  /* Read the file    &do loop to step through
  /*                   each line in the file
  /* Check read status

  /* Close the file

&end
```

This method of using [LISTFILE] could cause a problem if there's a large number of files in the directory. The 1,024-character limit restrains [LISTFILE], so if all the file names in a directory exceed this limit, the AML program fails.

To get around this problem, have [LISTFILE] create an output file. Nested reading loops could then be used to read all the files in the directory. The outer loop reads the file containing all the file names, while the inner loop reads each individual file. Modifying the last example, the code would look like this:

```
&s numobs = [listfile * -file file_list]
&s file_listunit = [open file_list openstat -read]
/* check to make sure the file was opened correctly

&S file = [read %file_listunit% readstat]
&do &while %readstat% = 0

  /* Check to see if the file exists
  /* Test whether we have read access to the file
  /* Open the file
  /* Test whether the file was opened correctly

  /* Read the file - &do loop to step through
  /*                   each line in the file
```

```
/* Check read status

/* Close the file

&S file = [read %file_listunit% readstat]
&end
```

In summary

You can use AML to create a system file containing a list of objects of a specified type, including coverages, AML programs, and INFO files. The [LISTFILE], [LISTITEM], and [LISTUNIQUE] functions are used to specify the objects in the list.

You can open, read, and close the system files created by [LISTFILE], [LISTITEM], and [LISTUNIQUE] using the [OPEN], [READ], and [CLOSE] functions. The value returned by [OPEN] sets an AML file unit number (i.e., 1 through 20) that you can use to reference the file. The value returned by [CLOSE] provides you with information about the completion status of the request. [READ] returns a string equal to a record in the system file. Status variables associated with the [OPEN] and [READ] functions provide you with detailed information about any errors that occurred during the process.

In the next lesson, you'll learn to write and delete system files using the AML functions [WRITE] and [DELETE].

Answer 10-1:
```
&sv numobs = [listfile /cannes/jack/modelst -grid grid.lis]
```

Answer 10-2:
```
&do &until %readstat% = 102 or %readstat% = 104
```

Exercises

Exercise 10.1.1

Write the syntax for a [LISTFILE] function that creates a file containing a list of all the names of lookup tables in a workspace (assume that .LUT is the filename extension).

Exercise 10.1.2

Write a program called DESCPOLY.AML that DESCRIBEs every polygon coverage in a workspace.

Exercise 10.1.3

Write a program called GETLINES.AML that allows the user to choose a workspace with the [GETFILE] function and then creates a file containing a list of all LINE coverages in the chosen workspace. The program should also list out the contents of the file to the user (&POPUP is sufficient).

Exercise 10.1.4

Write a program called READLINES.AML that reads the number of lines from a file that the user specifies and then displays them. Here's the usage for the READLINES command:

```
Usage: readlines <file> <num_lines | ALL> {start_num}
```

> `<file>` is the name of a system file. Make sure that this file exists. If it doesn't, inform the user and end the program.

> `<num_lines | ALL>` is the number of lines in the file you want to read. If the user specifies the -ALL option, make sure that all lines, beginning with the {start_num}, are displayed to the user.

> `{start_num}` is the position in the file from which to begin reading. If the user doesn't give one, assume that the file should be read from the beginning.

Check that the user gives you valid data (i.e., that the START_NUM is in the file, that from the given START_NUM you can actually read the number of lines asked for, and that the user actually typed `all` and not `boo`).

(Hint: You can load the file into an array of variables to determine how large it is and to access the data.)

Exercise 10.1.5

Write a program named MINMAXPOLY.AML that determines the minimum and maximum x- and y-coordinates for all of the polygon coverages in a workspace. The program should report these coordinates and name the coverages that contain them. Instead of having to set the MAPEXTENT to every coverage in your workspace, this program allows you to execute MAPEXTENT like this:

```
MAPEXTENT %xmin% %xmax% %ymin% %ymax%
```

(Hint: Use &DESCRIBE.)

Lesson 10.2—Writing system files

In this lesson

Now that you've learned everything there is to know about reading files with AML, you'll want to create some. You've already seen how to create files containing lists of specified objects using [LISTFILE], [LISTITEM], and [LISTUNIQUE]. In this lesson, you'll use the [OPEN] function to create a system file to which you can write using the [WRITE] function.

The AML code you've seen so far created and processed files but didn't delete any. This violates the AML programmer's creed: *pack out what you pack in.* In this lesson, you'll learn how to delete the temporary files that would otherwise litter the user's workspace.

Writing to a new file

To write to a new file, you must first create one. Use the [OPEN] function with the -WRITE option as shown here to create a file named MY.FILE:

```
&sv unit = [open my.file openstat -write]
```

If a file with this name already exists, it's overwritten. MY.FILE is now open and accessible though the [WRITE] function. As with the [READ] function, access the file using the file unit as follows:

```
&sv writestat = [WRITE %unit% 1,2567896AB78,RES]
```

The string that you want to write in the file follows the AML file unit. You must use a [WRITE] function for each line you write to the file. Quote the string if it contains blanks. The [WRITE] function strips these quotes before writing the string to the file.

The integer returned by [WRITE] indicates the completion status of the write request.

Returned value of [WRITE]	Indicates
0	Record written successfully
100	File not open for writing
101	Error during write
102	Unable to re-open file

As with [READ] and [CLOSE], these values help you manage the actions taken by your program during successful or unsuccessful processing.

The next example uses the file input/output (I/O) functions to create an AML program that restores environments in ARCEDIT. It's assumed that you set specific tolerances and environments in relation to an edit coverage. If you aren't currently editing a coverage, the program terminates.

```
/* savesettings.aml
&severity &error &routine bailout
&sv edit = [show edit]
&if [null %edit%] &then
   &return No edit coverage. Environments not saved.
&if not [exists %edit% -cover] &then
   &return [entryname %edit%] is not a coverage.
```

After determining that the edit coverage exists, the program creates an output file using the [OPEN] function. Name the file by appending the coverage name with the suffix _AESET.AML. For example, the program creates an output file called STREAM_AESET.AML for the coverage named STREAM.

```
&sv unit = [open [entryname %edit%]_aeset.aml openstat -write]
&if %openstat% ne 0 &then
   &return Unable to open file, error code: %openstat%
```

If the file opens successfully, the program writes the current environments and tolerances to it using the [SHOW] function. The file is actually an AML program that restores the current settings when it's executed. The file needs to contain not only the current settings, but also the commands that restore them. The following is an example of a file that might be created:

```
EDITCOVERAGE /RIDLAND/WORKBOOK/STREAM
DRAWENVIRONMENT ALL OFF, ARC ON
NODESNAP FIRST,59.912
DUPLICATEARCS YES
INTERSECTARCS OFF
EDITDISTANCE 599.12
DRAWSELECT ONE
SETDRAWSYMBOL 8 0
NODECOLOR DANGLE 1
NODECOLOR NODE 1
NODECOLOR PSEUDO 1
ARCSNAP ON,20
BACKCOVERAGE /RIDLAND/WORKBOOK/VEGETATION 3
BACKCOVERAGE /RIDLAND/WORKBOOK/GEOLOGY 1
BACKENVIRONMENT /RIDLAND/WORKBOOK/GEOLOGY ARC ON, LABEL ON
BACKCOVERAGE /RIDLAND/WORKBOOK/SOIL 4
&return
```

The code that creates this file follows. It uses the [WRITE] function for each line in the new program.

Examine the code that creates this line in the previous file: NODESNAP FIRST,59.912. The program uses the [QUOTE] function to quote the string because it contains spaces. (Using single quotes would have worked too, but code presented later in the program requires that [QUOTE] is used here.) Next, the program writes the NODESNAP command with the current settings (returned by [SHOW NODESNAP]). Like all lists returned by the [SHOW] function, this string is comma-delimited. The line writes to the file without having to manipulate it because ARC/INFO can accept strings in this format.

```
&sv writestat = [write %unit% [quote EDITCOVERAGE %edit%]]
&sv writestat = [write %unit% [quote DRAWENVIRONMENT ~
  [show drawenvironment]]]
&sv writestat = [write %unit% [quote NODESNAP [show nodesnap]]]
&sv writestat = [write %unit% [quote DUPLICATEARCS ~
  [show duplicatearcs]]]
&sv writestat = [write %unit% [quote INTERSECTARCS ~
  [show intersectarcs]]]
```

```
&sv writestat = [write %unit% [quote EDITDISTANCE ~
  [show editdistance]]]
&sv writestat = [write %unit% [quote DRAWSELECT ~
  [show drawselect]]]
&sv writestat = [write %unit% [quote SETDRAWSYMBOL ~
  [show setdrawsymbol]]]
&sv writestat = [write %unit% [quote NODECOLOR DANGLE ~
  [show nodecolor dangle]]]
&sv writestat = [write %unit% [quote NODECOLOR NODE ~
  [show nodecolor node]]]
&sv writestat = [write %unit% [quote NODECOLOR PSEUDO ~
  [show nodecolor pseudo]]]
```

Not all environments are so straightforward. For example, ARCSNAP expects something different from what [SHOW ARCSNAP] returns. If ARCSNAP should be ON, the following string is acceptable:

```
ARCSNAP ON,20
```

If, however, ARCSNAP should be OFF, and you use the results of [SHOW ARCSNAP], the following error occurs:

```
ARCSNAP OFF,0
End of line expected.
Usage: ARCSNAP ON {* | distance}
Usage: ARCSNAP OFF
```

The program anticipates the two ARCSNAP usages by checking to see if ARCSNAP is OFF. If it is, then the line `ARCSNAP OFF` writes to the file. If ARCSNAP is ON, the program writes to the file by accessing the data returned by the [SHOW] function.

```
&if [extract 1 [show arcsnap]] = OFF &then
  &sv writestat = [write %unit% [quote ARCSNAP OFF]]
&else
  &sv writestat = [write %unit% [quote ARCSNAP [show arcsnap]]]
```

Retrieving background coverages, their symbology, and their background environments presents another hurdle to overcome. The string returned by [SHOW BACKCOVERAGE ALL] appears like this: `cov1,sym1 cov2,sym2 cov3,sym3`, etc. It's difficult to access this list with a single &DO loop because pairs of data are needed at each pass through the loop. Instead, the following program uses [EXTRACT] to access each background coverage and symbol, then writes the BACKCOVERAGE command to the program. If there's a

BACKENVIRONMENT for the current background coverage (there should be, but the program checks, just in case), then the BACKENVIRONMENT command writes to the program. This avoids the problem of executing a statement like BACKENVIRONMENT GEOLOGY, which results in a usage error.

```
&do i = 1 &to [show number backcover]
  &sv bc = [extract 1 [show backcover %i%]]
  &sv bcsym = [extract 2 [show backcover %i%]]
  &sv writestat = [write %unit% [quote BACKCOVERAGE %bc% ~
      %bcsym%]]
  &if not [null [show backenvironment %bc%]] &then
      &sv writestat = [write %unit% [quote BACKENVIRONMENT ~
          %bc% [show backenvironment %bc%]]]
&end
&sv writestat = [write %unit% &return]
&call exit
&return
```

As always, the program closes the file when it finishes writing:

```
/*------------------
&routine EXIT
/*------------------
/* Close the new AML
&if [variable unit] &then &do
  &sv closestat = [close %unit%]
  &if %closestat% ne 0 &then
      &return FILE CLOSE ERROR %closestat%
&end
&return
/*------------------
&routine BAILOUT
/*------------------
&severity &error &ignore
&severity &warning &ignore
&call exit
&return &error An error has occurred in SAVSETTINGS.AML
```

STREAM_AESET.AML can be run the next time the user enters ARCEDIT to restore the settings needed to process the coverage.

Question 10-3: If you open a file in the -WRITE mode, can an AML program read the lines written to the file?

(Answer on page 10-25)

Writing to an existing file

As shown in the last example, the [OPEN] function with the -WRITE option creates a new file or overwrites an existing file. However, you may want to add to (rather than overwrite) an existing file.

Suppose you're using an application that captures point coordinates entered by the user. The program creates a file called POINT.FILE by using the [OPEN] function with the -WRITE option. You're busily entering points when an emergency calls you away. You quit the application even though you haven't finished entering all the points. When you come back and start the application again, the [OPEN] function deletes all the points you captured during the previous session and you have to start over. This isn't great for productivity, especially if you have already entered several hundred points.

To add to an existing file instead of overwriting it, use the [OPEN] function with the -APPEND option.

Instead of this:

```
/* OPEN the file to contain the points
&sv unit = [open point.file openstat -write]
```

code this:

```
/* OPEN the file to contain the points
&if [exists point.file -file] &then
  &sv unit = [open point.file openstat -append]
&else
  &sv unit = [open point.file openstat -write]
```

Managing system files

You won't always want to keep the files a program creates. Some programs produce intermediate files for processing, including system and INFO files. In most cases, you'll want the program to delete these intermediate processing files or give the user the option to do so. To accomplish this, use the [DELETE] function. [DELETE] gives you the ability to have AML delete ASCII files, directories, workspaces, and INFO files. The next example uses [DELETE] to delete temporary files created by AML:

```
&sv delstat = [DELETE tmp.xx]
```

[DELETE] returns a status of either zero or 100 as follows:

Returned value of [DELETE]	Indicates
0	Delete performed successfully
100	Couldn't delete file, directory, or INFO file

Use the [DELETE] function in an EXIT routine to ensure that your program deletes temporary processing files before the program finishes. Don't forget that a file must be closed before it can be deleted. The next example deletes an ASCII file stored in the variable TMP_FILE:

```
/*------------------
&routine EXIT
/*------------------
&if [variable unit] &then &do  /* If file has been opened.
  &sv closestat [close %unit%]
  &if %closestat% = 0 &then &do
    &dv fileunit
    &sv delstat = [delete %tmp_file%]
  &end
  &else
    &type Error closing unit %fileunit% Error code: ~
      %closestat%
&end /* &if [variable fileunit]
```

Other file management functions include [COPY] and [RENAME]. Use these functions in your applications rather than system commands to make your application portable across platforms.

The [SCRATCHNAME] function also allows you to create temporary file names that won't duplicate preexisting file names. Save the name of the temporary file early in the AML program using [SCRATCHNAME]:

```
&s tmp_file = [SCRATCHNAME]
```

Make sure that you create the file stored in TMP_FILE before using [SCRATCHNAME] again. If you use [SCRATCHNAME] a second time without creating the first file, then the second temporary name will equal the first:

```
Arc: &s tmp_file = [SCRATCHNAME]
Arc: &s tmp_file2 = [SCRATCHNAME]
Arc: &ty [CALC  %tmp_file% = %tmp_file2%]
.TRUE.
```

In this example, the [SCRATCHNAME] is used twice in a row to find a unique file name. As you can see, [SCRATCHNAME] wasn't successful because the first temporary file wasn't created before generating the second name.

```
Arc: &s tmp_file = [SCRATCHNAME]
Arc: &s unit = [OPEN %tmp_file% openstat -write]
Arc: &s tmp_file2 = [SCRATCHNAME]
Arc: &ty [calc %tmp_file% = %tmp_file2%]
.FALSE.
```

When the file is opened, it is created, so the second [SCRATCHNAME] function produces a unique name.

FYI

Use the file manipulation functions when dealing with files and directories. [COPY], [DELETE], and [RENAME] allow you to manipulate the chosen object (e.g., file, directory, workspace, or INFO file) from any subsystem, on any platform. Don't use platform-specific code here. The subsystems (ARCPLOT, ARCEDIT, etc.) won't pass unknown commands to the operating system like ARC does. Save yourself the trouble; use the [COPY], [DELETE], and [RENAME] functions.

In summary

The [OPEN] function with the -WRITE option creates a system file and opens it for writing. Using [OPEN] with the -APPEND option allows you to add information to an existing file. The [WRITE] function actually writes to the file, one record at a time. The value returned by the [WRITE] function indicates the completion status of the write request.

Use the [COPY], [DELETE], and [RENAME] functions to process files. Use the [SCRATCHNAME] function to generate unique file names for temporary files. Files must be closed using the [CLOSE] function before they can be manipulated.

Answer 10-3: An AML program can't read a file opened in the -WRITE mode until it's closed and then reopened with -READ access.

Exercises

Exercise 10.2.1

Write an AML program called SAVAPSET.AML that emulates this lesson's example of saving environments in ARCEDIT. Have your program save the ARCPLOT map-to-page transformation information. The program should also accept a coverage name as an argument and use it as the prefix for the name of the AML program. The program should execute the following commands:

```
MAPEXTENT
MAPANGLE
MAPLIMITS (in PAGE units)
MAPPOSITION
MAPSCALE
MAPUNITS
PAGESIZE
PAGEUNITS
```

Exercise 10.2.2

Write a program called DELFILES.AML that deletes a set of system files given a search string containing the asterisk wildcard character (*). Your program should provide the user with a list of the files to delete and ask for confirmation of the set—confirmation for *each* file isn't necessary.

Lesson 10.3—Using AML to write menu files

In this lesson

An effective application interface presents the user with an environment that's easy to recognize and easy to manipulate. In most cases, you design a single menu for a single purpose. There may be times, however, that you need the added flexibility of changing the appearance of a menu to perform the same tasks under slightly different circumstances.

The ability to dynamically alter an application's appearance is a powerful tool. In this lesson, you'll use the file-handling functions with which you're already familiar to create a custom interface that changes depending on the situation.

Creating menus dynamically from AML programs

Suppose you want to perform an IDENTIFY on a coverage. Instead of listing all attributes of a feature, imagine that you'd like to use a menu to select the attribute items you want to list. It's easy enough to create a menu that presents you with the items from a feature attribute table from which you can choose, then to apply your choice and see the list of items.

But what if you work with more than one coverage (a high probability if you're reading this book)? Do you want to code a menu for every coverage attribute table in the workspace? Not likely. Instead, have your program create the menu when you pass it a coverage and feature class. In this case, the menu is created automatically, as needed. You write the program; the program writes the menu.

AML form menus enable you to dynamically define an operation before execution, so operations vary depending on how the form is filled out. The form menus shown here contain check boxes from which selections are made (a check mark indicates a selection):

 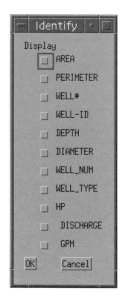

The OK and Cancel widgets are called *buttons;* they execute commands. In this case, OK executes the IDENTIFY command sequence and Cancel dismisses the menu. Don't worry if you don't fully understand form menus at this time; chapter 12 covers form menus in detail.

Here's the menu file for the menu on the left:

```
7
 Display
    %1 AREA
%1 checkbox QUERY_AREA
    %2 PERIMETER
%2 checkbox QUERY_PERIMETER
    %3 PARCEL#
%3 checkbox QUERY_PARCEL#
    %4 PARCEL-ID
%4 checkbox QUERY_PARCEL-ID
    %5 APN
```

```
%5 checkbox QUERY_APN
   %6 GENPLAN
%6 checkbox QUERY_GENPLAN
   %7 LANDUSE
%7 checkbox QUERY_LANDUSE
   %8 ZONING
%8 checkbox QUERY_ZONING
 %ok    %cancel
%ok button OK    &return; &s continue = .TRUE.
%cancel button 'Cancel' &return; &s continue = .FALSE.
```

You can use the file I/O functions that you already know to create files like this.
The rest of this lesson looks at the program that created these form menus,
IDENTIFY.AML. A detailed examination of the individual routines that create
the menus and identify features begins on page 10-32. First, IDENTIFY.AML in
its entirety:

```
/* identify.aml
&severity &error &routine bailout
&call getcov
&call getfeat
&call draw
&call create
&call menu

&if %continue% &then
  &call identify
&call exit
&return /* from AML

/*------------------
&routine GETCOV
/*------------------
&s source = [getcover *]
&if [null %source%] &then
  &return You must pick a coverage!
&return /* routine GETCOV

/*------------------
&routine GETFEAT
/*------------------
&s available_feat
&describe %source%
&if %dsc$pat_bytes% > 0 and not %dsc$qedit% &then
  &s available_feat = %available_feat% POLY POLYGONS
&if %dsc$aat_bytes% > 0 &then
  &s available_feat = %available_feat% LINE ARCS
&if %dsc$xat_bytes% > 0 &then
  &s available_feat = %available_feat% POINT POINTS
&if %dsc$nat_bytes% > 0 &then
```

```
      &s available_feat = %available_feat% NODE NODES
&if [token %available_feat% -count] = 0 &then
   &return No available features to identify.  Build or Clean coverage.
&s fclass = [unquote [getchoice -pairs %available_feat% -var ~
   draw_command -prompt 'Feature Class?']]

&if [null %fclass%] &then
   &return You must pick a Feature Class!
&return /* routine GETFEAT

/*------------------
&routine DRAW
/*------------------
CLEAR
MAPEXTENT %source%
&if %fclass% = NODE &then
   ARCS %source%
%draw_command% %source%
&return /* routine DRAW


/*------------------
&routine CREATE
/*------------------
/* Open the menu for writing
/*
&s menu_name = [scratchname]
&sv menuunit = [open %menu_name% openstat2 -write]
&s itlist = [listitem %source% -%fclass% -all]
/*
/* Create a form menu with checkboxes for each item, an
/* APPLY button to dismiss and execute, and a CANCEL button
/* to dismiss
/*
&sv wtstat = [write %menuunit% 7]
&sv wtstat = [write %menuunit% [format ' Display']]

&s i = 0
&do item &list %itlist%
   &s i = %i% + 1
   &sv wtstat = [write %menuunit% [format ~
      '    %%1% %2%' %i% %item%]]
   &sv wtstat = [write %menuunit% [format ~
      '%%1% checkbox %2%' %i% QUERY_%item%]]
&end

&sv wtstat = [write %menuunit% [format ' %ok     %cancel']]
&sv wtstat = [write %menuunit% [format '%ok button OK ~
   &return; &s continue = .TRUE.']]
&sv wtstat = [write %menuunit% [format '%cancel button ''Cancel''
&return; &s continue = .FALSE.']]
&sv close = [close %menuunit%]
&dv menuunit
&return /* routine CREATE
```

```
/*------------------
&routine MENU
/*------------------

&menu %menu_name% &position~
  &ul &display &ul &stripe Identify ~
  &pinaction '&return; &s continue = .FALSE.'
&return /* routine MENU

/*------------------
&routine IDENTIFY
/*------------------
/* Create an array of the chosen items and store it in
/* ITLIST. IDENTIFY until the user presses a key other
/* than 1
/*
&sv view_list
&do i = 1 &to [token %itlist% -count]
  &sv item = [extract %i% %itlist%]
  &if [value QUERY_%item%] &then
    &sv view_list - %view list% %item%
&end

units map
&flushpoints
&getpoint
&do &while %pnt$key% = 1
  identify %source% %fclass% %pnt$x% ~
    %pnt$y% %view_list%
  &getpoint
&end
units page
&return /* routine IDENTIFY

/*------------------
&routine EXIT
/*------------------
&if [variable menuunit] &then
  &sv close = [close %menuunit%]

&if [exists [value menu_name] -file] &then
  &s delstat = [delete %menu_name%]
&return/* routine EXIT

/*------------------
&routine BAILOUT
/*------------------
&severity &error &ignore
&call exit
&return &error /* routine BAILOUT
```

The main body of this AML program is simply an organized set of routines. The main body of the AML program is a series of &CALL directives, while the real work of the program is accomplished within the routines.

```
/* identify.aml
&severity &error &routine bailout

&call getcov
&call getfeat
&call draw
&call create
&call menu

&if %continue% &then
  &call identify
&call exit
&return /* from AML
```

The GETCOV and GETFEAT routines establish the coverage name and feature class. The coverage then draws within the DRAW routine. The CREATE and MENU routines generate a custom menu and display it. Finally, the identify routine performs the ARCPLOT IDENTIFY command.

The GETCOV routine uses the [GETCOVER] function to set the variable source with the coverage name. The AML program tests the value that returned from the user's selection to make sure they didn't dismiss the menu before choosing a coverage.

```
/*-----------------
&routine GETCOV
/*-----------------
&s source = [getcover *]
&if [null %source%] &then
  &return You must pick a coverage!
&return /* routine GETCOV
```

Once the GETCOV routine establishes the coverage name, the GETFEAT routine creates a list of available feature classes from which to choose. Rather than allow the user to choose any feature class, &DESCRIBE is used in conjunction with a series of &IF &THEN statements to establish which feature classes the coverage possesses. For instance, a STREET coverage may have LINE and NODE feature classes, but not POLY and POINT. In this case, the user would only be able to choose between the LINE and NODE feature classes.

The variable AVAILABLE_FEAT stores a list of the feature classes for the selected coverage. This variable is then used within the [GETCHOICE] function to allow the user to choose which feature class to query. The -PAIRS option sets the variable DRAW_COMMAND to contain the ARCPLOT command for drawing the coverage.

```
/*-----------------
&routine GETFEAT
/*-----------------
&s available_feat
&describe %source%
&if %dsc$pat_bytes% > 0 and not %dsc$qedit% &then
  &s available_fcat = %available_feat% POLY POLYGONS
&if %dsc$aat_bytes% > 0 &then
  &s available_feat = %available_feat% LINE ARCS
&if %dsc$xat_bytes% > 0 &then
  &s available_fcat = %available_feat% POINT POINTS
&if %dsc$nat_bytes% > 0 &then
  &s available_feat = %available_feat% NODE NODES

&if [token %available_feat% -count] = 0 &then
  &return No available features to identify. Build or Clean coverage.
&s fclass = [unquote [getchoice -pairs %available_feat% -var ~
  draw_command -prompt 'Feature Class?']]

&if [null %fclass%] &then
  &return You must pick a Feature Class!
&return /* routine GETFEAT
```

The DRAW routine uses the variable DRAW_COMMAND to draw the coverage on the ARCPLOT canvas. The only special case arises when the node feature class is chosen for query. In this case, the AML program also draws the arcs for the coverage as a point of reference for the nodes. Remember, the %SOURCE% variable contains the name of the coverage set in the GETCOV routine.

```
/*-----------------
&routine DRAW
/*-----------------
CLEAR
MAPEXTENT %source%
&if %fclass% = NODE &then
  ARCS %source%
%draw_command% %source%
&return /* routine DRAW
```

The CREATE routine creates a menu containing the items from the selected coverage and attribute table (feature class). First, the CREATE routine determines the name of the new menu file by using the [SCRATCHNAME] function. This file is then opened for writing with the unit number for this file stored in the variable MENUUNIT.

```
/*------------------
&routine CREATE
/*------------------
/* Open the menu for writing
/*
&s menu_name = [scratchname]
&sv menuunit = [open %menu_name% openstat -write]
```

The list of items for the selected coverage are stored within the variable ITLIST. The [LISTITEM] function generates this comma-delimited list. The -ALL option is used to retrieve all the items, including the redefined items.

```
&s itlist = [listitem %source% -%fclass% -all]
```

Now the menu needs to be created. In most cases, it's simply a process of using the [WRITE] function with the data required for each line in the menu. Assume that the menu shown on the left on page 10-28 is the menu being created. For review, the code for this menu is as follows:

```
7
 Display
    %1 AREA
%1 checkbox QUERY_AREA
    %2 PERIMETER
%2 checkbox QUERY_PERIMETER
    %3 PARCEL#
%3 checkbox QUERY_PARCEL#
    %4 PARCEL-ID
%4 checkbox QUERY_PARCEL-ID
    %5 APN
%5 checkbox QUERY_APN
    %6 GENPLAN
%6 checkbox QUERY_GENPLAN
    %7 LANDUSE
%7 checkbox QUERY_LANDUSE
    %8 ZONING
%8 checkbox QUERY_ZONING
 %ok    %cancel
%ok button OK    &return; &s continue = .TRUE.
%cancel button 'Cancel' &return; &s continue = .FALSE.
```

For now, don't focus on the meaning of the elements in the menu file, but on their format. Notice that some lines contain leading spaces and most lines include a single percent sign (%). Leading spaces aren't a problem for the [WRITE] function.

Additionally, the percent signs and variable values (item names) need to be on the same line. Neither single quotes nor the [QUOTE] function can accomplish this. Single quotes prevent the variable from being evaluated and the [QUOTE] function fails because of an unmatched variable delimiter error caused by the stand-alone percent signs.

The [FORMAT] function (discussed in chapter 7) provides a way of formatting the string in a way that the [WRITE] function can access. Consider the following code:

```
&sv string [FORMAT '    %%1% %2%' %i% %item%]
```

Here [FORMAT] places the value of the %i% variable in the location of the %1% placeholder and the value of the %item% variable in the location of the %2% placeholder. The string returned by the [FORMAT] function is quoted if it contains blanks. If %i% = 7 and %item% = LANDUSE, the contents of the STRING variable are as follows:

```
%string% = '    %7 LANDUSE'
```

Formatting the data this way enables the [WRITE] function to access it. [WRITE] removes the quotes and outputs the string to a file as shown below.

```
/* Create a form menu with checkboxes for each item, an
/* APPLY button to dismiss and execute, and a CANCEL button
/* to dismiss
/*
&sv wtstat = [write %menuunit% 7]
&sv wtstat = [write %menuunit% [format ' Display']]

&s i = 0
&do item &list %itlist%
  &s i = %i% + 1
  &sv wtstat = [write %menuunit% [format ~
    '    %%1% %2%' %i% %item%]]
  &sv wtstat = [write %menuunit% [format ~
    '%%1% checkbox %2%' %i% QUERY_%item%]]
&end
```

The previous code fragment contains an &DO &LIST loop that repeats once for each item contained in the variable %ITLIST%. Remember, this variable contains a comma-delimited list of items produced by [LISTITEM]. The code writes two lines in the menu for each item encountered. The first is the one shown in the example of using [FORMAT]. The second is a definition line for the first. Staying with the LANDUSE item, the output line is:

```
%7 checkbox QUERY_LANDUSE
```

FYI

> By allowing you to set a field width for each argument, [FORMAT] justifies the columns in the file to which you're writing. Use the syntax `%#,spaces%` (as shown in this example):
>
> ```
> &sv string = [FORMAT '    %%1,5% %2%' %I% %item%]
> ```
>
> In this case, the output line for the LANDUSE item (minus the quotes) is:
>
> ```
> '    %7    LANDUSE'
> ```

Note: When you check the box next to the text LANDUSE in the menu, the variable QUERY_LANDUSE is set to .TRUE.. The QUERY_%ITEM% variables are local because this menu will be launched from the AML program, so the menu is within the scope of local variables. The final two lines of the menu are written with the code below. The user presses these buttons when wanting to identify features or dismiss the menu. The menu file is then closed using the unit number stored in %MENUUNIT%.

```
&sv wtstat = [write %menuunit% [format ' %ok    %cancel']]
&sv wtstat = [write %menuunit% [format '%ok button OK ~
  &return; &s continue = .TRUE.']]
&sv wtstat = [write %menuunit% [format '%cancel button ''Cancel'' ~
&return; &s continue = .FALSE.']]
&sv close = [close %menuunit%]
&dv menuunit
&return /* routine CREATE
```

Notice the return string that's executed when the user presses the OK or Cancel buttons. First, &RETURN returns control to the AML program. Next, the variable CONTINUE is set to either .TRUE. or .FALSE. depending on which button is pressed. CONTINUE is used as a flag in the AML program to determine if you want to identify features or not. CONTINUE is tested within the main body of the AML program.

FYI

The following menu demonstrates some of the complexities of writing menus with the [WRITE] function. The menu file contains three lines of code:

```
1
Workspaces LISTWORKSPACES
'Available coverages' LISTCOVERAGES
```

Here's the pulldown menu:

The first line in the menu file is a single piece of text, the number 1. The code needed to write this line of code using the [WRITE] function follows:

```
&sv wt = [write %unit% 1]
```

The second line in the menu contains the text Workspaces and the command LISTWORKSPACES. This string must be quoted because of the space between the two words. [WRITE] strips a single pair of quotes before it outputs text to a file.

```
&sv wt = [write %unit% 'Workspaces LISTWORKSPACES']
```

The last line is more difficult because it contains both spaces and quotes. In this case, quote the string 'Available coverages' and the entire line 'Available coverages' LISTCOVERAGES. Use the [QUOTE] function to place both pairs of quotes so [WRITE] can strip them correctly.

Even though you could use quotes entered from the keyboard, it's a good idea to use the [QUOTE] function for consistency in your programming. Remember, the quote symbol suppresses AML interpretation, so you may not always get the result you want.

```
&sv wt = [write %unit% [quote [quote Available coverages] ~
    LISTCOVERAGES]]
```

At this point, the AML program has created the menu file. The MENU routine is now called to launch the menu that the CREATE routine produced.

```
/*-----------------
&routine MENU
/*-----------------
&menu %menu_name% &position~
  &ul &display &ul &stripe Identify ~
  &pinaction '&return; &s continue = .FALSE.'
&return /* routine MENU
```

In the &pinaction for this menu, the CONTINUE variable is set to .FALSE.. This means that if you close the menu by pulling the pin, the AML program skips the IDENTIFY routine and exits.

If the user clicks the OK button, the IDENTIFY routine is executed. The IDENTIFY routine first checks the values of all the QUERY_%ITEM% variables. If the variable evaluates to .TRUE. (check box is checked), that item is added to a list of item names contained within the variable VIEW_LIST.

```
/*-----------------
&routine IDENTIFY
/*-----------------
/* Create an array of the chosen items and store it in VIEW_LIST.
/* IDENTIFY until the user presses a key other than 1
&sv view_list
&do i = 1 &to [token %itlist% -count]
  &sv item = [extract %i% %itlist%]
  &if [value QUERY_%item%] &then
    &sv view_list = %view_list% %item%
&end
units map
&flushpoints
&getpoint
&do &while %pnt$key% = 1
  identify %source% %fclass% %pnt$x% ~
    %pnt$y% %view_list%
  &getpoint
&end
units page
&return /* routine IDENTIFY
```

The second part of the IDENTIFY routine uses the &GETPOINT directive that you saw in chapter 6. The value of %view_list% becomes an argument to the IDENTIFY command and lists only those items when the feature is identified.

The IDENTIFY command repeats as long as the user selects the 1 key. As soon as another mouse button is pressed, the IDENTIFY routine returns and the EXIT routine is called to clean up. The EXIT routine makes sure that the menu file closes and then deletes it.

```
/*-- --------------
&routine EXIT
/*-----------------
&if [variable menuunit] &then
  &sv close = [close %menuunit%]

&if [exists [value menu_name] -file] &then
  &s delstat = [delete %menu_name%]
&return/* routine EXIT
```

The AML program then finishes and returns control to the calling program.

The modular techniques used in this lesson use routines within a single AML program. A menu is launched and the AML program uses the values set in this menu to preform an IDENTIFY. This is one method for modular programming. Later, in chapters 12, 14, and 15, you'll learn more about forms, threads, and the ArcTools programming style. With this knowledge, IDENTIFY.AML can be rewritten in the same format as other ArcTools.

In summary

AML programs can write AML menus. The power to create form menus on demand gives you the ability to create a flexible interface that adjusts to changing situations. IDENTIFY.AML is a working model that demonstrates one style of programming for creating menus from programs. The [FORMAT] function used in conjunction with the [WRITE] function plays an important role in creating a menu file—or any file—from a program.

Exercises

Exercise 10.3.1

Write a program called TOPO_DESC that accomplishes the following tasks:

- Creates a list of coverages that were edited since their last clean or build.
- Creates a pulldown menu that allows you to DESCRIBE or view the LOG file for each coverage.

When complete, the menu should look like this:

The coverages displayed in these pulldowns should only be those that were edited since the last clean or build. The menu should be launched from the AML program that created it and have an Exit button that dismisses the menu when you're done making selections. When the AML program is finished, it should delete the menu file it created.

For more information on pulldown menus, see chapter 2, lesson 2, of this book.

(Hint: Use &DESCRIBE to find out if a coverage has been edited or not.)

11 Accessing your database

In chapter 6, you used AML for accessing and processing spatial data. But to create maps and reports you need attribute data.

This chapter teaches you how to access your INFO database with a variety of AML tools. This is an introduction to the suite of database-related functions found in AML. You'll learn when and why they're useful. You'll also explore the AML tools available for accessing other software. Finally, you'll see how using a tool called Cursors gives AML direct access to INFO and external database management systems (DBMSs).

This chapter covers the following topics:

Lesson 11.1—Accessing INFO tables

- AML functions that access INFO tables

Lesson 11.2—Using interactive software with AML

- Executing operating system commands from ARC/INFO
- Using &DATA with INFO
- Getting information from INFO that AML can use
- [TASK] it out

Lesson 11.3—Accessing attribute data

- Using Cursors to query and update attribute data
- Cursor variables
- Cursor descriptors

Lesson 11.1—Accessing INFO tables

In this lesson

This lesson examines the AML functions that access the attribute data stored in INFO tables. You'll find the ability of AML programs to access ARC/INFO attributes invaluable.

AML functions that access INFO tables

AML functions that include the keyword -INFO as an argument can assist you in working with INFO tables. This lesson describes how to use the following AML functions to access INFO:

- [LISTFILE] [listfile * -file]
- [EXISTS]
- [DELETE]
- [GETFILE]
- [GETITEM]
- [ITEMINFO]
- [GETUNIQUE]
- [GETCHOICE]

[LISTFILE], [EXISTS], and [DELETE]

You can use the -INFO argument with the [LISTFILE], [EXISTS], and [DELETE] functions introduced in chapter 10.

[LISTFILE] allows you to create an ASCII file containing a list of INFO file names (e.g., lookup tables) in the local workspace:

```
&sv numobs = [listfile *.lut -info -full lutnames.fil]
```

The [EXISTS] function checks for the existence of an INFO file and allows you to make programming decisions based on the outcome:

```
&if [exists soil.rel -info] &then
  RELATE RESTORE SOIL.REL
```

The [DELETE] function allows you to delete INFO files from any ARC/INFO subsystem:

```
&sv delstat = [delete tmp.dat -info]
```

Using these functions with the -INFO argument allows you to manage and query INFO files.

FYI

Some functions, including [LISTFILE] and [EXISTS], allow access to INFO tables through the ARC/INFO data model. For example, these two usages both access the INFO file soil.pat:

```
[exists soil.pat -info]

[exists soil -poly]
```

[GETFILE]

Like most of the [GET...] functions, [GETFILE] prompts the user to choose an object from a list. The -INFO option directs [GETFILE] to display data files from an INFO directory. The following code selects an INFO lookup table:

```
&sv table = [getfile *.lut -info]
```

If &LV is used on the TABLE variable, the following information is returned:

```
Arc: &lv table
Local: TABLE /carmel1/miker/amlworkbook/info!arc!res_date.lut
```

[GETITEM] 2-6

Another [GET...] function that accesses the INFO database is [GETITEM]. Suppose that you want to choose a coverage, an item, and a lookup table for the POLYGONSHADES command in ARCPLOT. This can be accomplished as follows:

```
&sv cov = [getcover * -poly 'Select a coverage to shade']
&sv item = [getitem %cov% -poly 'Select a lookup table item']
&sv lookup = [getfile *.lut -info 'Select a lookup table']
POLYGONSHADES %cov% %item% %lookup%
```

[ITEMINFO] 5-13

Suppose the user needs the option of shading polygons based on an item in the polygon attribute table (PAT). The program needs to make sure that the user chooses to shade by an integer item (type I). The [ITEMINFO] function returns a description of a specified INFO data file item, including its type. The syntax for [ITEMINFO] is as follows:

```
[ITEMINFO <specifier> <-type> <item>
          {-DEFINITION |-REDEFINED | -INDEXED | -EXISTS | -FULLDEF}]
```

Because of its numerous keyword options, [ITEMINFO] acts as four different functions. The keywords -REDEFINED, -INDEXED, and -EXISTS cause this function to return a logical value of .TRUE. or .FALSE., indicating whether the item is redefined, indexed, or exists.

The default option -DEFINITION causes the function to return the definition of the item (i.e., its width, output width, item type, and number of decimal places). For example, [ITEMINFO] used on the item COVER# returns an item definition such as 4,5,B,0, where 4 is the item width, 5 is the output width, B is the item type, and 0 is the number of decimal places. You can use [EXTRACT] on this string to determine if the item is type I (integer). The following example uses [ITEMINFO] with the -DEFINITION option to solve the hypothetical polygon shading problem:

```
/* polyshades.aml
&sv cov = [getcover * -poly 'Select a coverage to shade']
&if [query 'Do you want to use a lookup table'] &then &do
  &sv item = [getitem %cov% -poly 'Select a lookup table item']
  &sv lookup = [getfile *.lut -info 'Select a lookup table']
```

```
&if not [iteminfo %lookup% -info %item% -exists] &then
  &return %item% not present on table [entryname ~
  %lookup% -info]
POLYGONSHADES %cov% %item% [entryname %lookup% -info]
&end
&else &do
  &sv numitems = [token [unquote [listitem~
  %cov% -poly]] -count]
  &sv cnt = 0
&do &until %type% = I or %cnt% ge %numitems%
  &sv item = [getitem %cov% -poly ~
    'Please select an item to shade with']
  &sv def = [iteminfo %cov% -poly %item% -definition]
  &sv type = [extract 3 %def%]
  &sv cnt = %cnt% + 1
&end
 &if [extract 3 [iteminfo %cov% -poly %item% -definition]]~
   ne I &then &return No INTEGER item chosen
MAPEXTENT %cov%
POLYGONSHADES %cov% %item%
&end
&return
```

To restrict the types of items presented to the user, you can revise the previous code to eliminate [ITEMINFO]. Use [GETITEM] with the {item_type} to specify INTEGER as the item type, thus presenting only integer items to the user.

```
/* polyshades2.aml
&sv cov = [getcover * -poly 'Select a coverage to shade']
&if [query 'Do you want to use a lookup table'] &then &do
 &sv item = [getitem %cov% -poly 'Select a lookup table item']
  &sv lookup = [getfile *.lut -info 'Select a lookup table']
  &if not [iteminfo %lookup% -info %item% -exists] &then
  &return %item% not present on table [entryname ~
  %lookup% -info]
  MAPEXTENT %cov%
  POLYGONSHADES %cov% %item% [entryname %lookup% -info]
&end


&else &do
&sv item = [getitem %cov% -poly -integer ~
    'Select an item to shade with']
  &if [null %item%] &then &return No INTEGER item chosen.
MAPEXTENT %cov%
POLYGONSHADES %cov% %item%
&end
&return
```

[GETUNIQUE]

The functions you've seen so far address such structural issues as file names, item names, and item definitions. The [GETUNIQUE] function uses this information to enable the user to choose from a list of unique item values. In ARCPLOT, [GETUNIQUE] operates on the selected set of features. This enables you to select records with specific values and present these to the user.

FYI

[GETUNIQUE] doesn't limit its display to the unique values in the *selected set* unless the arguments of the [GETUNIQUE] function match the parameters used when the elements were selected. In the next example, [GETUNIQUE] displays the unique values from the fifteen reselected values:

```
Arcplot: RESELECT PARCEL POLY ZONE CN 'RES'
SOIL polys :15 of 173 selected.
Arcplot: RESELECT PARCEL POLY LANDUSE = ~
Arcplot: [getunique parcel -poly landuse]
```

In the next example, however, [GETUNIQUE] displays the unique values from every record in the INFO file because the function's arguments don't match the parameters set in the RESELECT statement:

```
Arcplot: RESELECT PARCEL POLY ZONE CN 'RES'
SOIL polys :15 of 173 selected.
Arcplot: RESELECT PARCEL POLY LANDUSE = ~
Arcplot: [getunique parcel.pat -info landuse]
```

ARCPLOT maintains selection information for each file. SOIL -POLY isn't the same type of selection as SOIL.PAT -INFO.

The [GETUNIQUE] function, a quick and easy interface to your data, is often used as shown here to prompt the user for data to use in selection statements:

```
&sv infofile = [getfile * -info]
&sv item = [getitem %infofile% -info]
&sv value = [getunique %infofile% -info %item%]
TABLES            ↝ 7·10
SELECT [entryname %infofile% -info]
RESELECT %item% = [quote %value%]
LIST                          ↳ 7-14
```

This code prompts the user for a file, an item, and then a value. It then goes to the TABLES subsystem, selects the file and the records, and displays them to the user. The selection process could be performed in any of the ARC/INFO subsystems that support attribute access (i.e., ARCPLOT, ARCEDIT, etc.).

[GETCHOICE]

→ 2-12

In chapter 2, you learned to use the [GETCHOICE] function to display user-defined choices in a menu. [GETCHOICE] can also use an INFO file or a system file to obtain its list of choices. The following example shows how you can use it to access an INFO file. Consider the following INFO items and their values:

Label   *class*

Soil_type	Soil_code
clay	1.1
loam	1.2

Label

Suppose you want to display the values of SOIL_TYPE in the menu but have the menu selection set a variable to the SOIL_CODE value. *class* For example, choosing CLAY from the menu would set a variable to 1.1. The [GETCHOICE] syntax needed to accomplish this is:

```
&sv soiltype [getchoice -info soil.exp -display soil_type ~
  -value soil_code -var soilcode -prompt 'Select a soil type']
```

Label

The -DISPLAY SOIL_TYPE option tells [GETCHOICE] to use the values for the SOIL_TYPE item from the named INFO file to display in the list. The -VALUE SOIL_CODE option tells [GETCHOICE] to return the values of the SOIL_CODE item to a variable named SOILCODE (specified with the -VAR option).

Using [GETCHOICE] in this way enables you to set two variables with one menu selection. One variable is set equal to the value of the selection and another is assigned the value of another item in the INFO file (e.g., SOILTYPE is set to CLAY and SOILCODE is set to 1.1).

In summary

Many AML functions enable you to access the tabular data in your INFO database. You can use the functions to access INFO tables, items, and the data in the tables for query. In the next lesson, you'll examine a more formal method of accessing INFO that you can also use to access other interactive software.

Exercises

Exercise 11.1.1

Write an AML program called RES.AML that performs a RESELECT in ARCPLOT. Enable the user to choose a file, an item, and a value. The program should perform the RESELECT and LIST the results. The program should also determine whether the user chooses a file or not and then exit with a usage message if no file is chosen.

Exercise 11.1.2

Rewrite the program you created in exercise 11.1.1 so that it offers the user a choice between these command tasks:

```
RESELECT
Clear the selected set
LIST
ITEMS
Get a new file
QUIT
```

Name the new program SELECT.AML. The program should continue until the user chooses QUIT. If the user chooses RESELECT, the selection process should be an intelligent one. If the item type is character, don't prompt the user for a logical operator; use EQ instead. If the item type is other than character, allow the user to choose from one of the following logical operators:

```
GT
GE
EQ
LE
LT
```

Exercise 11.1.3

Write a program called JOIN.AML that performs the JOINITEM command in ARC:

```
JOINITEM <in_info_file> <join_info_file> <out_info_file>
         <relate_item> <start_item> {LINEAR | ORDERED | LINK}
```

The program should prompt the user for the following data (e.g., with [RESPONSE]) and accept the <out_info_file> as an argument with &ARGS:

```
<in_info_file>
<join_info_file>
<relate_item>
<start_item>
relate_type: {LINEAR | ORDERED | LINK}
```

Exercise 11.1.4

Rewrite the program from exercise 11.1.3 to explicitly handle the following potential errors:

- Make sure that the program is run from ARC.
- Ensure that the <out_info_file> doesn't already exist.
- Check that the <relate_item> exists in the <join_info_file>.
- Be certain that the <relate_item> has the same name and definition in both the <in_info_file> and the <join_info_file>.
- Prompt the user for only valid relate items.
- Inform the user if there are no valid relate items.

Name the program JOINNEW.AML.

Lesson 11.2—Using interactive software with AML

In this lesson

You'll find times when you need to access external programs with AML. These include programs written in other languages such as C, statistical applications like SAS®, or other interactive software that doesn't directly interface with ARC/INFO. You may need to access these programs or simply have them produce output.

This chapter examines the AML methods for accessing other software. This discussion treats the database INFO as a separate application. Although there are other ways to access the INFO database (i.e., the functions presented in the previous lesson, TABLES, and Cursors, presented in the lesson to follow), the method presented here is the only way AML can fully interact with INFO.

Executing operating system commands from ARC/INFO

In some cases, you need your AML program to execute system commands. System commands issued from the Arc: prompt pass to the operating system and the following message appears on-screen:

Submitting command to Operating System...

ARC passes commands it doesn't recognize to the operating system, but the subsystems (ARCEDIT, ARCPLOT, etc.) don't. Instead they return the message Unrecognized command. To execute operating system commands from any ARC/INFO prompt, use the &SYSTEM directive.

↳ 1-13

This example shows how &SYSTEM initiates a subprocess that yields an operating system prompt at which you can enter commands. To return to ARC/INFO, type exit.

```
Arcplot: &system
% ls
Formedit      Mwm*          fmconsole.log  olvwm*  transfer/
Maker         dead.letter  motifsun/       olvwm*  proj1/
% exit
Arcplot:
```

If you use &SYSTEM in your application, be aware that the user may not know to type exit to return to the Arc: prompt. In this situation, a user might start ARC from the system prompt, an action that starts another ARC/INFO session. If this happens, multiple processes can cause problems with INFO file access. This can also create system resource problems if the user makes this mistake a number of times.

Issue &SYSTEM with an argument to execute an operating system command and return to ARC/INFO without having to type exit:

```
Arcplot: &system ls
Formedit      Mwm*          fmconsole.log  olvwm*  transfer/
Maker         dead.letter  motifsun/       olvwm*  proj1/
Arcplot:
```

FYI   Windows NT system commands aren't case sensitive, while UNIX system commands are all lowercase. For continuity, you should keep all system commands lowercase.

You may not realize that ARC is an operating system command. Its usage is as follows:

```
arc {arc_command} {command_arguments}
```

The ARC subsystem commands can also be issued from ARCPLOT. Just like the system command, it's done by prefixing the command with ARC. Suppose you're in ARCPLOT and want to draw a polygon coverage that doesn't have topology:

```
Arcplot: POLYGONSHADES LANDUSE LU_CODE LU_CODE.LUT
Cannot open PAL file
Arcplot:
```

If you're in the final stages of map production, you probably don't want to quit from ARCPLOT to build topology. Using ARC as shown here allows you to stay in ARCPLOT to perform the task:

```
Arcplot: arc build landuse poly
 Building polygons...
Arcplot:
```

Obviously, you can get the same result by opening a new window, starting an ARC session, and building the coverage. That would, however, be more difficult to incorporate into an AML program.

You can also access ARCPLOT commands from ARCEDIT by prefixing the commands with APC. Suppose you want to view a shaded polygon coverage (using a lookup table) while you're editing:

```
Arcedit: apc polygonshades landuse lu_code lu_code.lut
Arcedit:
```

> You can also access ARC commands from ARCEDIT by funneling the commands through ARCPLOT. The syntax would be APC ARC <command>. There isn't a direct link between ARCEDIT and ARC because you could corrupt data very easily if you're not careful. If you edit a coverage and then execute an ARC command to do some manipulations on that same coverage, the entire coverage could be lost. ARCEDIT commands often do the same thing as the ARC command. For example, the two different commands below give the same result, a list of coverages:
>
> ```
> Arc: LISTCOVERAGES
> ```
>
> ```
> Arcedit: DIRECTORY COVER
> ```

Accessing your database

→ can also go into another directory

Using &DATA with INFO

Sometimes it may be necessary to access other programs to add functionality or efficiency to your AML program. Use the &DATA directive to enable your AML program to interact with other programs. Place &DATA as the first line in a block of code that you want to pass to another software program as user input. &END terminates the interaction between the AML program and the other program.

The statements in an &DATA block are interpreted before the commands are sent to the operating system.

```
&DATA <operating_system_command>
{statement}
{...}
{statement}
&END
```

To use &DATA in your AML programs, you must know the correct operating system command and the arguments it expects from the user. &DATA must provide all the dialog the external program expects, including the command that terminates it.

Consider INFO as an example. There are two ways to execute the INFO database from the operating system. Use the `arc info` operating system command from a workspace, or use the `info` operating system command from an INFO directory. The INFO operating system command differs from the ARC/INFO command in that it doesn't automatically move the user in and out of the INFO subdirectory. You must move to the directory before executing the command.

&sv oldws = [show &wo]	&data arc info
&wo info	ARC
&data info	{...}
ARC	{...}
{...}	{...}
{...}	Q STOP
{...}	&end
Q STOP	
&end	
&wo %oldws%	

In either case, the AML program must provide the dialog INFO expects and quit from INFO before &ENDing the &DATA block. The statements in the &DATA block can contain directives, functions, and variables. AML executes these *before* accessing the external program. The following example demonstrates this:

```
/* reselect.aml
&sv infofile = [getfile * -info]
&sv item = [getitem %infofile% -info]
&sv value = [getunique %infofile% -info %item%]
&sv oldws = [show &workspace]
&wo info          4-30
&data info
ARC
SELECT [upcase [entryname %infofile% -info]]
RESELECT %item% = %value%
LIST
Q STOP
&end
&wo %oldws%
```

1-13

Q **Question 11-1:** In the previous example, why is the [UPCASE] function used on the SELECT line?

(Answer on page 11-31)

Examine what this code looks like when it's run with &ECHO &ON:

```
Arc: &echo &on
Arc: (thread0001, tty/7) &echo &on
Arc: &r reselect.aml
Arc: (thread0001, tty/8) &r reselect.aml
Arc: (thread0001, reselect.aml/1) &sv infofile = ~
/carmel1/miker/amlworkbook/info!arc!well.pat
Arc: (thread0001, reselect.aml/2) &sv item = HP
Arc: (thread0001, reselect.aml/3) &sv value = 200
Arc: (thread0001, reselect.aml/4) &sv oldws = ~
/carmel1/miker/amlworkbook
Arc: (thread0001, reselect.aml/5) &wo info
Arc: (thread0001, reselect.aml/6) &data info
(&DATA) Arc: (thread0001, reselect.aml/7) ARC
(&DATA) Arc: (thread0001, reselect.aml/8) SELECT WELL.PAT
(&DATA) Arc: (thread0001, reselect.aml/9) RESELECT HP = 200
(&DATA) Arc: (thread0001, reselect.aml/10) LIST
(&DATA) Arc: (thread0001, reselect.aml/11) Q STOP
(&DATA) Arc: (thread0001, reselect.aml/12) &end
Submitting command info
```

This display is interrupted to make a few points about &DATA. Notice that the &ECHO output for the code in &DATA...&END is displayed with a leading (&DATA). All of this code is interpreted *before* the command on the &DATA line executes (in this case, info). This means that all directives, functions, and variables are evaluated before the program ever executes the info command; therefore, the AML program must anticipate all actions needed by INFO beforehand. AML starts INFO, passes the content of the &DATA block as user input, and waits until INFO terminates.

Here's what happens next:

```
INFO  EXCHANGE  CALL
Copyright (C) 1994 Doric Computer Systems International Ltd.
All rights reserved.
Proprietary to Doric Computer Systems International Ltd.
US Govt Agencies see usage restrictions in Help files (Help
Restrictions)
ENTER USER NAME>ARC
ENTER COMMAND >SELECT WELL.PAT
       32 RECORD(S) SELECTED
ENTER COMMAND >RESELECT HP = 200
        4 RECORD(S) SELECTED

  ENTER COMMAND >LIST
                 8
  AREA              =          0.000
  PERIMETER         =          0.000
  WELL#             =       8
  WELL-ID           =      44
  WELL_LOC          =TEXAS ST. RES.
  DEPTH             =426
  DIAMETER          =24"
  WELL_NUM          =32
  WELL_TYP          =ELEC
  HP                =200
  GPM               =2700
  DISCHARGE         =TEXAS ST. RES.
                11
  AREA              =          0.000
  PERIMETER         =          0.000
  WELL#             =      11
  WELL-ID           =       7
  WELL_LOC          =WELL
  DEPTH             =674
  DIAMETER          =14"
  WELL_NUM          =37
```

```
WELL_TYP           =ELEC
MORE?Q STOP
HP                 =200
GPM                =1000
DISCHARGE          =1570
                   19
AREA               =          0.000
PERIMETER          =          0.000
WELL#              =     19
WELL-ID            =     46
WELL_LOC           =HIGHLAND AVE RES.
DEPTH              =600
MORE? <Now reading from terminal>
```

INFO expects either a carriage return, a YES, or a NO at the MORE? prompt, but all the commands needed by INFO aren't included in the &DATA block. Because the &DATA block doesn't include the correct dialogue, INFO accepts Q STOP as an answer to the MORE? prompt. INFO now wants input from the terminal because the &DATA block has been exhausted. When using &DATA, you must anticipate everything that the called program expects.

```
DIAMETER           =20"
WELL_NUM           =13
WELL_TYP           =ELEC
HP                 =200
GPM                =2500
DISCHARGE          =HIGHLAND AVE.RES.
                   20
AREA               =          0.000
PERIMETER          =          0.000
WELL#              =     20
WELL-ID            =     41
WELL_LOC           =MADEIRA WELL
DEPTH              =465
DIAMETER           =16"
WELL_NUM           =
WELL_TYP           =ELEC
HP                 =200
GPM                =1300
DISCHARGE          =1900
ENTER COMMAND >q stop
```

The user must enter the Q STOP to leave INFO. Control then returns to the program as it normally would.

```
Arc: (thread0001, reselect.aml/13) &wo
/carmel1/miker/amlworkbook
```

If you change the code by adding the line in bold, as shown below, it executes flawlessly:

```
&sv infofile = [getfile * -info]
&sv item = [getitem %infofile% -info]
&sv value = [getunique %infofile% -info %item%]
&sv oldws = [show &workspace]
&wo info
&data info
ARC
SELECT [upcase [entryname %infofile% -info]]
RESELECT %item% = %value%
LIST
YES
Q STOP
&end
&wo %oldws%
```

Inserting YES after the LIST command corrects the problem. But consider what happens if the file contains fifteen items with 10,000 records in the selected set and you insert the YES after the LIST command. It would take hours for the listing to scroll through record after record. Alternatively, you could write the program to insert NO after the LIST command. Of course, these decisions are based on having more than one screen of listed data. If less than one screen's worth of records are listed, then the YES or NO would be put on the command line. Since YES and NO aren't INFO commands, an error would be returned. Here, assume that there's going to be more than one screen's worth of data listed.

Examine another example that uses an &DATA block. In this case, there's a coverage called PROPERTY. The coverage contains parcel data, including land use and zoning. Suppose you're interested in leases on city-owned properties. The program enables you to select a target date by entering a number of days from today. All city-owned parcels whose leases expire before the date you specify should display on the screen.

&DATA is needed because AML by itself doesn't calculate dates by performing *date math*. Date math is the ability to add days to or subtract days from a date and obtain a new date as the outcome. Date math can also report the difference between two dates in number of days.

The following program subtracts days from a lease expiration date and compares it to today's date:

```
/* lease.aml
&severity &error &routine bailout
&if [show program] ne ARCPLOT &then
  &return This program must be run from ARCPLOT...
/* Set the MAPEXTENT to the PROPERTY coverage and display it.
MAPEXTENT PROPERTY
LINESET COLOR
POLYGONLINES PROPERTY 5
```

After displaying the coverage, the program selects all city-owned properties (specified by LANDUSE = 14) and shades them in red. If there are no city-owned parcels in the coverage, the program notifies the user and then terminates.

```
RESELECT PROPERTY POLY LANDUSE = 14
&if [extract 1 [show select property poly]] = 0 &then
  &return There are no city-owned parcels in the PROPERTY ~
coverage.
POLYGONSHADES PROPERTY 2
```

Next, the program asks the user for the number of days to search for from today's date. The program uses ARC to copy APN.CITYOWN to a temporary file, APN.TMP. This file contains the parcel numbers for all city-owned parcels. It also contains the dates that the leases expire in an item called LEASE_DATE. The program then uses ARC to add an item named TMP_DATE to the temporary file. After some calculation, this item is compared with today's date to see if a lease expired or will expire within the user-specified number of days. INFO calculates the value for the item TMP_DATE.

```
&sv numdays = [response 'Enter the number of days to ~
  search for leases that terminate']
&if [exists apn.tmp -info] &then
  &sv delstat = [delete apn.tmp -info]
arc COPYINFO APN.CITYOWN APN.TMP
arc ADDITEM APN.TMP APN.TMP TMP_DATE 8 10 D
```

The program now accesses INFO through an &DATA block. The program stores the current workspace and then enters the INFO subdirectory. &DATA INFO begins the INFO session. In INFO, the program selects the file APN.TMP and performs date math to calculate the TMP_DATE item to the result of LEASE_DATE (i.e., the number of days specified by the user).

The program compares the new date with today's date to determine whether the lease expires in the specified number of days.

As an example, suppose that today's date is March 25, 1997. The user wants to see all leases that expire within the next seven days. If a lease expires on April 1, 1997, then the program makes the following calculation:

```
TMP_DATE = LEASE_DATE - %numdays%
3/25/1997 = 4/01/1997 - 7
```

When the program compares the parcel to today's date, it calculates that today isn't within seven days of the expiration of the lease. It makes the comparison through the use of the INFO system item $TODAY. (For a complete list of INFO system items, search the online help index for "system items.") All parcels whose leases don't expire within the time specified are purged from the file. The program then quits from INFO and returns to the workspace.

```
[1]    &sv oldws = [show &workspace]
[2]    &wo info
[3]    &data info
[4]    ARC
[5]    SELECT APN.TMP
[6]    CALC TMP_DATE = LEASE_DATE - %numdays%
[7]    RESELECT TMP_DATE GT $TODAY
[8]    PURGE
[9]    Y
[10]   Q STOP
[11]   &end
[12]   &wo %oldws%
```

[1]   Store the current workspace
[2]   Go to the INFO directory
[3]   Start INFO in an &DATA block
[4]   User name ARC
[5]   Select temp file of city-owned parcels
[6]   Calc data = Lease date - number
[7]   Select records with date GT date chosen
[8]   Purge all records with expirations past chosen date
[9]   Confirm purge
[10] Quit from INFO
[11] End the data block
[12] Return to old workspace

As control returns to ARCPLOT, APN.TMP is used as a KEYFILE to select the parcels whose lease expires within the specified number of days. The selected parcels are then drawn.

```
RESELECT PROPERTY POLY KEYFILE APN.TMP APN
POLYGONSHADES PROPERTY 3
&call exit
&return
```

The exit routine deletes the temporary file:

```
/*------- -------------
&routine EXIT
/*-------------------
/* Delete temporary file
/*
&sv delstat2 - [delete apn.tmp -info]
&return
/*-------------- ---
&routine BAILOUT
/*-------------------
&severity &error &ignore
&call exit
&return &error An error occurred in (LEASE.AML)
```

This program performed all of its tasks without leaving ARCPLOT. The COPYINFO and ADDITEM commands are performed by prefacing the command with ARC, and the INFO processing completes using an &DATA block. Use these tools whenever possible to avoid the overhead of quitting from and starting ARC's subsystems.

These are the important things to remember about &DATA blocks:

• The command on the line with &DATA is an operating system command.

• You must anticipate all commands that the called program expects.

• All decision-making (&IF, &SELECT) or looping (&DO &WHILE, &DO &UNTIL, etc.) statements contained in the &DATA block execute *before* the other program executes. Thus, you can't base decisions on the results of that program.

• AML comments (/*) are treated as blank lines by the external program. Comments in the &DATA block should be in the language of the external program (e.g., REM when using INFO).

Getting information from INFO that AML can use

The &DATA block is invaluable when you must access INFO or any other external programs from your AML program. The &DATA block passes information from AML to external programs. When you need to get data from other software for AML to use, you can have the external program write an AML program containing the data needed to set to AML global variables.

The following program shows how INFO can pass data back to AML with the INFO file output tools. Any software that has a file output capability can use a file to pass data back to AML. The following example shows how an AML program accesses related tables in a database:

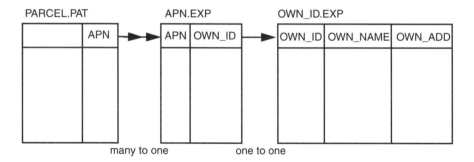

With this file structure, there are two ways to query a parcel for its owner's name and address and return this information to AML through AML variables. Employing the &DATA block is discussed here. The alternative, to use Cursors, is covered in the next lesson.

The following program allows the user to select a parcel by location or by assessor's parcel number (APN) and returns the owner's name and address through &TYPE. When the user doesn't pick the APN from a list (the choice is made spatially), the parcel number is retrieved from the APN item of the selected polygon. The program uses the [SHOW] function to retrieve the value of the item. This AML program also uses the [SCRATCHNAME] function to create a name that doesn't exist in the current directory.

```
/* owner.aml
&args cover
&severity &error &routine bailout
/* Check arguments
&if [NULL %cover%] &then
  &call usage
&if [exists %cover% -poly] &then &do
  &if not [iteminfo %cover% -poly apn -exists] &then &do
    &type Item APN  doesn't exist on [upcase %cover%].PAT
    &call usage
  &end
&end
MAPEXTENT %cover%
POLYGONS %cover%
&sv oldws = [show &workspace]
&sv tmp_file = [scratchname]
&sv choice = [getchoice APN LOCATION -prompt ~
  'Select a query method']
&select %choice%
  &when APN
    &do &until [extract 1 [show select %cover% poly]] = 1
      ASELECT %cover% poly
      &sv owner$apn = ~
      [getunique %cover% -poly apn 'Enter an APN']
      RESELECT %cover% poly apn = [quote %owner$apn%]
    &end /* &until
  &when LOCATION
    &do &until [extract 1 [show select %cover% poly]] = 1
      ASELECT %cover% poly
      RESELECT %cover% poly one *
      &sv owner$apn = [show select %cover% poly 1 item apn]
    &end /* &until
&end /* &select
&workspace info
```

 FYI   For more information about using the [SHOW] function with the SELECT option, search the online help index for "SHOW (ARCPLOT command)."

After determining the APN, the program uses an &DATA block to enter INFO and retrieve the owner information. After relating the files and finding the record, the program executes an OUTPUT command. This redirects all output to an ASCII file instead of the ARCNSP file, which enables the program to write the &SETVAR statements to the new file along with the data from the related files. Notice that the name of this new file is contained within the variable TMP_FILE. The new program assigns pairs of global variables for each owner of the selected parcel. These are retrieved by accessing them as indexed variables, the index being the value of OWN_ID.

```
&data info
  ARC
  COMO -NTTY /* turns messages off in INFO
  SELECT APN.EXP
  RELATE OWN_ID.EXP 1 BY OWN_ID
  RESELECT APN = [quote %owner$apn%]
  OUTPUT ../%tmp_file% INIT
  PRINT '&sv .owner$list'
  PRINT '&sv .owner$name[trim ',OWN_ID,'] = ',$1OWN_NAME
  PRINT '&sv .owner$add[trim ',OWN_ID,'] = ',$1OWN_ADD
  PRINT '&sv .owner$list = %.owner$list%', OWN_ID
  PRINT '&return'
  Q STOP
&end
&wo %oldws%
```

FYI

Notice that the &DATA block is indented. Indenting the &DATA block is the same as issuing spaces in front of your commands to the called program. Although it's not necessary, you may choose to indent in order to make your code more readable. If you do indent, make sure these leading spaces aren't going to affect the called program. In this case, the program is indented because INFO doesn't care about spaces in front of the command.

Now the new program runs. It assigns the data to the global variables and returns to OWNER.AML. OWNER.AML now determines the number of owners for the selected parcel by using [TOKEN] on the list of owner IDs. The program then loops through the IDs, printing the global variables containing the owner's name and address.

```
&run %tmp_file%
&type /&/&APN: %owner$apn%
&sv cntr = [token %.owner$list% -count]
&do i = 1 &to %cntr%
  &sv index = [extract %i% %.owner$list%]
  &type OWNER: [value .owner$name%index%]
  &type OWNER ADDRESS: [value .owner$add%index%] /&
&end
&call exit
&return
```

As always, the program cleans up any global variables and temporary files.

```
&routine USAGE
/*------------------
&type Usage: &run owner <parcel_cover>
&return &inform
/*------------------
&routine EXIT
/*------------------
/* Delete any global variables and temporary files
&sv delstat = [delete %tmp_file%]
&dv .owner$*
&workspace %oldws%
&return
/*------------------
&routine BAILOUT
/*------------------
&severity &error &ignore
&severity &warning &ignore
&call exit
&return &error An error occurred in (OWNER.AML)
```

OWNER.AML used INFO to write %TMP_FILE%. OWNER.AML then runs %TMP_FILE% to access the global variables it contains. This method is useful when you need to relate more files than ARC allows. The example contains two relates, but could be used to take advantage of any number of relates up to the nine allowed by INFO.

[TASK] it out

programing languages

The [TASK] function can return data to AML from a FORTRAN or C program. This can be helpful for adding more functionality or faster processing time to your AML program. If your AML program doesn't process large amounts of data fast enough, you may want to [TASK] out the processing to a compiled language like C or FORTRAN. You can use the [TASK] function to execute other programs as follows:

```
[TASK <program> <{FILE} argument_string>]
```

The <argument_string> contains the arguments for the program separated by spaces or commas. [TASK] can only pass one argument at a time, so if you want to pass multiple arguments separated by spaces, make sure that you quote the string. The quotes aren't necessary if you separate your arguments by commas.

The next AML program uses the [TASK] function to calculate the weighted (average) population center for the United States. The data contains 22,988 points that represent population centers throughout the United States. The x,y coordinates for each of these points are multiplied by their population size to find the weighted population center. This is an intensive task for AML, so these calculations are processed in a C program.

POP_CENTER.AML creates a display in ARCPLOT of all the population centers as well as the weighted population center for the country:

```
[1]   /* pop_center.aml
[2]   &severity &error &routine bailout
[3]   &s cover = population
[4]   clear
[5]   mape %cover%
[6]   markersymbol [getsymbol -marker]
[7]   points %cover%
[8]   arcs states
[9]   units map
[10]
[11]  &s coord_file = %cover%.indx
[12]
[13]  &if not [iteminfo %cover% -point x-coord -exists] &then
[14]     &do
[15]        arc addxy %cover% point
[16]     &end
[17]  arc tables
[18]  select [entryname %cover%.pat]
[19]  unload %coord_file% x-coord y-coord population
[20]  quit /* from tables
[21]
[22]  &s location = [task com [quote %coord_file% weight]]
[23]
[24]  markersymbol [getsymbol -marker]
[25]  marker [extract 1 %location%],[extract 2 %location%]
[26]
[27]  &call exit
[28]  &return
[29]
[30]  &routine exit
[31]  &return
[32]
[33]  &routine bailout
[34]  &severity &error &ignore
[35]  &call exit
[36]  &return &error An error occurred in (POP_CENTER.AML)
```

Lines [1]–[9] set up general environments like drawing and error checking.

Line [11] names the output file that is going to contain the list coordinates and population values for all the population centers.

Lines [13]–[20] create the text file containing three columns of data: x-coordinates, y-coordinates, and population. This is accomplished in two main steps:

- The items X-COORD and Y-COORD are added to the PAT using the ARC command ADDXY. Remember, you're in ARCPLOT, so you must preface ARC commands with arc.
- TABLES is then launched and a text file of the x, y, and population values is created using the UNLOAD command.

On line [22], the [TASK] function is used to launch the C program. Notice that there are two arguments passed: the name of the text file containing the x,y data, and the key word WEIGHT. The argument WEIGHT tells the C program to consider population in the calculations. The argument string uses [QUOTE] because there are multiple arguments. Finally, when the C program executes, the returned value is stored in the variable LOCATION.

In this example of [TASK], the name of the C program is given by itself. This only works if the C program resides in the current directory. The directories in your $PATH environment won't be searched, so if the C program resides elsewhere, a full pathname must be given.

Lines [24]–[25] plot the center of population for the country by extracting the x and y values from the string contained in %LOCATION%.

The exit and bailout routines follow to end the AML program.

When you compare the run times of the C program and an AML program using the above data, you find a significant time difference. The C program runs about 200 times faster than the equivalent AML code. POP_CENTER.AML uses the [TASK] function, but the weighted population center is also found using INFO within an &DATA block. INFO also processes this data considerably faster than AML.

The C program you create should be specifically designed to work with the [TASK] function. [TASK] wasn't designed to be used to access shell or systems programs.

FYI If you want an example of the C program used in the above example, look for COM.C in the chapter 11 database. If you want to run POP_CENTER.AML, then you'll have to compile COM.C to COM first. The AML program MASS_AML.AML also performs the same operation as COM.C. POP_CENTER2.AML uses MASS_AML.AML to do the calculations. To see the difference in performance, run POP_CENTER2.AML in the chapter 11 database.

The above example shows the functionality of [TASK] and how it's used. The next two pages deal with C programs and how they interact with [TASK]. These pages are meant as a reference for C programmers. In order to fully understand this section, you should have a working knowledge of the C programming language.

Both of the following programs receive an integer as an argument, then increment it by one and return this new value. The first example doesn't use the {FILE} option and reads the arguments from *stdin*:

```
/* nofile.c */
#include <stdio.h>

main()
{
        int     i;

        scanf( "%d", &i );
        i++;
        fprintf( stderr, "%d\n", i );
}
```

The arguments are read from *stdin;* there are no arguments passed to ARGV. The returned value (i), followed by a newline character (\n), is written to *stderr.* The newline is required in both the file and nonfile options of [TASK]. Using the [TASK] function with the program NOFILE produces the following output:

```
Arc: &type [task nofile 5]
6
```

With the {FILE} option, [TASK] creates a scratchfile and writes the arguments to that scratchfile. [TASK] then sends the scratchfile name to the C program as an argument. It's up to the C program to read the scratchfile and write the result back to that same file. The next example, FILEOPT.C, uses the {FILE} option to read in the arguments from the [TASK] function:

```
/* fileopt.c */
#include <stdio.h>
#include <stdlib.h>

int main( argc, argv )
int     argc;
char    *argv[];
{
        int     i;
        FILE    *fp;

        fp = fopen( *++argv, "r" );
        fscanf( fp, "%d", &i );

        i++;

        fclose( fp );
        fp = fopen( *argv, "w" );
        fprintf( fp, "%d\n", i );
        fclose( fp );

        exit( 0 );
}
```

One difference between this program and the last is the existence of the arguments passed to main (argc, argv). The scratchfile name is the argument that is passed to argv; nothing is read from *stdin*. The C program then follows these steps:

1. The scratch file is opened, the arguments are read, and the scratchfile is closed.

2. The scratchfile is reopened for writing, the result is written to the scratchfile, and the scratchfile is closed.

At the very end of the program you must exit with a zero (exit(0)). On Windows NT, you must also exit with a zero even with the nonfile option. Any other values returned produce the following error:

```
Arc: &type [task fileopt 'file 5']
AML ERROR - Problem in executing TASK function
```

After the C program finishes, [TASK] reads the result from the scratchfile. The *stdin* and *stderr* file pointers are never used.

When executing [TASK] with the {FILE} option, make sure that you include the keyword FILE within the single quotes along with the rest of the arguments. When FILEOPT is executed from the [TASK] function, the output should look like this:

```
Arc: &type [task fileopt 'file 5']
6
```

FYI

For the purpose of creating simple examples, the C programs shown above are very basic. They intentionally don't include any kind of error checking.

Of the two options, file and nonfile, the file option is more stable. It's sometimes possible to have an error with regard to your operating system when passing arguments through *stdin* and *stderr*. This is especially true if the file pointers are redirected through the OS.

If you want an example of the C programs used in the above examples, look for NOFILE.C and FILEOPT.C in the chapter 11 database.

In summary

You can use the &DATA block to access external software. The techniques demonstrated in this lesson using the INFO database apply to other interactive software. You can also [TASK] jobs out to other programs that might accomplish the jobs more efficiently. In the next lesson, instead of using an &DATA block or [TASK] to access the data contained in your database, you'll use a tool called Cursors.

A

Answer 11-1: The [UPCASE] function is used with the SELECT command because INFO expects all correspondence from the user to be in uppercase. Using [UPCASE] ensures that the user asks for the table.
Another option is to change the CASE FLAG inside of INFO, so that INFO accepts text written in any case.

Exercises

Exercise 11.2.1

Write statements to perform the following actions and return to the current environments using ARC:

Current environment: ARCPLOT
Action: Use MAPJOIN on coverages SOIL1 and SOIL2 to produce a new coverage, SOIL3.

Current environment: ARCEDIT
Action: Shade in the polygons for the coverage SOIL using the item TYPE with the lookup table SOIL.EXP.

Exercise 11.2.2

Write a program called LEASE2.AML that runs from ARCPLOT and allows the user to graphically select a city-owned parcel from the PROPERTY coverage (LANDUSE = 14). The program should report the lease date and number of days until the lease expires. Allow this to continue until the user presses the 9 key.

The program should draw the coverage, select the city-owned parcels, and display them in a different line color. The user should choose from the selected set (graphically). If the user doesn't select a parcel from the selected set, inform the user and present the selection crosshairs again.

Once the user selects a parcel, capture the parcel number (stored in the item APN), enter the INFO subsystem, and RELATE PROPERTY.PAT to APN.CITYOWN (the file containing the lease date) by APN. Create a temporary AML program to set global variables that return the lease date ($1LEASE_DATE) and the number of days until the lease expires (i.e., $1LEASE_DATE - $TODAY. You'll have to use the INFO numeric variable $NUM1 to accomplish this.) Print out this data for the user.

Exercise 11.2.3

Write a program called GENPOINTS.AML that loads an ASCII text file into the INFO database.
The file contains an ID, x- and y-coordinates, and attributes about WELL data. These are sample
records from the file:

```
 1,   9086.000,  48692.000,WELL          ,627,16"-20",GAS  ,250,1500,1350
37,  13166.000,  48064.000,MISSION WELL ,550,14"     ,ELEC ,150,1900,1350
28,  21418.000,  38031.000,LEE WELL     ,605,16"     ,ELEC ,100, 600,1570
24,  25158.000,  42941.000,NEW YORK ST. ,875,20"     ,GAS  ,250,1500,1570
```

Use this file to create a WELL coverage and an expansion file containing the WELL attributes.
Follow this six-step process:

1. Create an empty INFO file to receive the data. It should be named WELL.EXP and contain
   the following items:

```
DATAFILE NAME: WELL.EXP                                          2/14/1997
   10 ITEMS: STARTING IN POSITION     1
  COL  ITEM NAME       WDTH OPUT TYP N.DEC  ALTERNATE NAME
    1  WELL-ID           4    5  B   -
    5  X-COORD           4   12  F   3
    9  Y-COORD           4   12  F   3
   13  WELL_LOC         25   25  C   -
   38  DEPTH             3    3  I   -
   41  DIAMETER          7    7  C   -
   48  WELL_TYP          5    5  C   -
   53  HP                3    3  I   -          \
   56  GPM               4    4  I   -
   60  DISCHARGE             20   20  C   -
```

2. Import the data from the system file WELL.DAT using the INFO command ADD with the
   FROM option (i.e., ADD FROM ../WELL.DAT).

3. Create a new system file called GEN.FILE. Output the values of the WELL-ID, X-
   COORD, and Y-COORD items to this file. Also add the line END. (GENERATE requires
   this.)

4. GENERATE the coverage WELL using the data in the file GEN.FILE as input. BUILD
   topology for the coverage.

5. Drop the coordinate items from the file named WELL.EXP.

6. Finally, delete the file named GEN.FILE.

Exercise 11.2.4

Write a program called JUSTIFY.AML that left-justifies a character field in an INFO table. The program should accept as input the file name, item name, and item starting column as shown here:

```
&r justify <info_file> <character_item> <item_start_column>

&r justify own_id.exp own_name 5
```

Output should be the same file with the item left-justified. Make sure that the item is a character type. The file should create a copy of the INFO file to process in case of an error. As an example, examine the file structure of XXOWN_ID.EXP. The table appears as shown below during processing:

```
DATAFILE NAME: XXOWN_ID.EXP                               10/11/1996

    1 ITEMS: STARTING IN POSITION    1
  COL   ITEM NAME         WDTH OPUT TYP N.DEC  ALTERNATE NAME
    1   OWN_ID               4    5  B    -
    5   OWN_NAME            36   36  C    -
   41   OWN_ADD             28   28  C
        **  REDEFINED ITEMS  **
    5   FC                   1    1  C    -
   40   LC                   1    1  C    -
    5   LEFT                35   35  C    -
    6   RIGHT               35   35  C    -
```

The program needs to create these REDEFINEd items: FC, LC, LEFT, and RIGHT. Here is the structure of the file:

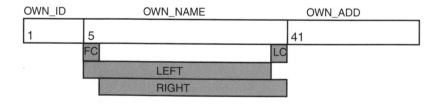

The following actions need to occur on a record-by-record basis:

1. Verify that data exists in the OWN_NAME item.

2. As long as the first character (FC) contains a space, move the data contained in the item called RIGHT to the item called LEFT. Then move a space to the last character (LC).

After the program uses REDEFINE on the file to create the new items, the program must create and execute an INFO program that performs the actions contained in the steps. You should first determine whether the program already exists. If it does, delete it before the code below occurs. The tilde (~) induces a line feed, and the RUN command compiles and runs the INFO program.

```
PROGRAM XXJUSTIFY.PG
SEL XXOWN_ID.EXP
PROGRAM SECTION TWO
IF OWN_NAME NE ' '
  DO WHILE FC EQ ' '
    MOVE RIGHT TO LEFT
    MOVE ' ' TO LC
  DOEND
ENDIF
PROGRAM END
~
RUN XXJUSTIFY.PG
DELETE XXJUSTIFY.PG
Y
```

Before exiting the AML program, remove the redefined items from the table and delete the INFO program, JUSTIFY.

Present the temporary file to the user after the INFO program executes successfully. If the user accepts the file, delete the original, copy the temporary file to the original name, and delete the temporary file. This program should work for *any* file and character item.

Note: INFO programs have even and odd sections, recognized by their even- and odd-numbered titles. Actions in the odd section are performed on the entire file, while actions in the even section are performed on a record-by-record basis. Most of your work should be done in an even programming section to reduce the number of records read. For more information on INFO programming, refer to chapter 14 in the *INFO User's Guide, "Programming with INFO."*

Lesson 11.3—Accessing attribute data

In this lesson

This lesson explores a tool set called Cursors. *Cursors* access data stored in external databases, as well as being able to access INFO datafiles beyond the constraints of the functions used in lesson 11.1, without using &DATA.

A mechanism for accessing individual elements in a selected set, Cursors are accessible from ARC, ARCPLOT, and ARCEDIT. Cursors step through each row, or *record,* in a selected set and access the attributes related to the record. Cursors aren't a method for establishing or manipulating a selected set, except when used in ARC. Although Cursors aren't AML tools, they're discussed here because they set AML program variables that can be used by programs and applications.

Using Cursors to query and update attribute data

With Cursors, the column values—or *item* values—of the current element in an INFO file or DBMS table are available for display or update through AML program variables. For example, you can write programs that use Cursors to access online data dictionaries to control your applications.

Cursors primarily serve as a mechanism for displaying and updating INFO files and DBMS tables through the form menu interface. With Cursors, data can be displayed, updated, inserted, and deleted from an INFO file or DBMS table using menu buttons, input fields, and widgets. Form menus are covered further in chapter 12.

When writing applications that use Cursors, consider the following:

- Cursors operate on the elements in the selected set.

  Cursors that process INFO file records use the ARCEDIT and ARCPLOT selection environments. In the ARC environment, the CURSOR command creates the selected set. The matching elements related to the current record can also be accessed one record at a time.

- Cursors can step through the records in the selected set.

  The current location of the cursor in the selected set is called the *current element*. The item values in the current element are available as AML program variables.

  The only difference between these variables and ordinary AML variables is their association with the cursor using a special naming convention. Use these variables just as you would any AML variable. Changing the value of the variable updates the value for the current record in the INFO file or DBMS table if the cursor is opened with RW (read/write) access. RW access is the default in ARCEDIT, while RO (read-only) is the default in ARC and ARCPLOT.

- Once a cursor is opened for a coverage feature class, INFO table, or DBMS table, subsequent ARC/INFO functions that operate on the selected set affect the current element only.

The steps for declaring and using a cursor differ slightly depending on whether you're in ARC, ARCPLOT, or ARCEDIT. The major difference is that ARC and ARCPLOT allow you to have ten cursors active, but ARCEDIT only allows one active cursor, named EDIT. This makes sense because when you're in ARC or ARCPLOT, you generally access many different geographic data sets or tables to accomplish a function. When in ARCEDIT, however, you edit one coverage feature class at a time, so you only need access to a single table.

The generic process follows:

1. Declare your cursor.

   In ARC and ARCPLOT, give the cursor a name and a file to operate on. This is accomplished automatically in ARCEDIT using the command EDIT or EDITFEATURE.

2. Select data.

   Use the selection commands in ARCPLOT and ARCEDIT (e.g., RESELECT, ASELECT, SELECT, etc.). In ARC, issue the selection statement when you declare the cursor.

STREAM-ID	STRM_NAM

Selected records:

3. Open the cursor.

A newly opened cursor points to the first record of the selected set and sets a number of AML program variables. These variables fall into two categories: cursor variables and cursor descriptors. *Cursor variables* list and update for the selected records. *Cursor descriptors* provide information about the status of the cursor. This lesson focuses on these two types of variables.

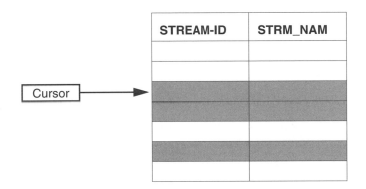

4. Query and manipulate data.

Tools within the CURSOR command allow you to step through the selected set on a record-by-record basis. These tools also operate on a related file, allowing you to access one-to-many relates. While accessing the records in the selected set, use AML directives to query and update the data in each record.

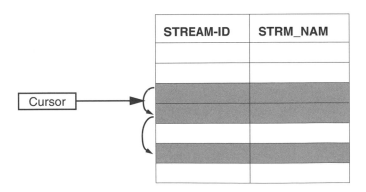

5. Close the cursor.

Close the cursor when you're finished processing records. You must close the cursor before changing the selected set to process another set of records. Close the cursor, change the selected set, and then reopen it for processing.

6. Remove the cursor.

When you finish using a cursor in ARC or ARCPLOT, remove it from the session to free up system resources.

Cursor variables

When you open a cursor, its variables are defined for each of the items in the selected table. Their values are the item values for the currently selected record. Cursor variables begin with a colon (:) as follows:

```
:<cursor>.<item>
```

where <cursor> is the name of a declared cursor and <item> is the name of one of the items from the feature attribute table or INFO file for which the cursor is declared.

For example, a cursor named TREECUR and an item called HEIGHT would result in the following cursor variable:

```
:treecur.height
```

Items in related tables are also available as cursor variables by using the relate name as shown here:

```
:<cursor>.<relate>//<item>
```

where <relate> is the name of a defined relate to an INFO file or external DBMS table. For example, a cursor named TREECUR, a relate called FORREL, and a related item called HARVEST_DATE would result in the following cursor variable:

```
:treecur.forrel//harvest_date
```

Cursor variables are AML *program variables*. AML program variables begin
with a colon and are the third type of variable listed when you issue the
&LISTVAR directive:

```
Arc: &listvar
Local:
    COVNAME         parcels
    COMPANY         esri
    CONAME          Environmental Systems Research Institute
    SCALE           1200
Global:
    No global variables are defined
Program:
    :PROGRAM                    ARC
    :TREECUR.HEIGHT              90
```

For more information on the syntax for the CURSOR command or cursor
variables, search for *cursor* in the online help index.

FYI

There are two ways to access the data stored in an external DBMS. One
method uses an ARC relate to the external DBMS and then accesses the data
with the last method shown.

Another method uses the DBMSCURSOR command, which accesses DBMS
data from ARC/INFO. DBMSCURSOR allows access to item values in *any*
DBMS tables, including data with no relationship to an INFO file or feature
attribute table.

The general process for operating the DBMSCURSOR is the same as
operating a cursor except for the following:

- You must CONNECT to the external DBMS before you can use the
  DBMSCURSOR command.
- You must issue an SQL select expression when you declare the
  DBMSCURSOR.

For more information about the DBMSCURSOR command and accessing an
external DBMS, search the online help index for *DBMS, cursor*.

Cursor variables aren't defined until the cursor is open. AML directives and functions operate on cursor variables just like any AML variable. As an example, consider a coverage called STREAM with feature class ARC and containing items named STRM_TYP and STRM_ORD. The cursor variables for a cursor called STRMCUR declared on the STREAM coverage are called :STRMCUR.STRM_TYP and :STRMCUR.STRM_ORD, respectively.

Consider a related file containing an item called STRM_NAM. If STRM_NAM is accessed with a relate called STRMREL, the cursor variable is called :STRMCUR.STRMREL//STRM_NAM. The following diagram shows this relationship:

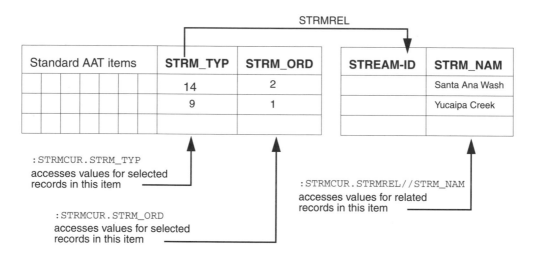

Look at an example in ARC for accessing the data in the STREAM coverage and the related table:

```
Arc: CURSOR STRMCUR DECLARE STREAM ARC RW
Feature cursor STRMCUR now declared using file
 STREAM.aat with Read Write access
Arc: CURSOR STRMCUR OPEN
Feature cursor STRMCUR now opened with
463 reselected records out of 463
Fetched record 1 for Feature cursor STRMCUR
```

After the cursor is opened, you can use directives like &TYPE and &LISTVAR to view the data as follows:

```
Arc: &type %:strmcur.strm_typ%
14
Arc: &lv :strmcur.strm_typ
Program: :STRMCUR.STRM_TYP
2
```

It's often useful to list multiple variables with one &LISTVAR statement. As shown in the following example, the first time you want these variables listed, they must be specified by name when the &LISTVAR directive is issued. Afterwards, they'll be shown with the default &LISTVAR listing (as long as the cursor is open). However, you only want to see the program variables, not the global and local variables. Here the &LISTPROGRAM (&LP) directive is used to see only the program variables. If you want to see just local or global variables, you can use &LISTLOCAL (&LL) or &LISTGLOBAL (&LG).

```
Arc: &lp :strmcur.strm_ord :strmcur.strmrel//strm_nam
Program:
:STRMCUR.STRM_ORD                            2
Program:  :STRMCUR.STRMREL//STRM_NAM         SANTA ANA WASH
Arc: CURSOR STRMCUR NEXT
Fetched record 2 for Feature cursor STRMCUR
```

Notice that the variables are displayed by default because they're specified by name above.

```
Arc: &lp
Program:
:PROGRAM                                     ARC
:STRMCUR.STRM_TYP                            9
:STRMCUR.STRM_ORD                            1
:STRMCUR.STRMREL//STRM_NAM                   YUCAIPA CREEK
```

After examining the current record, suppose you determine that the data in STRM_TYP is incorrect. To update the data, use the &SETVAR directive to set the variable to the new value.

```
Arc: &sv :strmcur.strm_typ = 1
Arc: &lp
Program:
:PROGRAM                              ARC
:STRMCUR.STRM_TYP                     1
:STRMCUR.STRM_ORD                     1
:STRMCUR.STRMREL//STRM_NAM            YUCAIPA CREEK
Arc: CURSOR STRMCUR CLOSE
Arc: CURSOR STRMCUR REMOVE
Feature cursor STRMCUR now removed
```

Question 11-2: What will you see if you type &lp now?

(Answer on page 11-51)

Cursor descriptors

The cursor descriptors are also program variables, set when you open a cursor. These are described below.

```
:<cursor>.AML$NEXT
:<cursor>.<relate>//AML$NEXT
```

AML$NEXT is a keyword. This variable is set to .FALSE. if you attempt to move the cursor past the end of the selected set of records or related records. The variable is set to .TRUE. if you successfully retrieved the next record.

```
:<cursor>.AML$NCOL
:<cursor>.<relate_name>//AML$NCOL
```

AML$NCOL is a keyword. This variable is set to an integer indicating the number of items in the INFO file and the related INFO or DBMS file.

```
:<cursor>.AML$NSEL
:<cursor>.<relate_name>//AML$NSEL
```

AML$NSEL is a keyword. This variable is set to an integer indicating the number of selected records. The number of records related to the current record is returned if the relate name is given.

To continue the previous example, the cursor descriptors for a cursor called STRMCUR and a relate called STRMREL are as follows:

```
:STRMCUR.AML$NEXT
:STRMCUR.STRMREL//AML$NEXT

:STRMCUR.AML$NCOL
:STRMCUR.STRMREL//AML$NCOL

:STRMCUR.AML$NSEL
.STRMCUR.STRMREL//AML$NSEL
```

FYI Cursors declared with the DBMSCURSOR command have additional cursor descriptor variables that contain item metadata information.

The cursor descriptors are very useful when using Cursors in AML programs. You can process the entire selected set by using <cursor>.AML$NEXT and <cursor>.<relate_name>//AML$NEXT as loop control variables.

The next example uses Cursors to transfer data from the polygon attribute table (PAT) to the arc attribute table (AAT). The program is used in a bathymetry application (depth measurement of water bodies) that assigns the minimum depth of two bordering polygons to the arc that separates them. An item called D-RANGE in the PAT stores the depth value, which is updated to the item called D-RANGE in the AAT. Before the program runs, two relates, RIGHT and LEFT, are defined from the AAT to the PAT, RIGHT on RPOLY# and LEFT on LPOLY#.

```
/* bath.aml
&severity &error &routine bailout
&if [locase [show program]] ne arcplot &then
  &return BATH.AML must be run from ARCPLOT.
  /* DECLARE a cursor on the coverage AAT, Read-Write access
needed
CURSOR DEPTH DECLARE BATH ARC RW
/* Get rid of the arcs bordering the universe polygon.
```

```
UNSELECT BATH ARC LPOLY# = 1 OR RPOLY# = 1
CURSOR DEPTH OPEN
RELATE RESTORE BATH.REL
/* Check to see if either polygon is coded 777 (canal). If
/* so calculate the D-RANGE item to equal 777. If not,
/* calculate the d-range item equal to the minimum value
/* of the d-range in the two bordering polygons.
```

An &DO &WHILE loop controls access to the table—as long as the cursor is pointing to a record the loop continues. The command CURSOR DEPTH NEXT drives the cursor through the file. Without this command, the cursor never moves and the loop can't finish. While in the loop, the program checks the value of the data contained in the PAT. This data can be handled like the data contained in any AML variable, with one difference: AML won't allow you to assign character data to numeric variables (items).

```
&do &while %:depth.aml$next%
  &if %:depth.left//d-range% = 777 or ~
    %:depth.right//d-range% = 777 &then
    &sv :depth.d-range = 777
  &else
    &sv :depth.d-range = ~
    [min %:depth.left//d-range% %:depth.right//d-range%]
  CURSOR DEPTH NEXT
&end  /* &DO &WHILE %DEPTH.AML$NEXT%
&call exit
&return
/*------------------
&routine EXIT
/*------------------
```

You can use the [SHOW] function with many Cursors options to retrieve information about the current cursor environment. The function [SHOW CURSORS] returns a comma-delimited list of cursor names:

```
  &do cursor_name &list [show cursors]
&if [locase %cursor_name%] = depth &then
    CURSOR DEPTH REMOVE
&end
&return
/*------------------
&routine BAILOUT
/*------------------
&severity &error &ignore
&severity &warning &ignore
&call exit
&return &error An error occurred in BATH.AML
```

FYI

Cursors are considered "noisy" because at each operation they report their status back to the user. You may find this annoying, as well as time-consuming, so it's a good idea to insert an &MESSAGES &OFF before declaring your cursors. If you choose to do this, remember to turn &MESSAGES &ON in your exit routine.

As you become comfortable using the techniques shown in this and the previous lesson, you'll be able to make informed decisions about which method to use to access/update your database. As a rule, using INFO is faster, though more difficult, while using Cursors is slower, but more versatile in its relationship to AML. The needs of your application should dictate your choice.

The next example solves a mapping problem in ARCPLOT: accessing a one-to-many relate to put multiple-related item values on a polygon. This presents a problem because the ARC relate doesn't specifically support one-to-many relationships. Cursors can access the related file and walk through all related records to collect the values of the item. The procedure in this case is to nest the related loop in the main cursor loop. This allows all related records to be processed before accessing the next polygon record.

```
/* stacktxt.aml
&severity &error &routine bailout
&args cover relate_name related_item
/* Test arguments.
MAPEXTENT %cover%
POLYGONS %cover%
RELATE RESTORE parcel.rel
&if [null %related_item%] &then  &call usage
&if not [exists %cover% -POLY] &then &do
  &type %cover%  isn't a polygon coverage.
  &call usage
&end
&if [show PROGRAM] ne ARCPLOT &then  &return You must be ~
  in ARCPLOT

/* Declare the cursor STACK_CUR for polygons
/* in coverage %cover%.
CURSOR STACK_CUR DECLARE %cover% POLY
/*  Open the cursor STACK_CUR.
CURSOR STACK_CUR OPEN
/* Use the CURSOR command with the NEXT option to skip the
/* cursor STACK_CUR past the universe polygon.
CURSOR STACK_CUR NEXT
```

After skipping the universe polygon, the outer loop is set up to step through the rest of the polygons, labeling each one with its related item value(s). This loop repeats as long as there are more polygons to label (i.e., the cursor variable %:STACK_CUR.AML$NEXT% is .TRUE.).

```
&do &while %:STACK_CUR.AML$NEXT%
/* Set a variable QFIRST_TEXT_LINE = .TRUE. for the first
/* pass through the inner loop.  This 'flags' the first
/* related item value.
&sv qfirst_text_line = .TRUE.
```

Now the inner loop is set up to step through all the related item values for one polygon. This loop repeats as long as there are more related item values for the polygon.

The cursor variable :STACK_CUR.%RELATE_NAME%//AML$NEXT is .TRUE. as long as there are more related item values for the polygon in the related file, accessed using the %RELATE_NAME% variable (which was passed as an argument). Notice the use of the [VALUE] function to nest the evaluation of the variable.

```
&do &while [value :stack_cur.%relate_name%//aml$next]
```

The following section of code produces a string that appears with line breaks and contains all of the related values for the current record. The backslash (\) between the variable and the related item value causes each value to print on a separate line.

```
    &if %qfirst_text_line% &then
      &s conc = [value :stack_cur.%relate_name%//~
      %related_item%]
   &else
   &sv conc = ~
      %conc%\[value
:stack_cur.%relate_name%//%related_item%]
  /* Set QFIRST_TEXT_LINE to .FALSE. for rest of the
  /* related item values of this polygon, the first one
  /* is already processed
    &sv qfirst_text_line = .FALSE.
```

The cursor STACK_CUR moves to the next related record in the file, which is accessed using the relate %RELATE_NAME%. This action steps to the next related item value for the current polygon. When the cursor moves, it changes the value of :STACK_CUR.%RELATE_NAME%//AML$NEXT based on whether or not there's another related record. If not, the inner &DO &WHILE loop exits, the polygon is labeled, and the next polygon is accessed by the outer loop.

```
    CURSOR STACK_CUR RELATE %relate_name% NEXT
&end   /* &DO &WHILE inner loop
```

The section of code in the inner loop builds a string of related item values. For example, suppose that the related information is owner names for a parcel (e.g., Theodore, Ridland, and Boyd). The goal is to label the polygon with all of these names. You can accomplish this using the LABELTEXT command as follows:

```
LABELTEXT <cover> '''string'''
```

The triple quotes allow you to put any text string on a polygon label. In this case, you want to stack the names so that the polygon appears like this:

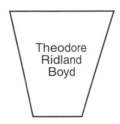

You can stack text by building a string with backslashes (\) to create line breaks like this:

```
Theodore\Ridland\Boyd
```

After collecting all the item values, the polygon is labeled using the text string syntax for LABELTEXT (three quotes surrounding the string). The [QUOTE] is used so that the variable CONC is evaluated.

```
    LABELS %cover% NOIDS
    LABELTEXT %cover% ''[quote %conc%]'' UC
```

Get the next polygon from %COVER%.PAT. This changes the value of
:STACK_CUR.AML$NEXT based on whether there's another polygon to label:

```
CURSOR STACK_CUR NEXT
&end  /* &DO &WHILE outer loop
&call exit
&return
/*------------------
&routine USAGE
/*------------------
&type Usage: STACKTXT <coverage> <relate_name> <related_item>
&return
/*------------------
&routine EXIT
/*------------------
/*  Remove the cursor: STACK_CUR if it was declared.
&do cursor_name &list [show cursors]
  &if [locase %cursor_name%] = stack_cur &then
    CURSOR STACK_CUR REMOVE
&end
&return
/*------------------
&routine BAILOUT
/*------------------
&severity &error &ignore
&call exit
&return &error Bailing out of stacktxt.aml...
/* End of AML.
```

This program demonstrates the automated method of processing a selected and
related set of records. After selecting the records and opening the cursor, use an
&DO &WHILE loop to process the entire set, querying or modifying as you go.
If you need to process related data, nest another &DO &WHILE loop in the first.
This ensures that you process all related records before proceeding to the next
selected record.

FYI

> The true power of Cursors is unleashed when they're used in conjunction with
> form menus. By using Cursors with forms, application builders can easily
> develop an AML interface to select, display, and update the data associated
> with a coverage. A form's fields are AML cursor program variables. Typing in
> the fields updates the data file. (Data integrity and security can be maintained
> according to the user's access rights.) A form can also contain buttons and
> icons to move through the selected set of features or reestablish the selected
> set based on logical expressions or spatial queries.

In summary

Cursors access, query, and update data contained in INFO or an external DBMS. Using Cursors is an asset when you need direct access to query and modify the data in INFO and related tables. It's also needed when &DATA can't perform the necessary tasks.

Answer 11-2: Now when you type &lp, you see the following:

```
Program:
:PROGRAM                              ARC
```

The cursor variables no longer exist.

Exercises

Exercise 11.3.1

Fill in the missing parts of the following AML program. BOOK.AML uses Cursors to access an INFO file that stores the locations of images associated with buildings on selected lots. The user is prompted to select a parcel, then two IMAGEVIEW windows are created and a menu (BOOK.MENU) is executed. The user can turn the pages of the "book" to see other scanned documents describing the selected building. By turning the pages, the program walks the cursor through a selected set of images. The comments above the missing parts give you hints for what you should fill in.

```
/* book.aml
&severity &error &routine bailout
&args routine turnto
/* Check to see if the program is running from the menu
&if not [null %routine%] &then &do
  &call %routine%
  &return
&end
/* Display the image and lot polygons. Shade the buildings
MAPEXTENT LOTS
SHADESET COLOR
IMAGE WOODLAND
LINECOLOR 1
ARCS LOTS
CLEARSELECT
POLYGONSHADES BUILDINGS 7
POLYGONS BUILDINGS
/* Allow the user to select a building
&type Select a building
&getpoint &map &push
RESELECT BUILDINGS POLY ONE *
&if [extract 1 [show select buildings poly]] = 0 &then
   &type No building located.
&else &do
   /* Make sure that the file ATTRIB exists.  This file contains the
   /* locations of the appropriate images. If it does, select all of
   /* the images associated with the selected parcel.
   &if ^ [exists attrib -info] &then
      &return Cannot open attrib.
   POLYGONSHADES BUILDINGS 2
   RESELECT ATTRIB INFO BUILDINGS-ID = ~
```

```
      [show select buildings poly 1 item  buildings-id]
/* Create imageview windows
IMAGEVIEW CREATE 'The Other' SIZE 300 400 POS LR SCREEN LR
IMAGEVIEW CREATE 'One Page' SIZE 300 400 POS LEFT WINDOW 'The Other'
/* Use threads to create the menu that allows the user to turn pages
/* in the "book" of images.
&sv origthread = [show &thread &self]
&thread &create bookthread &menu book &stripe 'Turn Page' &pos &ll
  &window 'One Page' &ul &pinaction 'imageview ~
destroy ''One Page''; imageview destroy 'The Other'; ~
cursor bookcursor remove; &thread &delete bookthread'
/* Declare and open a cursor named BOOKCUR on the INFO file
/* ATTRIB. Call the routine that gets the images and allow the user
/* to interact with the menu.
&sv .page = 1
&call getimg
&thread &focus &off &self
&return
&routine trnpg
/* Controls two imageview windows, emulating page turning.
/* Variable used: turnto; Value: next or previous;
/* Purpose: displays images corresponding to next and previous page.
/* If the user chooses to go to the next page, and the current page
/* is less than the number of selected images (in ATTRIB) -1 then
/* get the image. If there are no more pages, inform the user.
&if %turnto% = next &then &do
   &if %.page% < [calc _____ - 1] &then &do
      &sv .page = %.page% + 1
      &call getimg
   &end  /* if
   &else &sv .bookmes = No more pages.
&end
&else &if %.page% > 1 &then &do
        &sv .page = %.page% - 1
        &call getimg
     &end
     &else &sv .bookmes = No more pages.
&return
&routine getimg
/* Gets an image from an info file and displays it and the next one
&sv messvar = [show &messages]
&message &off
/* If the page isn't the first one, move cursor to the correct page.
&sv pageno = %.page% - 1
&if ^ %pageno% = 0 &then
   &do i = 1 &to %pageno%
   &end
/* Now that the cursor is at the correct page, use the IMAGEVIEW
```

```
/* command with the image stored in the item IMAGE in the file
/* ATTRIB.
IMAGEVIEW _____ # # 'One Page'
/* Move the cursor to the next page. Use IMAGEVIEW to display the
/* image stored in the item IMAGE.
IMAGEVIEW _____ # # 'The Other'
&message %messvar%
&return
&routine bailout
&severity &error &fail
/* If the program fails, determine if BOOKCURSOR exists. If so, remove it.
&if _____ &then
   _____

&return &error Bailing out of BOOK.AML
```

12 Form menus

Form menus are the most sophisticated type of menu that AML supports. Like all AML menus, form menus are ASCII files you can create with a text editor. By interacting with form menus, you can dynamically define an operation before it executes. This ability to execute different operations depending on how the user fills out the menu is what sets form menus apart from the other menu types. Form menus create a powerful graphical user interface for completing complex ARC/INFO tasks.

Form menus can present choices, verify input, and provide online help. Form menus accept interactive input in several ways. Information can be typed at the keyboard, selected from a list of choices, checked off a list, or set by moving a sliding bar. These various ways of interacting with form menus are supported through graphical objects called *widgets*. The user provides information to complete an operation by interacting with the widgets.

FormEdit is a sophisticated tool for designing and implementing form menus. Using FormEdit, you can select, position, arrange, and apply properties to the menu widgets to create the appearance and functionality you want. FormEdit allows you to lay out the widgets for your interface quickly and easily.

This chapter builds on the basic menu information presented in chapter 2. If you haven't familiarized yourself with this preliminary material, do so before continuing.

This chapter covers the following topics:

Lesson 12.1—Introducing form menus

- Form menu basics
- Introducing FormEdit

Lesson 12.2—Creating form menus with FormEdit

- FormEdit tutorial
- Using variables in form menus
- Managing variables in form menus

Lesson 12.3—Beyond FormEdit

- Designing form menus
- FormEdit and menu files

Lesson 12.1—Introducing form menus

In this lesson

Like paper forms, form menus gather information from the user. A form menu provides an interactive interface for conducting a lengthy transaction with the user. The user sets all the parameters required on the form to execute a complex operation, then presses a button to initiate the task. This lesson describes how and why to use form menus and introduces the ARC/INFO FormEdit utility.

Form menu basics

When you complete a form on paper, you use a pencil to interact with the form. You provide information by answering questions, filling in blanks, circling choices, and checking boxes. With a form menu, the user provides information by interacting with visual objects, called *widgets*, that collect the data. The user's interaction with a widget results in assigning a value to an AML variable. The variable is then used in an ARC/INFO command to perform a task.

Examine the following form menu. It consists of a scrolling widget, which displays a list of coverages to choose from, and a choice widget, which offers a choice between polygon, line, or point feature types. Two variables are set, one to the selected coverage name and one to the selected feature type. When the button widget Build is pressed, it uses the assigned variables to execute the BUILD command and creates topology for the selected coverage and feature type.

Each kind of widget that makes up a form menu consists of *properties*. These widget properties dictate its appearance, the type of information it accepts, and any action it performs. One widget might accept only a symbol number, another a true/false response, and another only an integer number in a specific range. Each widget is specifically designed to accept one type of data.

Form menus can provide the user with online help about a menu or menu option. Help can come in three forms: a single-line message, the display of a text file describing an operation, or a list of valid choices from which the user can select.

Form menus can reduce errors by testing whether the information the user enters is the correct type (e.g., numerical or character), is in a specified range of acceptable values, and is complete.

Introducing FormEdit

You can use FormEdit to design, implement, and modify form menus. Its graphical interface allows you to select and lay out widgets quickly, apply properties and actions to widgets, and establish general menu parameters.

Invoke FormEdit from any ARC/INFO subsystem as follows:

```
Arc: formedit
```

FormEdit displays the two windows illustrated here:

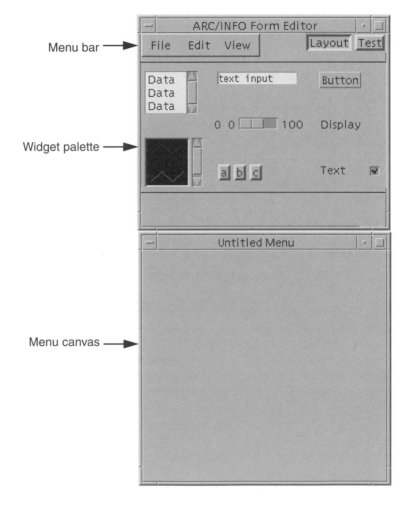

The top window contains the menu bar and the widget palette. The bottom window is the blank menu canvas where you design and create menus.

FYI

Note that these graphics and examples were created using the UNIX Common Desktop Environment (CDE). If you're using Windows NT, the FormEdit utility looks and behaves slightly different. Consult the online help for tips on using the Windows NT version of FormEdit. Note that FormEdit on both UNIX and Windows NT creates menu files and the ASCII text files that go behind the menus. You can create a menu on UNIX and use it on Windows NT, or vice versa.

The menu bar

The three buttons on the left side of the menu bar perform file management, widget editing, and menu canvas display, respectively. The View button also offers online help. The two buttons on the right side of the menu bar are used to toggle between the two modes of operation for FormEdit (i.e., layout mode or test mode).

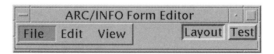

The File button displays the File menu, which contains the commands you need to create a new form menu, open an existing menu, save changes to the current menu, or save a copy under another name. The Quit option closes FormEdit and returns you to the command-level prompt. The File menu looks like this:

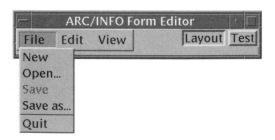

The Edit button on the menu bar displays the Edit menu, which you can use to modify widget properties on the menu canvas, change the widgets' positions, or even convert one type of widget to another.

The Edit menu looks like this:

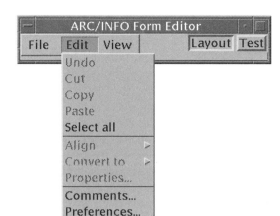

The last two lines on the Edit menu perform operations that affect the entire menu rather than selected widgets. Use the Comments option to include information like the author's name, date written, basic functions performed, and other pertinent information about your menus. Use the Preferences option to define overall menu properties.

The View button on the menu bar displays the View menu, which allows you to display the snapping grid on the menu canvas or refresh the canvas. The snapping grid is always active, but you can have it visible or hidden. The View menu is shown here:

The last option on the View menu displays online documentation for menu bar commands and options. For example, choose Help to obtain information on the properties defined with the Preferences option in the Edit menu.

Use the two buttons at the right of the menu bar to indicate whether you want to see your menu in layout or test mode. Layout is the default; use it as you add and arrange widgets on the menu canvas. Test allows you to see the actual size of your menu as it appears in an ARC/INFO session. Placement of the widgets dictates the size of the menu, not the size of the menu canvas in layout mode. See the online help for a more detailed discussion of the menu bar.

The widget palette

Widgets are graphical objects that appear on form menus. Widgets present information and receive user input. All widgets have two characteristics: a unique graphical appearance and a set of properties.

Nine types of form menu widgets are found on the widget palette, which is in the top window of FormEdit, below the menu bar. The widgets are arranged on the palette as shown here:

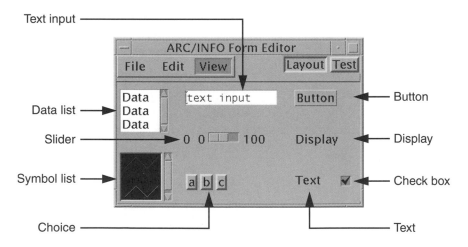

When you build a form, you can choose any widget from the widget palette and drag it to the menu canvas. All widgets can be repositioned and aligned with other widgets. Most widgets (except for buttons, check boxes, text, or choices) can be enlarged or reduced. Placing the widgets in the order in which they should be used makes it easier for the user to follow the form. You'll practice manipulating widgets in the next lesson.

You can incorporate enhanced pulldown menus (menu type 8) *within* your form menus. See "Embedding a menu in a form" on page 14-24 for more information.

A form can contain one, several, or all nine types of widgets. Each type of widget collects a specific kind of information; you choose which type of widget to use depending on which information is needed. The following table describes the information each widget accepts:

Widget name	Description
Button	Executes commands using the variables set by other widgets.
Check box	Toggles a setting between .TRUE. and .FALSE..
Choice	Presents a set of choices from which one can be selected.
Data list	Presents a scrolling list of choices (coverages, INFO or system files, TINs, grids, libraries, items, unique values, etc.).
Display	A display-only widget that displays a variable as text, an icon, or a symbol (e.g., displays the current symbol chosen from the symbol list).
Slider	Presents a horizontal bar with a handle. Slide the handle to set a numeric value (e.g., a buffer distance or snap tolerance).
Symbol list	Displays a scrolling list of symbol choices (line, shade, marker, or text).
Text	Displays background text for descriptions or messages.
Text input	A blank line where the user types input (e.g., a coverage name).

Question 12-1: Which type of widget would be most appropriate to use for each of the following tasks: (a) retrieving an INFO file, (b) retrieving a shade symbol number, (c) displaying a coverage name, (d) specifying whether or not to display node errors, and (e) applying all retrieved information to create an ARCPLOT display?

(Answer on page 12-11)

Widget property sheets

The properties of a widget dictate which information it displays, which it accepts from the user, and which actions it performs. After you place a widget on the menu canvas, you need to define its properties to activate it.

Each widget type has a unique set of properties. For example, the properties of a slider widget include a range of acceptable values, an increment for the movement of the slider, a default value, and a help message. Properties for a text input field include the type of data (e.g., character or integer), the acceptable length for an input string, and a flag that stops data from echoing on the screen as it's typed (e.g., a password).

You define a widget's properties on its property sheet, displayed by double-clicking on the widget in the menu canvas (not in the widget palette).

The slider widget and its property sheet are displayed here:

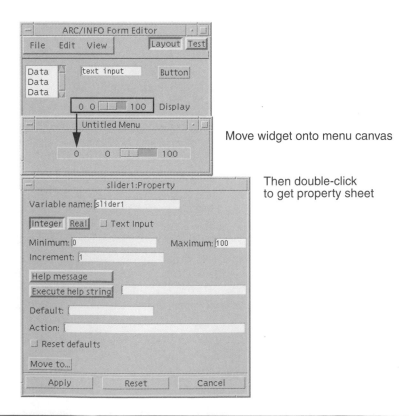

Move widget onto menu canvas

Then double-click
to get property sheet

Except for buttons and text, all widgets assign a value to a variable. The first entry on most property sheets is a place to enter the widget variable name. Form menus accept user input through the widget variable. For example, when the user moves the slider bar defined in the menu above, the value is assigned to the variable `slider1`. An ARC/INFO command can use this variable to accomplish such tasks as setting a tolerance or setting a buffer distance.

Other properties for the slider widget include online help, data verification parameters, a default value, text that appears on the widget, and the action, if any, that occurs when users manipulate the widget.

Rather than assigning a variable, the buttons use the variables set by other widgets to perform an action. On a button property sheet, you name the button and define the action it performs.

A text widget has no properties and therefore no property sheet. Double-clicking on a text widget allows you to edit the text string.

In the next lesson, you'll experiment with widgets and their property sheets as you create a form menu.

In summary

Form menus are used to build customized interfaces to ARC/INFO functionality. Form menus are ASCII files, but are easily constructed through a graphical interface tool called FormEdit. Widgets are the graphical components of the menu and accept user input. Widget properties dictate what information the widget displays and how the widget acts when it's manipulated.

Answer 12-1:
(a) Data list (of INFO files)
(b) Symbol list (of shade symbols)
(c) Display
(d) Check box
(e) Button

Exercises

Exercise 12.1.1

What are some advantages to using form menus instead of other screen menus, such as pulldown, sidebar, or matrix?

Exercise 12.1.2

In FormEdit, reference widgets are located on the menu _____ and dragged to the menu _____ to create a new form menu.

Exercise 12.1.3

In FormEdit, what's the difference between the preferences set from the Edit menu and properties set from a widget's property sheet?

Exercise 12.1.4

Can comments be added to form menus? If so, how?

Exercise 12.1.5

Explain how the snapping grid works in FormEdit.

Lesson 12.2—Creating form menus with FormEdit

In this lesson

This lesson presents an online tutorial so you can practice selecting, moving, and arranging widgets on the menu canvas and defining widget properties using FormEdit.

Form menus are dynamic because widgets can share data. This lesson shows how variables set in one widget can affect the appearance and operation of another widget.

Note that the following tutorials use FormEdit graphics and examples from the UNIX CDE windowing system. If you're using FormEdit on Windows NT, refer to the online help for further instructions.

FormEdit tutorial: Part 1

The following tutorial allows you to experiment with the widgets presented in the previous lesson. To get the full benefit from this tutorial, you should have access to a machine that has ARC/INFO loaded. In this lesson, you'll create a form menu to perform the BUILD command.

This tutorial refers to the mouse buttons. Roughly speaking, you use the 1 (left) button to do something and the 3 (right) button to find something. The 2 (middle) button isn't used in this tutorial. The layout of a typical three-button mouse is shown below:

FYI If you're using Windows NT with a two-button mouse, the mouse shouldn't give you any trouble in the tutorial. If you do run into problems, consider acquiring a three-button mouse.

To begin building your first form menu, start FormEdit:

Arc: **formedit**

The FormEdit interface appears featuring the menu bar, widget palette, and menu canvas introduced in the previous lesson. If you prefer to see the snapping grid while you lay out the menu, choose the Show grid option from the View menu.

Lay out the menu widgets

In ARC/INFO, the form menu you'll create looks like this:

Begin by selecting widgets from the widget palette and arranging them on the menu canvas. Follow the example menu shown above to determine where to place the widgets. There are five widgets you need to place on the menu canvas. From the top of the menu, they're text, data list, choice, and two button widgets.

- Move the five widgets you need, one widget at a time, from the widget palette. Use the 1 button on your mouse to click on and drag it to a position on the menu canvas.

  Reposition widgets on the canvas the same way: click, hold, drag, release.

Your initial menu should look like this:

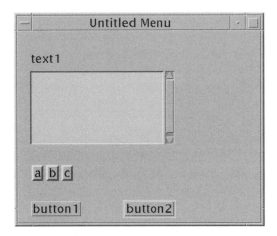

Don't worry if the widgets aren't placed exactly as shown in the example; you can adjust them later. After placing a widget, you may notice that its position changes slightly because it snaps to the grid. If you select a widget that you don't want, delete it using the Delete key on your keyboard.

Now, try dragging a widget so it falls partially outside the menu canvas on the bottom or on the right. On the right side, between the widget palette and the menu canvas, the *margin icon* appears. The margin icon indicates that a widget falls outside the menu canvas; an arrow indicates direction. If you need more room on the menu canvas, enlarge the window.

Creating static text

The text string text1 (at the top of the menu) should tell the user what the data list widget is for. You need to modify this text so it reads appropriately.

1. To modify the text string, use the 1 button on the mouse to double-click on text1.

   Now the text string appears in a larger type-in box (i.e., text1), indicating it can be modified. Double-click again and text1 appears highlighted.

---

2. Type `Coverage to build:` and press Return.

   The type-in box disappears and the operation is complete.

Creating a data list

A scrolling data list is the most effective way to present available objects (e.g., coverages, items, or grids). Use the property sheet for the data list widget to define which objects to display.

1. Double-click the data list widget to display its property sheet.

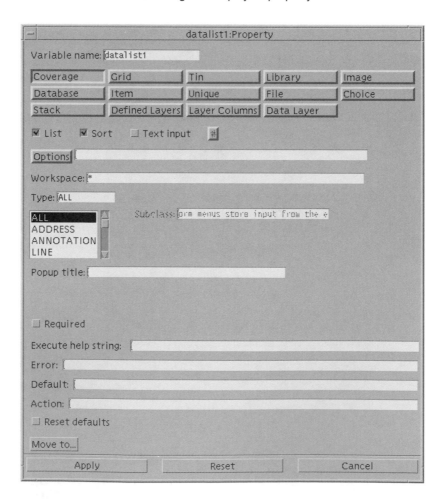

The properties on this sheet allow you to define how the data list looks and what happens when the user interacts with it. Don't be alarmed by the length of the property sheet. Not every data list requires that you define all of these properties. You only need to define a couple of properties now. The other data list properties are discussed later in this lesson.

2. Double-click the variable name datalist1 and rename it cover.

When the user selects a coverage name from the list, it's stored in the variable named cover. You can name the variable anything you want, but the name should describe the type of information the variable stores, in this case a coverage name.

3. Select the kind of information you want to display in the list from the choices following the input field for the variable name.

Notice that the Coverage choice appears pressed in. This is the default value and specifies that the data list will contain coverages.

4. Apply the properties you defined for the widget by pressing the Apply button.

All other properties remain unchanged at this time. The top portion of your property sheet should look like this:

5. Press Cancel to dismiss the property sheet.

Notice that the data list doesn't yet contain coverages. When the menu runs in ARC/INFO, the coverages appear. (Note: The scroll bar appears on the right side of the data list in the menu canvas, but during an ARC/INFO session it appears on the side defined by your windowing environment.)

Creating choices

To build topology, you need to specify the type of features a coverage contains. The BUILD command can build only one type of feature at a time; therefore, the choices must be mutually exclusive.

Choice widgets provide a set of mutually exclusive choices from which the user can select. (A choice widget provides the same visual feedback to the user as a push-button radio does.) The button representing the selected choice appears pressed in while all other buttons appear raised.

The choice widget property sheet allows you to define the text and set the variable value for each choice. (Using the radio example, you may want the button to read KLOS, but set the dial to the value 95.5.)

1. Double-click on the choice widget to display its property sheet.

2. Double-click the variable named `choice1` and rename it `feat`.

Choosing a feature type with the choice widget sets the variable `feat` with the name of the feature type.

3. Edit the names of the choices currently named a, b, and c. First, click on choice a, then click on Edit (located below the text field).

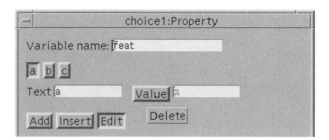

The Text and Value fields are displayed and, by default, contain the value of a. To edit the Text field, double-click on the lowercase a and rename it Polygon. Then press Return. Next, click on Value, double-click on the lowercase a in the value field, replace it with polygon, and press Return. Polygon now appears as the text for the first widget choice.

Repeat this process to change the second choice, b, to Arc in the Text field and arc in the Value field. Arc now appears as the text for the second widget choice. Repeat this process to change the third choice, c, to Point in the Text field and point in the Value field. Point now appears as the text for the third widget choice.

Your property sheet should now look like this:

 FYI The choice widget isn't limited to three choices. You can use the Add button to add more choices after the current choice. Delete choices by making them active and then pressing the Delete button.

As you construct the choices at the top of the property sheet, the choice widget replicates in the Default: field.

4. Move the mouse to the Default: field and choose Polygon. This means that the value polygon will be stored in the variable FEAT when the menu is initialized.

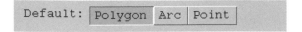

5. Press `Apply` to apply the properties to the widget.

6. Press `Cancel` to dismiss the property sheet.

The choice widget now appears with the new text in the menu canvas.

Creating a button to execute an action

Now that you have a way to capture the coverage name and feature type and store them in variables, you need a way to use them as arguments to execute the BUILD command. Button widgets execute an action defined by one ARC/INFO command, several commands (separated by semicolons), an AML program or menu (using &RUN or &MENU), or system commands (using &SYSTEM).

1. Double-click on the widget named `button1` to display its property sheet.

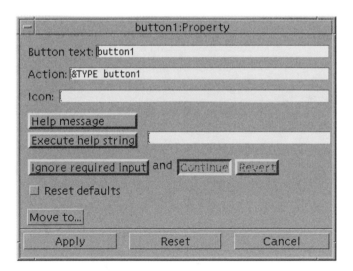

2. Double-click on the button text `button1` and replace it with the word `BUILD`.

3. Define a new Action: field. To highlight the action `&TYPE button1`, triple-click inside the type-in box to highlight the whole line. Next, type `BUILD %cover% %feat%`.

4. Press `Apply` to apply the properties to the widget.

The top of the property sheet should look like this:

5. Press Cancel to dismiss the property sheet.

The button widget now appears with the new text in the menu canvas.

Creating a button to cancel an action

Menus should always provide an escape when the user doesn't want to continue. Define the action for this situation in another button.

1. Double-click on the widget named button2 to display its property sheet.
2. Double-click on the button text button2 and replace it with the word CANCEL.
3. Define a new Action: field.

   Highlight the action &TYPE button2 and replace the text by typing &return.

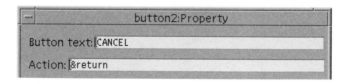

4. Press Apply to apply the properties to the button widget.
5. Press Cancel to dismiss the property sheet for the button widget.

The second button widget now appears with the new text in the menu canvas.

Saving the menu

The Apply button updates the property sheet, but the changes aren't saved to the menu file. You must save the menu to save your changes.

1. From the File menu on the menu bar, select Save as.

2. The default menu name (untitled1.menu) appears in the Path: field.
3. Double-click untitled1.menu and replace it with build.menu. After you hit Return, the new text appears as the title of the menu canvas window.

Viewing the menu in test mode

Thus far, you've seen the menu canvas in Layout mode. The size of the canvas has remained constant. Test mode displays the size the menu appears when it's executed in an ARC/INFO session.

1. From the menu bar, select Test mode.

The size of the menu in test mode is usually different because it's based on the number and placement of the widgets. "Window layout" on page 12-49 offers more detail about sizing menus.

2. Return to Layout mode.

Using the menu in ARC/INFO

Now see how the menu looks and operates in ARC/INFO.

1. From the File menu, select Quit.

FYI

> You don't have to quit FormEdit to run the menu. Instead, you can open another window to execute ARC/INFO and run your menu from there. When launching the menu, make sure you're in the proper workspace or have used &MENUPATH.

2. Making sure there are coverages in your current workspace, invoke your form menu as follows:

```
Arc: &menu build.menu
```

If the current workspace doesn't contain any coverages, move to another workspace, specify an &MENUPATH to the directory containing your form menu, and invoke the menu as follows, substituting your pathnames and terminal specification:

```
Arc: workspace /carmel3/miker/covers
Arc: &menupath /carmel3/miker/menus
Arc: &terminal 9999
Arc: &menu build.menu
```

Remember, you must set the &TERMINAL directive before you can display screen menus. Note that it only needs to be set once during an ARC/INFO session.

FYI

> The form doesn't look exactly the same in FormEdit test mode as it does in ARC/INFO because they use two different widget tool kits.

3. Select a coverage and the appropriate feature class, then press BUILD

The BUILD command executes. While this occurs, the menu is inactive, indicated by the pointer becoming a clock symbol. As control returns to the menu, you may see the text Updating scrolling list data. This occurs because the menu must update itself—conditions may have changed while the menu was busy (e.g., if you were running a CLEAN, there would be a new coverage to display in the scrolling list).

FYI

ARC/INFO commands display messages to the user in the window from where the menu was invoked. In this case, the BUILD command sends messages like Building polygons. You can turn these messages off by issuing &MESSAGES &OFF before invoking the menu. Remember to turn &MESSAGES &ON after you're done.

4. Press CANCEL to dismiss the menu.

Your first form menu is complete.

FormEdit tutorial: Part 2

In this part of the tutorial, you'll enhance BUILD.MENU by making the following modifications:

- Provide online help as a form menu option.
- Add widgets to allow the user to specify an output coverage.
- Resize widgets to accommodate information.

The revised menu looks like this:

If you changed workspaces after the last step in part 1, return to the workspace containing the form menu and invoke FormEdit as shown in the example:

```
Arc: workspace /carmel3/miker/menus
Arc: formedit build.menu
```

Alternatively, you can invoke FormEdit and use the File menu to Open the menu file. In this case, the FormEdit windows are displayed, but the menu canvas is smaller than before. Its size is the minimum allowed by the arrangement of the widgets. For now, leave the menu canvas this size.

Providing HELP

Even though your menus should be usable with little or no instruction, always provide online help for new users.

Form menus can offer help in two ways: as a one-line message appearing across the bottom of the menu, or as an entire page of information. Help is presented only when the user explicitly asks for it.

Follow steps one through six to add online help to BUILD.MENU.

1. Double-click on the choice widget to display its property sheet.

Assume the user doesn't know how to interact with the widget. Notice the two choices in the middle of the property sheet, Help message and Execute help string.

2. Click on Help message and then on the line to the right. Type the following message on the line:

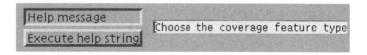

The message now appears whenever the user places the mouse on the widget and presses the right mouse button.

3. Press Apply and then Cancel.

To provide more help, display a customized text file or ARC/INFO help file.

4. Double-click on the Build button to display its property sheet.
5. Click on Execute help string and then on the line to the right.
    Type the following command on the line:

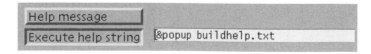

6. Press Apply and then Cancel.

If you create a text file called BUILDHELP.TXT, then when the user places the mouse on the Build button and presses the right mouse button, the text file appears in another window to describe what happens when the Build button is selected.

Adding a check box

Many users like to create a new coverage as a backup when they perform a BUILD on a coverage. Add this option to BUILD.MENU with a check box for indicating whether or not a new coverage should be created.

First you need to make room on the menu for the check box.

1. Enlarge the canvas and move the Build and Cancel buttons down.

The default width for the data list accommodates names up to seven characters. You can widen the data list so long names don't appear truncated.

2. Click and hold on the data list and stretch it to accommodate coverage names.

   After you click on the data list, selection handles appear at the corners and at the middle of each side to facilitate resizing the widget.

   Move the mouse to the handle in the middle of the right side. The cursor turns into an arrow and vertical bar icon.

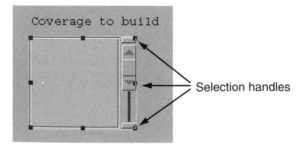

Selection handles

   Press and hold the left (1) button on the mouse and drag to the right, stretching the data list until it's as wide as the text above it.

3. Select a check box widget and text widget from the palette above BUILD.MENU and drag them into position.

   Use the check box to indicate whether the user wants a new output coverage. Double-click on the text and change it to new output coverage.

4. Double-click the check box to display its property sheet.

   Change the variable name to qnewcover, a logical variable with a value of either .TRUE. or .FALSE..

5. Press the Un-checked box in the Default: field.

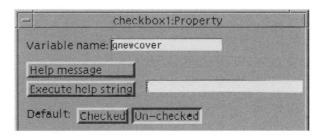

6. Press Apply and then Cancel.

Adding a text input field

Complete the following four steps so that the user can specify a new output coverage name through a text input field.

1. Select a text input widget from the palette and place it on the menu in the location shown on page 12-26.

2. Double-click the widget to display the property sheet.

   Update the variable name to newcover.

You're giving the user a blank slate on which to write, so you should validate the data that's entered to make sure it's acceptable for the operation. For example, some operations require a numeric value, others may require an x,y coordinate, and still others require a character string no longer than twenty-five characters.

3. In this form menu, the input field requires a coverage name. The maximum length for a coverage name is thirteen characters. Update the Maximum number of the Characters: field to thirteen.

In the series of choices below that field, notice that Character is already set as the default type of input.

To examine some additional verification tools, click on Integer or Real and notice that the Range: field is activated. This allows you to verify that numeric values are in an acceptable range for the operation. You can also specify that input is required for a text input field by checking the Required box. If the text input field isn't filled in, an error message displays at the bottom of the form.

Change the type back to Character before you continue.

4. Press Apply and then Cancel.

Changing the BUILD button

Now modify the action performed by the Build button to create an output coverage if the user wants one. The original action specified on the property sheet looks like this:

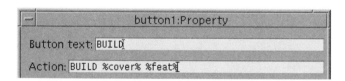

Before the BUILD command executes, you need to check whether the user wants a new output coverage and, if so, copy the original coverage to the new name. The following AML program performs this task:

```
/* buildcov.aml
&args cover feat qnewcover newcover
&if %qnewcover% &then
  &do
    &if not [null %newcover%] &then
      &do
        &if not [exists %newcover% -cover] &then
          &do
            COPY %cover% %newcover%
            BUILD %newcover% %feat%
          &end    /* do if not exists newcover
        &else
          &type Coverage %newcover% already exists!
```

```
     &end      /* do if not null newcover
   &else
     &type You must specify an output coverage.
&end  /* do if qnewcover
&else
  BUILD %cover% %feat%
&return
```

1. Double-click the `Build` button to display its property sheet.
   Replace the text in the Action: field with the following:

```
&r buildcov.aml %cover% %feat% %qnewcover% %newcover%
```

Now that you've made all the necessary changes to the widgets, go ahead and save those changes.

2. From the `File` menu (on the menu bar), select `Save as`. Replace `build.menu` with `buildcov.menu`, then `Quit`.

Using the menu in ARC/INFO

Test how the menu looks and operates in ARC/INFO as you did before (see page 12-23). Examine the following enhancements:

- The look and operation of the check box
- The look of the text input field
- How the thirteen-character limit affects the operation of the text input field

Question 12-2: What are the ways to invoke a menu that's not in your current workspace?

(Answer on page 12-39)

Congratulations on completing the FormEdit tutorial. Refer to the online help for detailed information about each widget and its properties. You can also get online help from FormEdit by pressing the right (3) mouse button over any widget in the menu canvas.

Using variables in form menus

The remainder of this lesson focuses on using variables in form menus. Variables store the value set by a widget. Variables also define properties for individual widgets and set preferences for the entire menu. Variables make form menus dynamic by allowing the widgets to change appearance as the variables that define them change their values.

Although variables are an integral part of form menus, there are some places where you can't use variables. The following lists outline where you can and can't use variables.

Variables can be referenced in:

- A display widget
- Options for the data list and symbol list widgets
- The action property of any widget
- The execute help string option of any widget
- The initialization string on a preferences sheet

Variables can't be referenced in:

- The text property of the choice widget
- Range or increment values on a slider widget
- Range values for a text input widget
- Button text
- Button icon
- Background text
- Modifying a widget's layout

 FYI

You can use the [WRITE] function to create dynamically changing form menus, even where variables aren't allowed. See lesson 12.3 for a discussion of the form menu file and chapter 10 for an example of creating this file with the [WRITE] function.

The following example uses a display widget to illustrate how variables set in one widget affect others. A display widget can be a text string, an icon, or a symbol. A variable named markersym, set with a symbol list widget, then displays in the display widget.

The relevant parts of the two property sheets and the resulting menu are shown below:

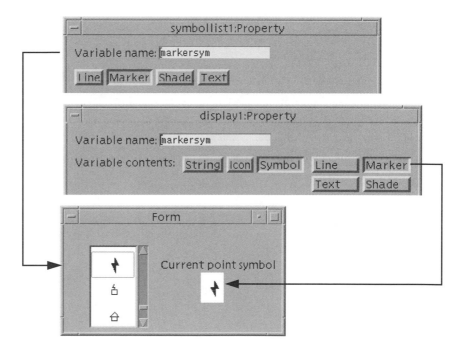

 FYI You can store icon files used for display in other directories and access them with the &MENUPATH directive. See chapter 3 for more information about &MENUPATH.

Using variables to define a scrolling data list can also dynamically change the look of a menu. Parameters for a data list are defined in one of two ways on the property sheet: either using the Options: properties, which can contain a single string with many parameters, or using the Workspace:, Type:, Subclass:, and Popup title: fields. These methods are mutually exclusive.

The following examples use both of these methods for setting data list variables to make the list dynamic. The first example shows how to set parameters using the Options: field.

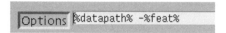

One benefit of using the Options: field is that you can assign an entire string to one variable. If you set a variable like this:

```
&sv options = %datapath% -%feat%
```

it can be used in a property sheet like this:

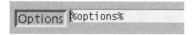

The next example sets the same parameters using the Workspace: and Type: fields.

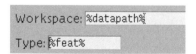

Assuming `feat = polygon` and `datapath = /carmel3/miker/data`, the above example displays all the polygon coverages in the `/carmel3/miker/data` directory.

A scrolling data list is similar to using a [GET...] function for the specified type (e.g., if the data list type is ITEM, it creates the same list as the [GETITEM] function does). The Options: field can contain the optional parameters available with the [GET...] functions, except for the NONE, OTHER, and PROMPT options.

See chapter 2 if you aren't familiar with the [GET...] functions in AML.

You can dynamically change the help messages by using a variable for the file name in the Execute help string: property.

Managing variables in form menus

Many variables are associated with a form menu, so it's important to keep track of exactly when each variable is initialized, updated, reset, and deleted. There are three ways to manage form menu variables: AML programs (external to the menu), the property sheet of each widget, or the preferences sheet (selected from the Edit menu).

Initializing form menu variables

The following examples demonstrate the three methods for initializing variables.

Creating an AML program to initialize the form menu variables is an easy way to manage variables. The following AML program initializes the variables FEAT and DISTANCE before invoking the menu:

```
/* some AML that initiates form menu variables
&sv feat = arc
&sv distance = .25
&menu editcov.menu
&return
```

These variables don't need to be global because local variables set in an AML program are available to every menu the program calls. However, if a menu subsequently runs another AML program, local variables are *not* available to the called program. Use the &ARGS directive or global variables in this situation. (If you aren't familiar with the use and scope of local and global variables, see chapters 1 and 8.)

The second way to initialize variables is on a widget property sheet by placing a value in the Default: field. A choice property sheet with the `feat` variable set to `arc` looks like this:

The third way to initialize variables is with the preferences sheet, accessed through the Preferences option on the Edit menu from the menu bar. Variables set in the preferences sheet are initialized when the form menu executes.

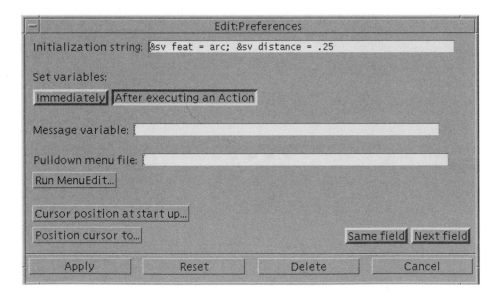

Alternatively, you could run an AML program to initialize global variables from the Initialization string on the preferences sheet.

Initialization string: &r init.aml

Updating form menu variables

After they're initialized, variables constantly update as the user interacts with the form menu. The Set variables: field on the preferences sheet contains two alternatives for specifying how variables are updated. Set variables: immediately updates variable values as soon as the user interacts with a widget. Set variables: after executing an Action updates variables after the specified action executes.

For an application that displays multiple menus simultaneously, using Set variables: immediately ensures that a variable used in more than one menu updates in every menu at the same time. Use this option.

If you use the Set Variables: after executing an Action option, other menus aren't updated until the action executes. This results in the display of inconsistent values.

FYI

After executing an Action is the default choice for the Set variables: field. Before AML supported multiple menus, this choice was used in many applications and remains the default choice for downward compatibility.

The Ignore required input option is unique to the button property sheet. It allows the button action to execute without verifying that the user entered all required input in text input fields. This button option is used when the button action doesn't rely on the settings of other widgets. As an example, most Cancel and Help buttons use this option, allowing the user to dismiss the menu or get help without needing to fill in required arguments.

The Ignore required input option has two settings that affect variable assignment, Continue and Revert.

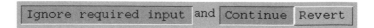

The Continue setting *retains* any changes made to the form after the button action is completed. Alternatively, Revert *resets* the form to the settings that were last saved and doesn't retain changes made by the user. Regardless of which option you use, if the button action dismisses the menu, all variables are initialized when the menu reactivates after the button action executes.

Resetting form menu variables

Reset defaults is an option available for all widgets except display. This option establishes whether the Default: value is reset after an Action: is performed. It's difficult to manage this option because it resets *all* widgets with their Default: values, no matter how their individual default is set.

Instead of using Reset defaults with Default: values, it's better to use AML programs to reset variables. Initialize your variables through an AML program in the Initialization string (or before running the menu) and update them in an AML program executed during the Action: as shown here:

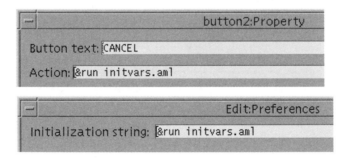

Deleting form menu variables

It's a good programming practice to delete variables when they're no longer needed. If you don't, outdated values can show up in the menu.

Consider a Cancel button whose Action: is to dismiss the menu with &return. To delete all the variables, you could modify the Action: to include deleting variables as shown here:

```
Action: &return; &dv .xpoint .ypoint .size
```

FYI

Using the menu name followed by a dollar sign ($) to name the variables that are associated with the menu makes them easier to manage. For example, if the names of all variables used with BUILDCOV.MENU begin with .buildcov$ (e.g., .buildcov$cover), then &dv .buildcov$* deletes them all.

In summary

FormEdit allows you to arrange widgets graphically on the menu palette and define properties and actions on their property sheets. Form menus accept user input through the widget. Form menus are dynamic because each widget stores information in a variable that other widgets can use to modify their own appearance and operation.

Variables constantly update as the user interacts with the form menu, so keeping track of when variables are initialized, updated, reset, and deleted is important. Using AML programs is an easy way to control variables. You can run these programs from the menu or before the menu executes.

A

Answer 12-2: To invoke menus that aren't in your current workspace, you can do the following:
(a) Physically move to where the menu is located, then launch the menu
    (i.e., first type w /carmel/ridland/project1
    and then type &menu main.menu).
(b) Specify a path to where the menu is located when you launch the menu
    (i.e., type &menu /carmel3/ridland/project1/main.menu).
(c) Use the &MENUPATH directive first, then launch the menu
    (i.e., first type &MENUPATH /carmel3/ridland/project1
    and then type &menu main.menu).

Exercises

Exercise 12.2.1

(a) Explain how to assign a variable *as soon as* a form menu is invoked.

(b) Explain how to assign a variable *before* a form menu is invoked.

Exercise 12.2.2

Follow the guidelines listed below to create a menu called AE.MENU that sets up a working environment in ARCEDIT. Begin by laying out all the widgets in the proper location. Next fill out the property sheets needed to perform the following tasks:

- The scrolling list displays all coverages in the current workspace and sets the MAPEXTENT and the EDITCOVERAGE.
- The choice widget sets the EDITFEATURE.
- The slider bar sets the tolerance for ARCSNAP and NODESNAP.
- The series of check boxes set the features for the DRAWENVIRONMENT.
- The Draw button calls an AML program named DRAWCOV.AML. The menu stays displayed on the screen. The user can then choose another coverage to draw.
- The Cancel button dismisses the menu and calls an AML program named CLEANUP.AML that deletes the global variables and resets any environments it created.

Lesson 12.3—Beyond FormEdit

In this lesson

Just knowing the mechanics of creating form menus doesn't ensure a good user interface. Designing an interface involves combining the available tools with a set of user requirements and good design principles. Your goal as an applications programmer should be to create an interface that performs the desired tasks easily and efficiently, requiring a minimal amount of training. Toward this end, this lesson presents you with some guidelines for designing an interface.

Like all AML menus, a form menu is an ASCII file. Your use of FormEdit actually writes the menu file. This lesson looks at the relationship between the widgets and widget properties as displayed in FormEdit and the ASCII menu file FormEdit writes.

If you haven't read the section "Introducing menu design" in chapter 2, lesson 2.3, do so before continuing with this lesson.

Designing form menus

The extensive set of widgets and functions that form menus utilize enables you to perform a task in a variety of ways. Whether you design one form menu or an entire application, you need to address such design considerations as organization, logical flow, visual appearance, ease of use, error checking, and online help. The following sections specifically address the design issues of menu appearance, labels, use of icons, and consistency.

Designing the visual appearance of menus

Users base their first impressions of an application on the appearance of the interface. Although you're providing an interface to sophisticated software tools and a complex application, your menu should appear easy to use. If users think they'll succeed, they're more inclined to experiment and therefore learn more quickly.

Something familiar is easier to use. It's your job to discover how a task was accomplished in the past and attempt to mimic the operation. For example, if you design a form menu to replace a paper form, create the form menu as a replica of the paper version. Although minor improvements are usually well received, avoid making users restructure the procedures they've known for years.

Presenting unambiguous and intuitive labels

Always use terminology that users understand. Unambiguous labels for objects on your form menus make them easier to learn. Text on widgets, as well as any background text you use on the form, should help clarify operations for the user.

Operations like CLEAR, SAVE, and SELECT can have several meanings in an ARC/INFO session. On the following menu, it's difficult to determine the purpose of the choice widget (One, Many...), and it isn't obvious whether the CLEAR button clears the screen or clears the selected set.

Adding informative text as shown here eliminates these problems:

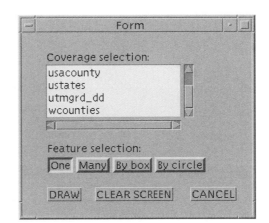

You should determine the educational and professional background of those who will use your application to define the appropriate terminology. Don't fall into the trap of representing ARC/INFO commands and terminology on buttons. If users aren't familiar with ARC/INFO, placing an ARC/INFO term on a button won't help them understand the process better.

The type of widget you use influences how quickly the user learns. Take advantage of natural, intuitive, and spatial analogies. Consider designing a layout of appropriate widgets to show values of high, medium, and low, or up, down, left, and right. The horizontal slider and choice widgets shown on the left may not work as well as the set of intuitively positioned buttons shown on the right.

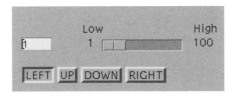

You've used check marks to select an item from a list or to indicate that a task on a list has been completed. Similarly, you can use check-box widgets for making nonexclusive choices, such as showing that an operation is complete, indicating coverages you want to display, or selecting features to include in a drawing environment. Because check boxes are nonexclusive, checking one doesn't affect other check boxes. For setting exclusive choices like ON/OFF settings, choosing a single feature type, or choosing a value to insert in a database, use choice widgets.

Using icons

An *icon* is a small graphic representing a function that is performed by the application. Buttons and display widgets can appear as icons in form menus. Using icons doesn't necessarily ensure a more intuitive application. Icons are effective for operations that are easy to represent graphically. Icons that aren't ambiguous and follow a real-world metaphor are easiest for users to recognize and understand.

You encounter many well-designed icons every day. For example, controls on the dashboard of a car, like headlight, windshield wiper, and ventilation controls, are presented with easily recognized icons. Many road signs and international symbols convey information without using language.

Take advantage of culturally accepted standards on your form menus. For example, you might consider using the international symbol for information to represent the IDENTIFY command, which retrieves information about a feature. An icon is effective in this situation because IDENTIFY is a common, repetitive operation and this icon is widely recognized.

Icons are best used for frequently repeated tasks. When used properly, icons create an appealing interface. Icon interfaces commonly suffer from two problems: either too many icons are used, overwhelming the user, or icons are poorly designed, confusing the user. Consider the following questions before deciding to use icons:

- Can a simple graphic represent the task?

  Some tasks can be depicted with a simple graphic, some can't. Consider the difference between joining two tables and building topology. The first is conceptually easy, the second involves many steps and the input/output (i.e., a coverage) is difficult to represent with a picture.

- Which simple graphic best represents the task?

  Icons that depict the user's conceptual model of the operation are most successful. Think about a graphical representation for a zoom-in function. The following examples illustrate some possibilities:

- Is the same operation performed on different objects (e.g., listing coverages, listing items, and listing records)?

  The icon for listing a selected object should be consistent. In this case, it's difficult to create a simple graphic to depict the operation, while a button labeled LIST gets the job done.

- Are icons distinguishable from one another?

  An interface is easier to use when its icons are easily distinguishable. Icons are small graphics; too much detail is difficult to see. Icons aren't efficient if it takes a user too long to figure out which one to use.

- Does the user prefer working with icons instead of text?

  The important point here is the preference of the user, not the preference of the designer. The user may want buttons that display familiar terminology instead of icons. A simple graphic may be worth a thousand words, but they must be words the user understands. Don't allow your desire to use icons override the needs of the user.

- Are the icons understandable without text?

  Text can be part of an icon, but if text is needed, consider using only the text.

After you answer these questions, you have a good basis for deciding whether or not to use icons in your interface.

Considering graphical and functional consistency

When you design multiple menus to use in an application, make sure that both the graphical appearance of the menus and the presentation of functionality are consistent in all the menus. Some examples:

- Always use the same widget to perform the same function.

  For example, if you use a choice widget to define a feature type, use a choice widget every time and in every menu the user needs to perform this operation.

- Buttons with the same name should act the same.

  For example, every OK button executes an operation and erases the menu from the screen; every APPLY button executes the operation, but doesn't erase the menu.

- The same icon should perform the same operation every place it appears.

- A widget should appear in the same relative position on every menu on which it's used.

  For example, a list of coverages always appears in the upper left corner, and the EXIT and OK button are always side by side at the bottom center of the menu.

- Change the method of user input only when necessary.

  Switching between the keyboard and the mouse slows down the user and disrupts the flow of the menu.

Providing help and feedback

The more information the user needs to memorize or look up in a manual, the longer it takes to become proficient using your application interface. Place the information and instructions needed to complete the form directly on the menu or make them accessible with background text or online help.

Provide as much feedback to the users as possible, and make it meaningful. Most operations either create something or set parameters. Always indicate the result of the operation to the user. A display widget can indicate the result of an operation (e.g., the new coverage or the current text symbol).

Don't leave the user guessing about how long an operation will take. If you display a completion status for commands that take a long time, then users won't get frustrated and can make more efficient use of their time. The slider and display widgets icon can display a completion status that looks like this:

Avoiding and handling errors

No matter how well you design an interface, errors still occur. Try to conceive of every possible mishap and protect against it, or at least handle it gracefully. Take advantage of the widget property sheets to set appropriate values, present only valid choices, and check user input. These precautions prevent many errors and make the application easier to use. The following example illustrates how to establish the Range: for a text input field to check user input. You can use this technique to prevent the user from entering inconsistent or illegal values in your database.

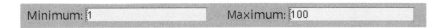

When a program encounters errors, provide the user with meaningful feedback. Indicate what the user did incorrectly instead of what happened to the execution of the program. Provide the user with solutions whenever possible.

Keep error messages, warning messages, and confirmation messages consistent and distinguishable from one another. If the program stops executing, an exclamation point with a descriptive message like the one shown here might be appropriate.

Notice how the heading on the previous menu changes on the next menu to reflect its purpose. This is achieved using the &STRIPE option on the &MENU directive, discussed under "Window layout" on page 12-49.

For a confirmation message, the exclamation point could be changed to a question mark. These symbols are easy to distinguish and recognize, and follow a real-world metaphor.

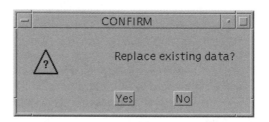

When designing your applications, try to anticipate and prevent user errors. For example, provide a confirmation message (e.g., Are you sure?) to prevent users from deleting the wrong coverage or to verify that edits made in a lengthy ARCEDIT session were saved.

FYI

The ArcTools menu interface that comes with ARC/INFO has four standard message tools. You can use these "prebuilt" menus to provide information to the user. To see a sample of these tools, type the following commands:

```
/* set your aml and menu paths to the ArcTools directory
&run $ATHOME/lib/setpaths
/* make sure your terminal is set properly
&term 9999

/* to get the previous 'ERROR' message (page 12-47), type:
&run msconfirm 'Cannot complete operation' ~
'Did you enter a number between 1 and 100?' ~
# 'ERROR'

/* to get the previous 'CONFIRM' message (page 12-47):
&run msconfirm 'Replace existing data?' # # 'CONFIRM'
```

You can have these menus automatically launch when your AML program or menu application encounters an error. Consult the online help for more information on these and other ArcTools standard message tools.

Window layout

Always consider the design and layout of the windows as you develop a graphical interface. Many windows can be displayed simultaneously, so their spatial relationship, graphical consistency, and organization are important.

It's not uncommon for your screen display to include a graphics canvas, a form menu, and a popup error message. Attempt to organize these windows where they should appear, minimizing the distance between them without overlap. If the windows must overlap, consider the order in which the user interacts with them.

Each window has a frame that defines its boundary. This window can be sized and positioned as necessary for the application. Use the &MENU directive with the &SIZE and &POSITION parameters to manage the window layout. The following example displays the upper left corner of CHANGECOV.MENU at the same location as the upper right corner of the display canvas.

```
DISPLAY 9999 1
&menu changecov.menu &position &ul &display &ur
```

If the font size and style used by FormEdit differs from those defined for your working ARC/INFO session, you may need to modify menus you create using the &SIZE and &POSITION parameters.

It's easier to work with a consistent font so that what you see in FormEdit is what you get in ARC/INFO. You can create a file named Formedit to contain the specifications of the font and other window environment parameters. The Formedit file might look like this for systems running the UNIX OPEN LOOK® or OSF/Motif® user interfaces:

```
Formedit*Background:        gray80
Formedit*Font:              courier-14
```

FYI

When designing multiple menus, the &THREAD directive used with &MENU supports the same optional parameters. Invoking a menu with &THREAD also allows you to use [SHOW &POSITION <thread>] and [SHOW &SIZE <thread>]. Managing multiple menus with threads is discussed in chapter 13.

In addition to a frame, every window has a title bar. Use the &MENU directive with the &STRIPE option to place a title on each window. The title should announce the purpose of the menu, or which button it was called from, and should distinguish it from other menus. The following codes invoke the BUILDCOV menu:

```
&menu buildcov.menu
```

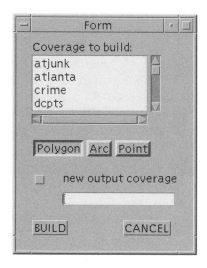

```
&menu buildcov.menu &stripe 'Build a coverage'
```

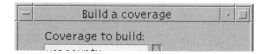

When form menus and messages are displayed using the Windows NT or UNIX CDE window manager, a staple symbol appears in the upper left corner. (Other window managers use other symbols.) The staple is an icon that represents a CANCEL function. If the user double-clicks the staple with the mouse, the window is dismissed. An accidental dismissal could be disastrous for your application, so control it using the &PINACTION parameter of the &MENU directive.

If you want the main menu to reappear (i.e., the application to restart) when the user double-clicks the staple, issue the menu directive in this way:

```
&menu myform.menu &pinaction &run startapp.aml
```

The user knows best

In the world of application development, the user is the boss. Understanding what makes sense to the users of your application is vital to its success. You need to understand their conceptual models, how they perform tasks, and most importantly, their terminology. Keep in mind that the designers, or programmers, usually don't use the application on a daily basis.

Interviewing users before development builds a good foundation. These interviews often generate ideas for design specifications from which you can begin to build prototype menus. *Prototypes* are partially working versions of an application that can be tested for usability (e.g., visual appearance, functionality, consistency, etc.). As soon as the menus are usable, have the users test them.

Incorporate the feedback you receive from your users into the interface and test again. Prototyping is a cyclical process; several iterations aren't uncommon. Don't consider prototyping a nuisance—it provides the information you need to build a successful application.

FormEdit and menu files

FormEdit is a pictorial version of an AML menu file that lets the programmer design and create form menus without having to know AML form menu syntax. Whether created with a text editor or using FormEdit, all form menus are stored as ASCII files.

Displayed below is BUILDCOV.MENU (created in the tutorial) and the menu file from which it's generated:

```
[1]    7buildcov.menu
[2]    /*
[3]        Coverage to build
[4]        %datalist1
[5]
[6]
[7]
[8]
[9]        %choice1
[10]       %x1  New output coverage
[11]           %input1
[12]
[13]       %button1            %button2
[14]   %datalist1 INPUT COVER 19 TYPEIN NO SCROLL YES ROWS 4 COVER ~
           * -ALL -SORT
[15]   %choice1 CHOICE FEAT PAIRS ~
           'Polygon' 'polygon' 'Arc' 'arc' 'Point' 'point'
[16]   %x1   CHECKBOX QNEWCOVER INITIAL .FALSE.
[17]   %input1  INPUT NEWCOVER 33 TYPEIN YES SCROLL NO SIZE 13 ~
           CHARACTER
[18]   %button1    BUTTON KEEP 'Build' &r  buildcov.aml %cover% %feat% ~
[19]       %qnewcover% %newcover%
[20]   %button2  BUTTON KEEP CANCEL 'Cancel' &return
```

The number on the first line indicates the type of menu: "7" indicates a form menu. Lines [3] through [13] are called *field references*—they define the layout of the form. Field references must never be in column 1. Lines [3] and [10] include *background text*. Notice the similarity of their relative positions to the widgets in the graphical display.

Lines [14] through [20] are called *field definitions* and indicate all the widget properties in AML menu file syntax. All field definitions must start in column 1. Each widget, property, and preference defined in FormEdit translates to an AML keyword or term that appears in the menu file.

The following examples show how the information entered on property sheets appears in the menu file.

The data list property sheet and the text it generates in the menu file are shown below:

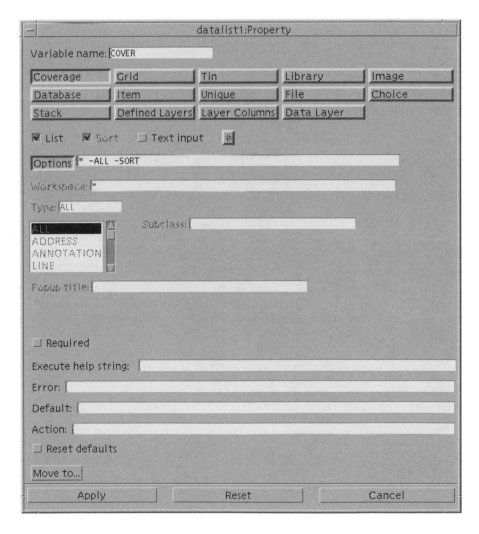

```
%datalist1 INPUT COVER 19 TYPEIN NO SCROLL YES ROWS 4 COVER ~
  * -ALL -SORT
```

Everything except for 19 and ROWS 4 are explicitly defined in the property sheet. 19 and ROWS 4 are defined when the user graphically resizes the widget on the menu canvas.

The choice property sheet and the text it generates in the menu file are shown here:

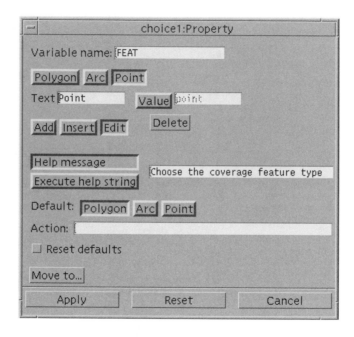

```
%choice1 CHOICE FEAT PAIRS~
    'Polygon' 'polygon' 'Arc' 'arc' 'Point' 'point'
```

The following reference table lists AML syntax found in form menu files and the corresponding FormEdit terminology:

AML syntax	Property sheet terminology	FormEdit widget
CANCEL	Ignore required input and revert	Button
<choice_text>	Text	Choice
ERROR	Error	Data list
-EXTENSION	Extension	Data list
HELP	Help message	Button; Check box; Choice; Slider; Text input
KEEP not used	Reset defaults	Button; Check box; Choice; Data list; Slider; Symbol list; Text input
ICON	Variable contents: Icon	Button; Display
INITIAL	Default	Check box; Choice; Data list; Slider; Symbol list; Text input
<max_value>	Maximum	Slider
<min_value>	Minimum	Slider
NEXT	Move to...	Button; Check box; Data list; Slider; Symbol list; Text input
NOECHO	Password	Text input
-NUMBER	Display symbol number	Symbol list
PAIRS	Text and value don't match	Choice
-PROMPT	Popup title	Data list; Symbol list
QUERY	Execute help string	Button; Check box; Choice; Data list; Slider; Symbol list; Text input
RANGE	Range	Symbol list; Text input
RETURN	Ignore required input and continue	Button
RETURN <return_string>	Action	Button; Check box; Choice; Data list; Slider; Symbol list; Text input
REQUIRED	Required	Data list; Symbol list; Text input
ROWS	Size widget	Button; Data list
SCROLL YES	List	Data list; Symbol list
<set_value>	Value	Choice
SINGLE	Text and value match	Choice
SIZE	Maximum number of characters	Text input
STEP	Increment	Slider
-SORT	Sort	Data list
SYMBOL	Variable contents: Symbol	Display
TYPEIN YES	Text input	Data list; Slider; Symbol list
VALUE	Variable contents: String	Display

The preferences sheet sets parameters for the entire menu. This table shows AML syntax for form menus and the corresponding FormEdit terminology used in the preferences sheet.

AML syntax	Preferences sheet terminology
%FORMINIT	Initialization string
%FORMOPT MESSAGEVARIABLE	Message variable
%FORMOPT SETVARIABLES IMMEDIATE	Set variable: Immediately
%FORMOPT SETVARIABLES RETURN	Set variable: After executing an action
%FORMOPT STARTFIELD	Cursor position at startup...
%FORMOPT NEXTFIELD	Position cursor; Same field; Next field
%FORMOPT MENU <menu_file>	Embeds enhanced pulldown menu (type 8) in form

FYI

Many who know the syntax for AML form menus like to use FormEdit and the text editor together. They use FormEdit to lay out and define the widgets and finish by editing the form menu file in the text editor. Others like to first type the menu file in the text editor, then open the file in FormEdit to arrange and align the widgets in a graphically pleasing layout.

The AML form menu syntax for each widget is discussed in the online help under *Customizing ARC/INFO with AML >> FormEdit >> FormEdit controls*.

Below are some tips for creating understandable menu files:

- Use comments.

  Comments should state the author's name, the purpose of the menu, variables used, other menus and AML programs the menu runs, and so forth.
- Don't mix field references and field definitions.

  Place all field references at the top of the file and field definitions below them.
- Use descriptive field reference names.

  If you prefer, use the type of information stored (e.g., coverage) rather than the name of the widget type (e.g., data list).

In summary

A graphical user interface should provide users with a functionally consistent, visually appealing, and easy-to-use set of tools to perform their work. Before you design an interface, you should interview prospective users to determine their needs. Always have users test menu prototypes and then use their feedback to improve the interface.

You can create form menus using FormEdit, a text editor, or both. In any case, an ASCII menu file is the final result. You can modify the appearance of a menu using options available with the &MENU and &THREAD directives.

Exercises

Exercise 12.3.1

Discuss four ways you can implement online help for the user.

Exercise 12.3.2

Considering the design guidelines presented in this chapter, which improvements would you make to the following ARCPLOT menu interface?

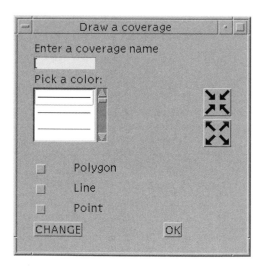

Exercise 12.3.3

Given the following property sheet, what's the line it creates in the menu file?

Exercise 12.3.4

Given the following line in a menu file, fill out the property sheet:

```
%slider1 SLIDER bufdist 21 TYPEIN YES QUERY '&popup bufhelp.txt'
     STEP 50 INTEGER 100 5000
```

13 / *Threads: Sewing your application together*

A robust application interface allows the user to interact with more than one menu at a time. This chapter shows how to display and manage multiple menus in an application using AML *threads*. Threads are the mechanisms that deliver input to the AML processor.

In this chapter, you'll progress from working with one active menu to managing multiple menus. You'll explore a working ARCPLOT display and query application that looks like this:

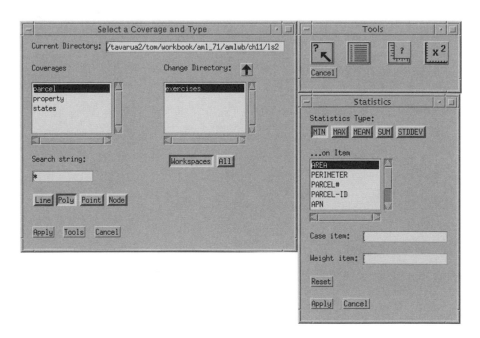

This application includes three menus for selecting and displaying coverages in ARCPLOT, and for performing queries using the commands IDENTIFY, LIST, MEASURE, and STATISTICS.

This chapter covers the following topics:

Lesson 13.1—Introducing threads

- Using a single thread
- Using multiple threads

Lesson 13.2—Managing threads

- Focusing in on threads
- Thread modality
- Synchronized threading

Lesson 13.3—Managing a real-world application

- Examining the display and query application

Lesson 13.1—Introducing threads

In this lesson

In ARC/INFO, a thread carries input to the AML processor. Threads accept input from three sources: AML programs, TTYs (typing at the keyboard), and AML menus. To build the multithreaded application shown on page 13-1, you must first understand how a single thread works.

This lesson examines how input sources (programs, TTYs, and menus) act in a single-thread environment and explores how multiple threads affect them. As an application designer, you'll find that using threads gives you more flexibility in deciding where to position menus, when to present them, and when to make them accessible to the user. This allows you to create an interface that guides the user through the application logically.

Using a single thread

The programming you've done so far used a single thread. With a single thread, only one menu can be active at a time and the following statements are true:

- Programs that invoke menus stop executing until the menu is dismissed.
- Menus that execute programs become busy until the program finishes.
- Menus that invoke menus disappear until the second menu is dismissed.

Visualize input sources stacked on the thread, with the user having access only to what's at the top of the stack. Start with a TTY input source (i.e., a prompt):

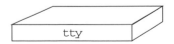

Now suppose you run an AML program from the command line:

```
Arc: &run getdrawcover.aml
```

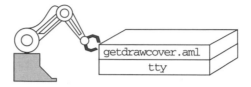

Now GETDRAWCOVER.AML is on top of the stack. There's no access to ARC/INFO through the command line prompt as long as the AML program runs.

GETDRAWCOVER.AML executes GETDRAWCOVER.MENU, which moves to the top of the stack:

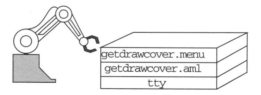

GETDRAWCOVER.MENU appears on the screen:

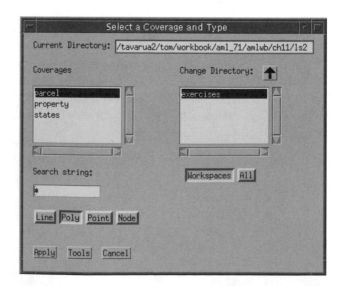

GETDRAWCOVER.MENU tops the stack so GETDRAWCOVER.AML pauses until the menu is dismissed.

Finally, suppose the user invokes TOOLS_ICON.MENU by pressing the Tools button on the menu. GETDRAWCOVER.MENU disappears from the screen as TOOLS_ICON.MENU is placed on top of the stack and appears on the screen.

Even though GETDRAWCOVER.MENU isn't visible on the screen, it's still on the stack:

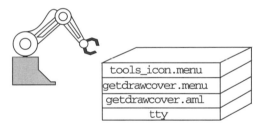

There are now four input sources on the single-thread stack. Only the input source on top of the stack is visible and active. When TOOLS_ICON.MENU is dismissed with the Cancel button, it's removed from the top of the stack, and GETDRAWCOVER.MENU moves to the top, reappearing on the screen.

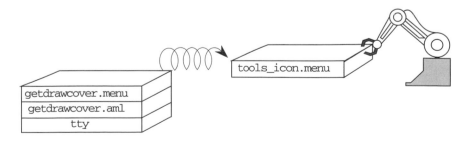

When GETDRAWCOVER.MENU is dismissed, it's removed from the stack and GETDRAWCOVER.AML continues executing where it left off:

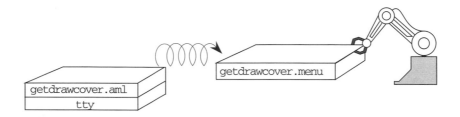

When GETDRAWCOVER.AML finishes executing with &RETURN, control returns to the user through a prompt (TTY input source):

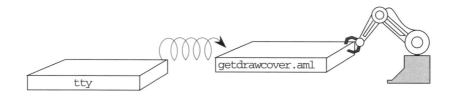

As you can see, everything you do in AML goes on the thread. The &TB directive shows what's on the current thread. (TB is an abbreviation for *trace back*.) &TB lists the contents of the current thread. An asterisk (*) marks the thread name. When using a single thread, the most current process (i.e., the top of the stack) appears as the last item in the &TB display.

If you execute &TB when TOOLS_ICON.MENU is on top of the stack, as in the previous example, you see the following list:

```
* Thread thread0001
Unit 0    Line 11   Terminal      tty
Unit 1    Line 2    AML File      /amlwb/getdrawcover.aml
Unit 0    Line 1    Form Menu     /amlwb/getdrawcover.menu
Unit 0    Line 1    Form Menu     /amlwb/tools_icon.menu
```

All input sources appear on the &TB list, but you can only interact with the input source at the bottom of the list (i.e., TOOLS_ICON.MENU) because it's on top of the stack.

Question 13-1: If a menu on top of the stack executes an AML program, does the menu disappear until the program finishes and then reappear?

(Answer on page 13-14)

Using multiple threads

Because AML handles multiple threads, you can generate multiple stacks containing AML programs, menus, and TTYs as sources of input. Having more than one thread means you can have more than one active menu with which users can interact. You can design a full-screen application instead of the hide-and-seek interfaces you've already seen (i.e., the menu appears, disappears, reappears, etc.).

Although ARC/INFO allows twenty active threads, the AML processor handles only one source of input at a time. In addition, AML only processes one action at a time for any of the three types of input sources—TTYs, AML menus, and AML programs (e.g., enter a command at the keyboard, execute a menu widget, or run an AML program that supplies commands as if you were typing them). Consider, for example, what would happen if two AML programs could run at the same time. The AML processor couldn't know which program supplied a line of code or where to send the output. The result would resemble placing your code in a blender and pressing the puree button.

AML avoids this by processing only a single input source at a time. In the case of multiple threads where multiple input sources are invoked, remember this rule:

- An AML program or TTY on top of any thread stack takes precedence over all other threads.

Therefore, if an AML program on one thread creates a new thread that runs another AML program, the first program pauses until the second one finishes (like the single-thread model). Knowing this process allows you to create applications that present multiple menus and input sources to the user at one time.

Returning to the example, the goal is to get both GETDRAWCOVER.MENU and TOOLS_ICON.MENU on the screen at the same time. This requires creating a new thread using the &THREAD directive. The usage for &THREAD &CREATE is as follows:

```
&THREAD &CREATE {thread} {&MODAL} <&RUN <args> | &MENU <args> |
&TTY {args}>
```

AML gives default names to threads if you don't name them. Although naming your threads is optional, it's recommended. You'll find threads with user-specified names easier to manage than those with default names.

When ARC/INFO starts, AML always provides a thread, usually named THREAD0001:

GETDRAWCOVER.AML is executed from the command line as follows:

```
Arc: &run getdrawcover.aml
```

Instead of invoking GETDRAWCOVER.MENU on the existing thread, in this example GETDRAWCOVER.AML creates a new thread called tool$getdrawcover:

```
&thread &create tool$getdrawcover &menu getdrawcover.menu
```

Now GETDRAWCOVER.MENU appears on the screen, but it's busy (indicated by the pointer turning into a watch) so the user can't interact with it:

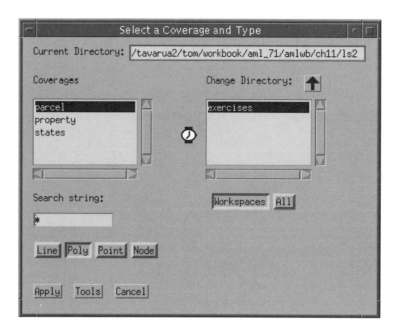

Examining the existing threads reveals why the user can't interact with the menu. The thread `tool$getdrawcover` has GETDRAWCOVER.MENU on top of its stack and `thread0001` has GETDRAWCOVER.AML on top. Because an AML program or TTY on top of a stack takes precedence over other threads, GETDRAWCOVER.AML continues to run and GETDRAWCOVER.MENU can't be accessed. For this reason, control returns to GETDRAWCOVER.AML, which launches another thread, `view$tools`:

```
&thread &create view$tools &menu tools_icon.menu
```

The new thread carries TOOLS_ICON.MENU as its input source. There are now two menus on the screen, but because GETDRAWCOVER.AML is still on top of its stack, both are busy. The result is nice for show, but lacks functionality.

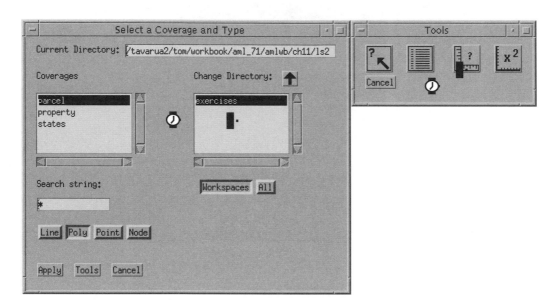

GETDRAWCOVER.AML needs to move from the top of its stack to make these menus active. Using an &RETURN doesn't work because a TTY replaces the AML program and the two menus remain busy. One solution is to have GETDRAWCOVER.AML launch a menu so each of the three threads have menus on top. This allows the user to interact with all three menus. This works until the user dismisses the menu on thread0001, returning control to GETDRAWCOVER.AML, and making the menus on the other two threads busy again.

A better solution, illustrated below, is to have GETDRAWCOVER.AML delete the thread it currently occupies. This leaves two threads with a menu on top of each.

```
&thread &delete &self
```

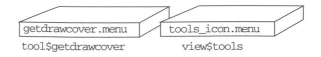

Because the two remaining threads have menus on top, they both can accept user
input:

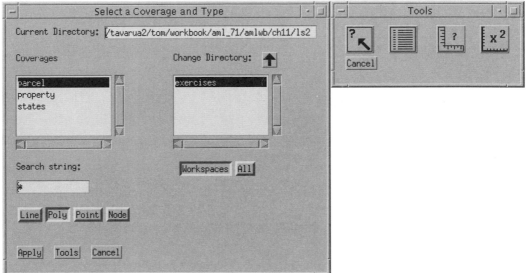

Selecting the Cancel button from the TOOLS_ICON.MENU removes the menu
from its thread with &RETURN and AML deletes the thread. AML automatically
deletes a thread created with the &THREAD directive when it no longer contains
an input source.

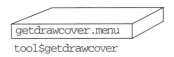

tool$getdrawcover

Finally, pressing the Cancel button on the GETDRAWCOVER.MENU deletes
the last thread, meaning that no input sources now exist. Fortunately, AML
knows this and creates a new thread with a TTY as input:

thread0002

Examine the usage of &THREAD &DELETE:

```
&thread &delete <thread | &SELF | &ALL | &OTHERS>
```

You can delete threads using the four options as follows:

- &THREAD &DELETE THREAD deletes a named thread.

  You can use this option to delete any thread by name, including the one executing the &THREAD &DELETE statement. EXIT routines often delete threads by name. The following example shows how to use the [SHOW] function to determine the names of program-created threads and then delete them:

  ```
  &routine EXIT
  &if [show &thread &exists edit$select] &then
    &thread &delete edit$select
  &return
  ```

  Knowing thread names helps you manage them. Be sure to clean up your application's environment. If you start with one thread, there should be only one thread remaining when the program finishes.

- &THREAD &DELETE &SELF deletes the thread that executes the &THREAD &DELETE statement.

  LAUNCH.AML gives an example of how this option is often used in the AML program that launches an application:

  ```
  launch.aml
  &thread &create browse &menu browse
  &thread &create ident &menu ident
  &thread &create disp &menu disp
  &thread &delete &self
  &return
  ```

Invoking all the menus at once and deleting the original thread is just one way to launch an application. It's not necessarily the best way to perform this function. How you execute this task depends on your application.

- &THREAD &DELETE &ALL deletes all threads and creates a TTY thread.

  This option is useful in the debugging stage as well as when an application finishes. You might find it convenient to combine this usage with the &PINACTION option of the &MENU directive as follows:

  ```
  &thread &create browse &menu browse ~
     &pinaction '&thread &delete &all; &run start cleanup'
  ```

  When used this way, launching the menu BROWSE.MENU creates the thread named BROWSE. When the menu is dismissed, the &PINACTION argument terminates the application by deleting all threads and runs the START.AML program, passing it the routine name CLEANUP.

- &THREAD &DELETE &OTHERS deletes all threads *except* the one that executes the &THREAD &DELETE statement.

  This option is useful for debugging an application because during development you're often left with busy menus and a prompt. In this case, you have two options:

  ```
  &thread &delete &self
  ```

  or

  ```
  &thread &delete &others
  ```

  &THREAD &DELETE &SELF deletes the TTY and activates your menus (unless other problems exist). &THREAD &DELETE &OTHERS removes the menus and leaves the TTY so you can repair the application and run it again.

The other options of &THREAD, &FOCUS, and &SYNCHRONIZE (discussed in lesson 13.2) have the same keyword options discussed here.

FYI

> In addition to the AML limit of twenty threads at one time, ARC/INFO limits to twenty the number of open display canvases allowed at one time. Each scrolling list of symbols created by AML is considered a display canvas. Also included in this limit are the ARCPLOT and ARCEDIT display canvases, and the canvases created with the WINDOW and IMAGEVIEW commands.

In summary

A *thread* carries input from an AML program, a TTY, or an AML menu to the AML processor. Input from these sources stack on the thread with the most current at the top of the stack. With a single thread, the user can interact with only one input source at a time (i.e., the one at the top of the stack).

When you start ARC/INFO, AML automatically creates a thread with a TTY as its input source. You must use the &THREAD directive to create and delete any additional threads. Multiple threads allow you to display and access many menus at one time. A thread with an AML program or TTY at the top of a stack takes precedence over all other threads.

Multiple threads need to be managed. In the next lesson, you'll learn more about multiple threads and how to manage them to obtain the results you want.

Answer 13-1: No. The menu remains on the screen, but it's busy until the AML program finishes.

Exercises

Exercise 13.1.1

Write a command to launch a menu called EDIT_ARC.MENU on a thread called EDIT_ARC.

Exercise 13.1.2

Write an &DO loop that tests whether the threads EDIT_ARC, EDIT_SELECT, and DRAWENV exist and, if so, deletes them.

Exercise 13.1.3

An AML program contains the following code. Draw a picture of the resulting thread stacks and identify the threads with which the user can interact.

```
&thread &create displ &menu displ.menu
&thread &create ident &menu ident.menu
&thread &create zoom &menu zoom.menu
&thread &delete &self
```

Exercise 13.1.4

If your application bails out and leaves you with two busy menus and a prompt, which two commands could you type to clean up the environment?

Lesson 13.2—Managing threads

In this lesson

Thread management becomes vital when you use more than one thread. This lesson covers an important thread management technique known as focusing.

When only one menu at a time is accessible to the user, it's easy for the user to know what to do. When multiple menus appear on the screen, however, it's hard for the user to know which menu to use. Focusing ensures that the user interacts with the menus in the appropriate order.

Focusing in on threads

Focusing a thread allows the user to interact with the input source on top of that thread. By default, every thread is focused when you create it, but you may not be able to interact with it because of the current thread environment. For example, a new thread is busy, or *unfocused,* if you create it when there's another thread with an AML program or TTY on top of its stack—a situation you saw in the previous lesson.

Consider a situation where several threads have menus at the top. You want the user to interact with some, but not all, of them. For example, you don't want a user to interact with a menu that performs an IDENTIFY function without first selecting a coverage and setting a MAPEXTENT. Focusing and unfocusing menus can direct the user as to what to do next. If only one menu among several is focused, the user easily recognizes it as the one to use.

The following example applies focusing to the display and query application developed in the previous lesson. So far, the application looks like this:

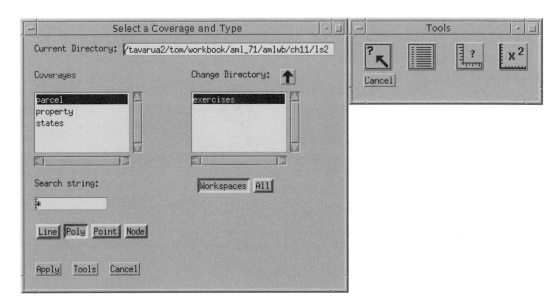

The menu on the left (GETDRAWCOVER.MENU) allows the user to select a coverage and a feature class. Pressing the Apply button draws them. The menu on the right (TOOLS_ICON.MENU) contains the tools IDENTIFY, LIST, MEASURE, and STATISTICS, from left to right. As the application stands, the user could cause an error by selecting the tool to execute an IDENTIFY without first defining a coverage and feature class.

It's the menu designer's job to foresee this kind of error and prevent it from occurring. Keeping the TOOLS_ICON.MENU unfocused (i.e., busy) until the user selects a coverage and feature class eliminates this error.

In this case, the application looks like this when it's invoked:

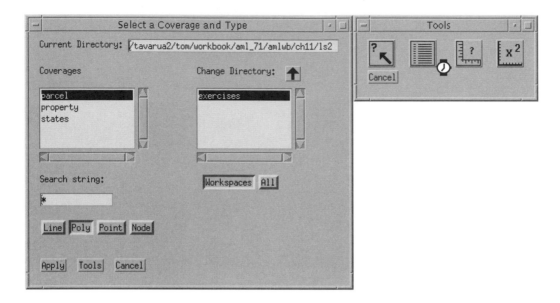

Take another look at the application beginning at the point when
GETDRAWCOVER.AML creates the thread view$tools:

```
&thread &create view$tools &menu tools_icon.menu
```

Next, GETDRAWCOVER.AML deletes its own thread, causing the other two
threads to focus and giving the user access to both menus.

The user, however, shouldn't gain access to the TOOLS_ICON.MENU until choosing a coverage and feature class from GETDRAWCOVER.MENU. You can correct this situation using &THREAD &FOCUS as follows:

```
&thread &focus <&ON | &OFF> <thread | &SELF | &ALL | &OTHERS>
```

As you can see, the options `<thread | &SELF | &ALL | &OTHERS>` are also available for &THREAD &DELETE. See page 13-12 if you need to review these options.

GETDRAWCOVER.AML needs to execute the following statement to turn the focus off the `view$tools` thread:

```
&thread &focus &off view$tools
```

Next, GETDRAWCOVER.AML deletes its own thread:

```
&thread &delete &self
```

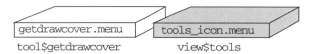

The environment shown in the graphic on page 13-18 is now set. The menu on the right, TOOLS_ICON.MENU, remains unfocused while the user interacts with GETDRAWCOVER.MENU.

In this version, pressing the Apply button on GETDRAWCOVER.MENU runs GETDRAWCOVER.AML on its own thread, passing a routine name (i.e., apply) as an argument. (See chapter 8, lesson 3, if you need to review passing arguments in the &RUN statement.)

```
&run getdrawcover apply
```

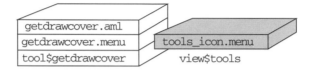

GETDRAWCOVER.AML sets the MAPEXTENT and displays the chosen coverage and feature class. With these defined, it's all right to give the user access to the TOOLS_ICON.MENU. GETDRAWCOVER.AML refocuses the view$tools thread and &RETURNs GETDRAWCOVER.MENU to the top of the stack.

```
&thread &focus &on view$tools
&return
```

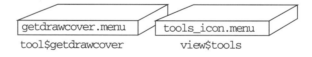

The user can now interact with both menus as needed.

FYI

> If you don't refocus other threads before deleting the currently active thread (&SELF), you receive the following message:
>
> ```
> AML WARNING - Can't select a thread, creating new one
> Arcplot:
> ```
>
> At this point, you have the following options: (1) issue &thread &delete &others and start over, or (2) issue &thread &focus &on &others and &thread &delete &self to return to the rest of the application.

Thread modality

Consider a menu that allows the user to enter a data source in a text input field. Needing assistance, the user presses the 3 button on the mouse to display a browse menu that allows the user to look around the system to find and select the needed information (e.g., a coverage name).

To ensure that the user defines the data source with the browse menu before doing anything else, you can make it *modal*. When a menu is modal, all other threads in the application remain unfocused until the interaction with the modal menu is finished. Conditions return to their previous state when the modal thread is deleted.

Modal threads are created using the &THREAD &CREATE with the &MODAL option. The usage is as follows:

```
&thread &create {thread} {&MODAL} <&RUN <args> |
                &MENU <args> | &TTY {args}>
```

The &MODAL option forces all input to the named thread. Regardless of what's on top of other thread stacks, no interaction is allowed with the rest of the application until the user deals with the modal thread. Therefore, if an AML program is on top of one thread and it creates another modal thread containing a menu, the AML program pauses until the modal thread is deleted.

In the application, the user can't access the TOOLS_ICON.MENU until he or she chooses a coverage and a feature class from GETDRAWCOVER.MENU. After the TOOLS_ICON.MENU is focused, the user can select the STATISTICS button (the button on the far right) to invoke the STATISTICS tool shown here:

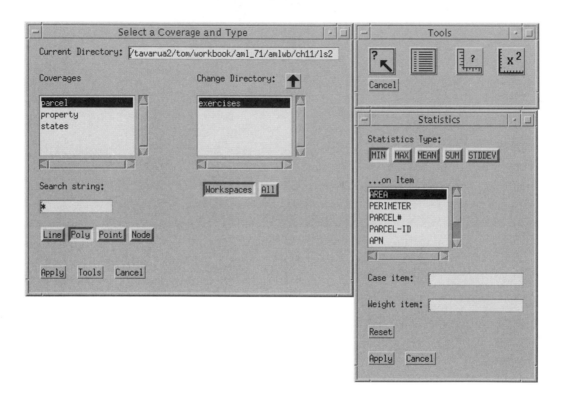

STATISTICS.AML accepts the following arguments:

```
&args routine coverage fclass itemlist position stripe
```

When STATISTICS.AML launches STATISTICS.MENU, the data in the variable ITEMLIST is displayed in a scrolling list. The fact that data isn't passed globally from GETDRAWCOVER.MENU to STATISTICS.MENU (i.e., they don't share local variables) can lead to a problem of consistency between menus. This means that when the user selects a new coverage, the STATISTICS.MENU isn't updated. This is illustrated below:

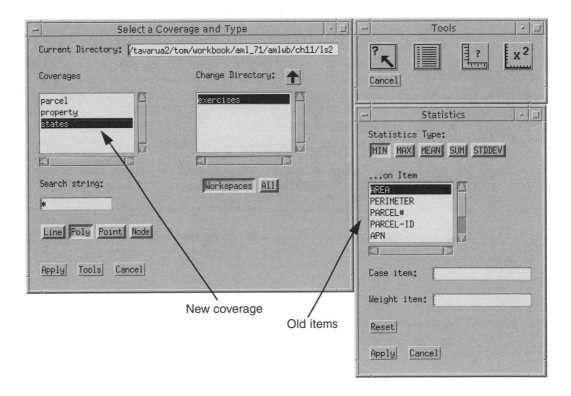

New coverage

Old items

Using &THREAD &CREATE with the &MODAL option allows you to unfocus GETDRAWCOVER.MENU and TOOLS_ICON.MENU until the user finishes interacting with the tool selected from the TOOLS_ICON.MENU. This is accomplished as follows:

```
&thread &create tool$statistics &modal &menu statistics
```

This produces the desired result, illustrated here:

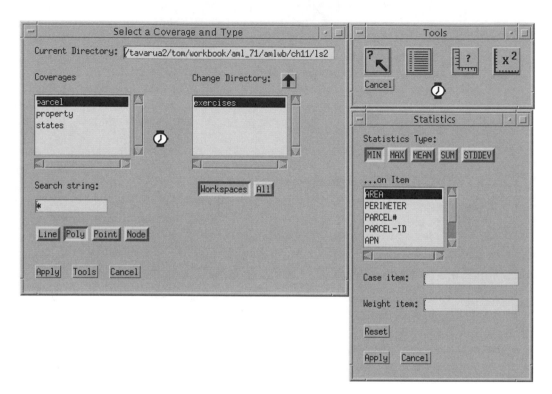

When the Cancel button deletes the thread tool$statistics, AML restores the previous thread environment. Previously focused menus are refocused and previously unfocused menus stay that way.

The previous example demonstrates how you can manage the threads of interdependent menus. Manage the focus for the TOOLS_ICON.MENU by requiring that the user define a coverage and feature class before using the tools. The example also demonstrates how to focus threads to force the user to interact with a chosen tool before selecting new data. Focusing is one way of managing threads to ensure the user interacts with the application properly. Although this application could use global variables to remove the necessity for restricting the user's interaction with the menus, you should realize that thread management is always needed in complex applications.

Synchronized threading

Synchronization is another thread management tool. As you saw in the last example, menus often share data. A focused menu updates its display as data changes; unfocused menus don't. An unfocused menu isn't updated until it's focused. When focused and unfocused menus share variables, your application can display current data in one menu and outdated data in another. In addition, when a menu invokes an AML program that changes its data, the new data isn't displayed until the program finishes executing because the menu is unfocused while the program runs.

Synchronizing a thread updates the data in an unfocused menu without having to refocus it. This allows you to update your menu based on what's happening in your AML program. One example would be a slider bar that moves as an AML program progresses:

```
/* timer.aml
&r $ATHOME/lib/setpaths /* to access arctools default menus

&thread &create tool &r msworking init ~
'One moment...processing data'
&thread &create slider &m slider.menu ~
  &position &ul &thread tool$msworking &ll ~
  &size [show &thread &size tool$msworking] ~
  &stripe 'Percent Done:'

&do count = 1 &to 100 &by .5
&thread &synchronize slider
&end
```

Below is an example menu with only one widget—a slider bar with a variable called count.

```
7 slider.menu
  %list
%list SLIDER COUNT 70 TYPEIN NO STEP .1 REAL 0 100
```

At each pass through the loop, the menu SLIDER.MENU is synchronized with TIMER.AML. Because a menu that's launched from an AML program shares local variables, the menu can illustrate the progression of the loop.

Executing TIMER.AML yields these results:

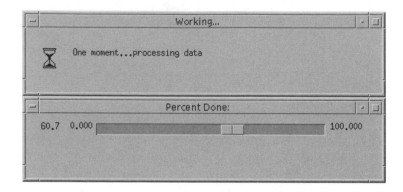

In the next example, you'll see an AML program that updates a form menu with status messages. The form sets up the buffering environment and executes BUFFERCOV.AML to do the actual buffering:

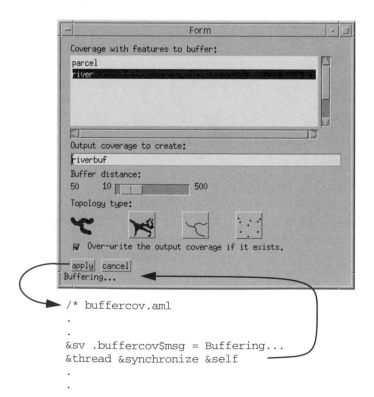

```
/* buffercov.aml
  .
  .
&sv .buffercov$msg = Buffering...
&thread &synchronize &self
  .
  .
```

Here you have BUFFERCOV.MENU running BUFFERCOV.AML. In this case, local variables won't work. You have to use the global variable .buffercov$msg to communicate between the AML program and the menu.

```
7 buffercov.menu
 Output coverage to create:
 %2
 Buffer distance:
 %3
 Topology type:
 %4          %5          %6          %7

 %8 Over-write the output coverage if it exists.
 %b4    %button5
%1 INPUT INCOV 60 TYPEIN NO SCROLL YES ROWS 6 COVER *
%2 INPUT OUTCOV 60 TYPEIN YES SCROLL NO CHARACTER
%3 SLIDER DISTANCE 30 TYPEIN NO ~
   STEP 10 ~
   INTEGER 10 500
%4 DISPLAY .buffercov$iconname 3 ICON

/* Other field definitions go here.

%formopt messagevariable .buffercov$msg
```

Other variables that you want to retain while the menu is being synchronized must also be global variables. The display field is displaying an icon name stored within the variable .buffercov$iconname. If this variable isn't global, the icon disappears the first time the menu is synchronized.

In summary

Use the &THREAD directive with the &FOCUS option for managing threads. The user can intcract only with focused threads. By controlling the focus of threads, you can ensure that the user performs tasks in the order your application requires.

Like focusing, thread modality allows you to control how the user interacts with the application. Create modal threads using the &THREAD directive with the &MODAL option. The user must interact with a modal thread first, regardless of what's on top of the other thread stacks.

Exercises

Exercise 13.2.1

If an AML program executes the following code, what happens if you dismiss both menus?

```
/* edit.aml
.
.
&thread &create select &menu select.menu
&thread &create drawenv &menu drawenv.menu
&thread &focus &off &self
```

Exercise 13.2.2

How can you ensure that a menu on a new thread is the only one focused?

Exercise 13.2.3

Why would you use the &MODAL option to &THREAD &CREATE instead of managing thread focusing with &THREAD &FOCUS?

Exercise 13.2.4

Assume that you have two menus, MENU1 and MENU2, each on its own thread. The threads share the same names as their menus. MENU1 is focused and MENU2 isn't. MENU1 allows the user to type the name of a coverage to create (stored in a variable called .CREATE$NEWCOV). MENU1 runs CREATECOV.AML when an OK button is selected. When the user types the coverage name, CREATE$NEWCOV, MENU2 should be focused. Write the code that CREATECOV.AML needs to execute to do this.

Lesson 13.3—Managing a real-world application

In this lesson

The previous lessons showed how a single-thread application can evolve into a multiple-thread application with threads managed through the technique of focusing. To give you a real-world view of how to use &THREAD in your applications, this lesson examines the AML code used in the routines found in the example application's AML programs.

While viewing the examples in lessons 13-1 and 13-2, you may have noticed that the executable statements didn't exactly match the result shown in the menu displays (i.e., menu stripe, positioning, etc.). This was done intentionally to simplify the application for use as a teaching aid. This lesson examines the actual AML code needed to use and manage threads in a real-world application.

 FYI

The CD–ROM included with this workbook includes the complete code for the programs and menus shown in this lesson. You many want to examine it to see where the code pieces shown in this lesson fit.

Examining the display and query application

The ARCPLOT display and query application you saw in the previous lessons consists of a number of tools made of pairs of AML programs and menus, shown in the following table:

AML program	AML menu
getdrawcover.aml	getdrawcover.menu
tools.aml	tools_icon.menu
statistics.aml	statistics.menu
measure.aml	measure.menu

This is the ArcTools programming style introduced in chapter 8 and is examined fully in chapter 15. Each of the AML programs in the application contain many routines that the menus access.

For review, the application interface looks like this:

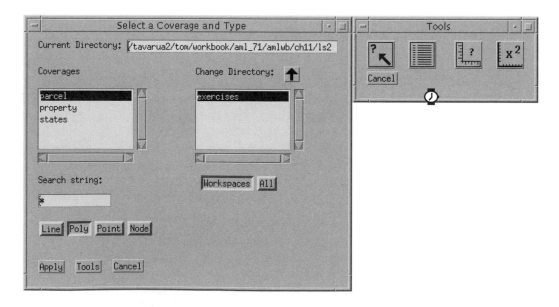

All of the tools in the application accept both a routine name and an optional position and stripe as arguments from the &RUN statement. For example, the following code makes up the body of GETDRAWCOVER.AML:

```
&args routine wildcard type position stripe
&severity &error &routine bailout
/* Check arguments
&if [NULL  %routine%] &then
  &call usage
&call %routine%
&return
```

The rest of the program is routines. GETDRAWCOVER.AML is initially run by passing a routine named INIT. This routine initializes variables, including those that specify the menu position and stripe (if they weren't initialized with the &RUN statement):

```
&if [NULL %position%] or %position%_ = #_ &then
  &set position - &ul &screen &ul
&if [NULL %stripe%] or %stripe%_ = #_ &then
  &set stripe = Select a Coverage and Type
```

INIT uses these values to create a thread:

```
&thread &create tool$getdrawcover ~
  &menu getdrawcover.menu &position [unquote %position%] ~
  &stripe [quote [unquote %stripe%]] ~
  &pinaction '&run getdrawcover cancel'
```

GETDRAWCOVER.AML runs TOOLS.AML, passing the appropriate routine name (i.e., init), positioning argument (i.e., &ul &thread tool$getdrawcover &ur), and menu stripe (i.e., Tools) as follows:

```
&run tools init '&ul &thread tool$getdrawcover &ur' Tools
```

Like GETDRAWCOVER.AML, TOOLS.AML contains an &ARGS statement to accept a routine name, menu position, menu stripe, and the code needed to call the named routine:

```
&args routine position stripe
&severity &error &routine bailout
&if [NULL %routine%] &then
  &call usage
&call %routine%
&return
```

The INIT routine in TOOLS.AML creates a new thread called view$tools:

```
&thread &create view$tools ~
    &menu tools_icon ~
    &position [unquote %position%] ~
    &stripe [quote [unquote %stripe%]]
```

TOOLS.AML then &RETURNs to GETDRAWCOVER.AML, which unfocuses the view$tools thread before deleting its own thread.

```
&thread &focus &off view$tools
&thread &delete &self
```

Choosing a coverage and feature type from the GETDRAWCOVER.MENU and pressing the Apply button runs GETDRAWCOVER.AML, which executes the following routine:

```
/*-------------------
&routine APPLY
/*-------------------
/* Display coverage and focus tools
&if ^ [null %.getdrawcover$cover%] and ~
    ^ [null %.getdrawcover$type%] &then &do
  CLEAR
  MAPEXTENT %.getdrawcover$cover%
  %.getdrawcover$type%S %.getdrawcover$cover%
  &if [show &thread &exists view$tools] &then
    &thread &focus &on view$tools
&end
&return
```

This code displays the chosen coverage, determines if the view$tools thread exists, and, if so, allows the user to interact with the TOOLS_ICON.MENU by refocusing the thread. TOOLS_ICON.MENU allows the user to perform four functions. Each of the buttons runs TOOLS.AML passing a routine name. TOOLS.AML executes the commands IDENTIFY and LIST. TOOLS.AML executes other AML programs that, in turn, create new threads to perform the MEASURE and STATISTICS operations.

The following diagram illustrates this process:

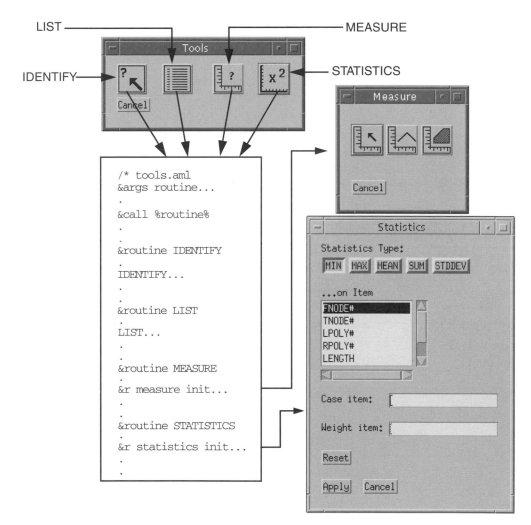

Pressing the statistics icon on TOOLS_ICON.MENU executes TOOLS.AML.
The program receives STATISTICS as the routine name along with positioning
and stripe information to pass to STATISTICS.AML with an &RUN statement.

Before running STATISTICS.AML, TOOLS.AML translates the global variable set by GETDRAWCOVER.MENU to local variables as follows:

```
/*-----------
&routine TRANSLATE
/*-----------
&sv source = %.getdrawcover$cover%
&sv class = %.getdrawcover$type%
&return
```

STATISTICS.AML calls its INIT routine to initialize the local variables and calls the MENU routine to launch a modal thread named tool$stats. This thread contains the menu STATISTICS.MENU as the input source. The following diagram shows what happens when the statistics icon is chosen from the Tools menu:

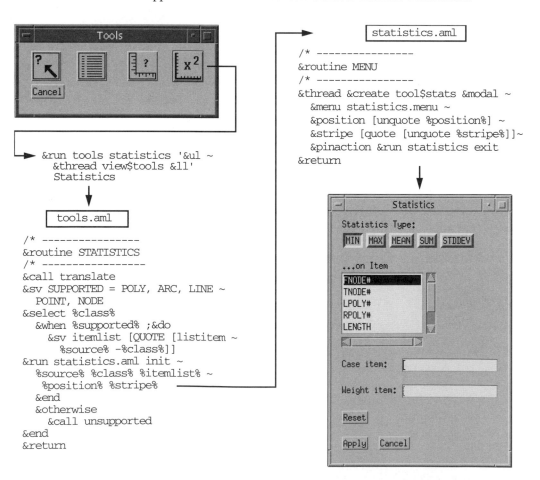

The data that passed to TOOLS.AML and then STATISTICS.AML (i.e., &ul &thread view$tools &ll) positions the tool$stats thread. This illustrates how you can position menus relative to other threads. The new menu is positioned relative to the view$tools thread as shown here:

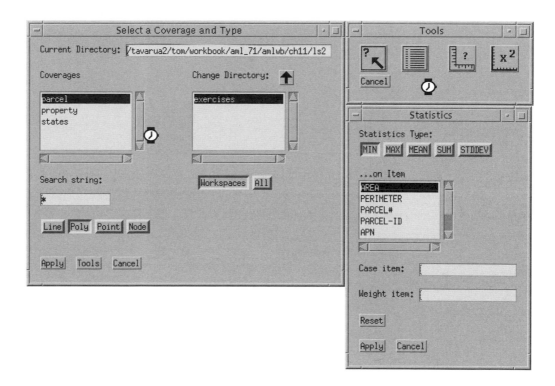

Pressing the Cancel button dismisses the Statistics menu. The Cancel button executes the following return string:

```
&return; &run statistics exit
```

STATISTICS.MENU was the only input source on the thread named tool$stats, so the &RETURN causes the thread to be deleted. Because tool$stats was launched as a modal thread, removing it causes AML to restore the previous focusing environment. In this case, AML refocuses the threads named tool$getdrawcover and view$tools. STATISTICS.AML runs and the EXIT routine deletes all global variables associated with the tool.

The Cancel buttons in the other menus perform similarly; they &RETURN to remove the thread and delete global variables.

In summary

Developing applications using threads adds a level of complexity to managing the user's interaction with the application. The &THREAD directive is used to invoke and remove AML programs and menus. This directive is also useful for managing the order in which a user interacts with the menus. The ability to focus and unfocus menus allows you to guide the user through your application in a way that avoids errors and ensures the result you intend. Use &THREAD in conjunction with routines and &ARGS in multiple programs to develop your applications.

Exercises

Exercise 13.3.1

Assume you have an existing threaded menu called `disp`. Write the command that invokes SELECT.MENU on a new thread called `select` and positions it on the right side of the existing menu with their tops aligned.

Exercise 13.3.2

Choosing `Attribute` from the Polygon Display Properties menu runs GETATTRIBUTE.AML. The program creates a thread named `tool$getattribute` and launches the focused menu shown below. The menu is for selecting an item and/or a lookup table with which to shade the polygons. This thread needs to be modal to force the user to interact with it before using any of the widgets on the Polygon Display Properties menu.

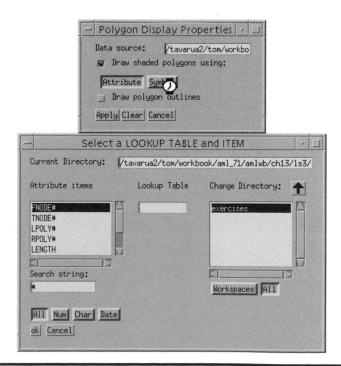

Examine the end of the INIT routine shown below. Fill in the missing code to determine whether the thread `tool$getattribute` exists and, if it does, delete it. Launch GETATTRIBUTE.MENU on the modal thread `getattribute`.

```
/*-------------------
&routine INIT
/*-------------------
/*    .
/*    .
/*    .
&if _____ &then

    _____
    _____ ~
    _____ ~
    &position [UNQUOTE %position%] ~
    &stripe [QUOTE [UNQUOTE %stripe%]]
    &pinaction '&run getattribute cancel'
&return
```

Exercise 13.3.3

If an AML program includes the following statements, what's the resulting thread environment?

```
/* some.aml
    .
    .
    .
&thread &create menu1 &menu menu1.menu
&thread &create menu2 &menu menu2.menu
&thread &create &modal menu3 &menu menu3.menu
```

What happens if the user dismisses MENU3? Which statement should be present in the AML program to ensure that the user can interact with MENU1 and MENU2 after MENU3 is dismissed?

 **Enhanced pulldown menus**

Pulldown menus perform an important role in application development. Many applications allow you to interact with the software solely through the use of menus. As you saw in chapter 2, AML provides a wealth of simple menus like pulldown, sidebar, and matrix menus. These menus allow you to make one selection from a set of choices.

Enhanced pulldown menus include and expand upon the functionality of pulldown, sidebar, and matrix menus. Enhanced pulldown menus allow you to pull choices down from the main bar (as in Type 1 menus), and also to pull choices to the right (as in Type 2 menus). Choices on an enhanced pulldown menu can have accelerator (shortcut) keys and can be grayed out if not applicable. These menus can also be incorporated into form menus, allowing you to combine the functionality of form menus with the added capabilities of enhanced pulldown menus.

All of the menu building in this chapter is demonstrated with the enhanced pulldown menu editor, MenuEdit. MenuEdit, like FormEdit, is another powerful tool for building application components graphically.

This chapter relies on information presented in chapter 12. Before continuing, you should be familiar with form menus. Most of lesson 14.3 assumes that you have a working knowledge of threads, which were covered in chapter 13.

This chapter covers the following topics:

Lesson 14.1—Introduction to enhanced pulldown menus

- Enhanced pulldown menu basics
- Introduction to MenuEdit

Lesson 14.2—Creating enhanced pulldown menus with MenuEdit

- Building a simple enhanced pulldown menu
- Looking at the code

Lesson 14.3—MenuEdit power tools

- Using and creating accelerator keys
- Graying out choices
- Embedding a menu in a form

Lesson 14.1—Introduction to enhanced pulldown menus

In this lesson

Enhanced pulldown menus resemble the pulldown and sidebar menus covered in chapter 2. They combine the functionality of pulldown and sidebar menus into a powerful menu type that can be used in a wide range of applications.

In chapter 2, you learned that menus have a strict syntax and coding procedures that must be rigidly followed. Enhanced pulldown menus can be completely constructed using an editor called *MenuEdit*. MenuEdit makes the generation of enhanced pulldown menus much simpler and faster than using a text editor to write the menu file.

Enhanced pulldown menu basics

Enhanced pulldown menus have three kinds of items: Menu Bar Items, Pull Right Items, and Menu Items. The *Menu Bar Items* are the main items across the top of the menu. Unlike pulldown menus that can execute an action directly from the Main Menu Bar, enhanced pulldown menus have a submenu associated with each Menu Bar Item choice. *Pull Right Items* are submenus underneath each Menu Bar Item or other Pull Right Item. When a Pull Right Item is selected, it displays a submenu of choices to its right. *Menu Items* execute a return string and can be placed beneath Menu Bar Items or Pull Right Items. The diagram below shows the three different items found in an enhanced pulldown menu:

Introduction to MenuEdit

MenuEdit is the easiest and fastest way to start creating enhanced pulldown menus. MenuEdit provides a graphical interface where, just as in FormEdit, the code for a menu is created as you build it. The following diagram points out some of the key parts of MenuEdit:

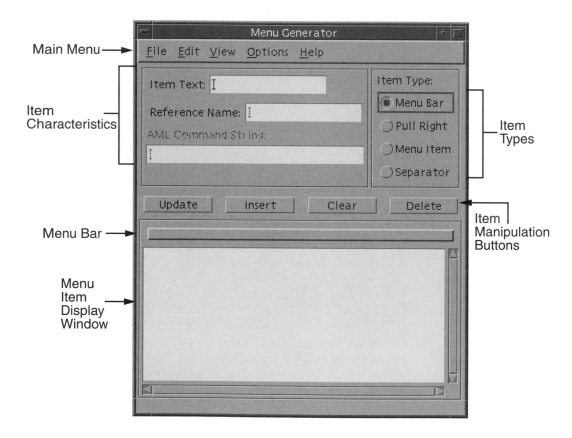

The Menu Item Display window at the bottom of the form lists each element (Menu Bar Items, Pull Right Items, and Menu Items) included in the enhanced pulldown menu. The Menu Bar shows the menu as it will appear on your screen. The type and characteristics of the item selected in the Menu Item Display window are given at the top of the form. In the next lesson, you'll learn how to use each part of MenuEdit to create an enhanced pulldown menu.

 FYI

The Menu Bar is not present on the Windows NT version of MenuEdit. To view your menu in Windows NT, press the Test button. A sample menu displays. This is similar to the Test option in FormEdit.

In summary

Enhanced pulldown menus provide more functionality than pulldown, sidebar, and matrix menus. MenuEdit allows you to create enhanced pulldown menus quickly through a graphical user interface.

Lesson 14.2—Creating enhanced pulldown menus with MenuEdit

In this lesson

Like form menus, enhanced pulldown menus have an editor, called MenuEdit, that allows you to create your own menus. In this lesson, you'll use MenuEdit to create a menu (ACTION.MENU) that includes some of the functionality of an enhanced pulldown menu. You'll also take a quick look at the code MenuEdit creates.

The focus of this lesson is on the creation of the menu layout, not on the commands that the menu choices execute. The completed ACTION.MENU will look like this:

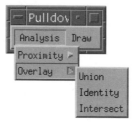 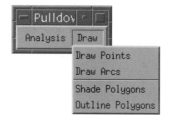

Building a simple enhanced pulldown menu

MenuEdit is used throughout the rest of this lesson to create ACTION.MENU. To start MenuEdit, simply type `menuedit` at the `Arc:`, `Arcedit:`, `Arcplot:`, or `Grid:` prompt.

The first step in making an enhanced pulldown menu is to create the Menu Bar Items. These are the main headings across the top of the menu that lead to more choices. The following diagram illustrates how Menu Bar Items can be added to an enhanced pulldown menu:

1. In the Item Type box, check `Menu Bar`.
2. In the Item Text input field, type `Analysis`. (In this example, this will be the text of the Menu Bar Item.)
3. Click the Insert button to display the new Menu Bar Item in the Menu Item Display window and on the Menu Bar.

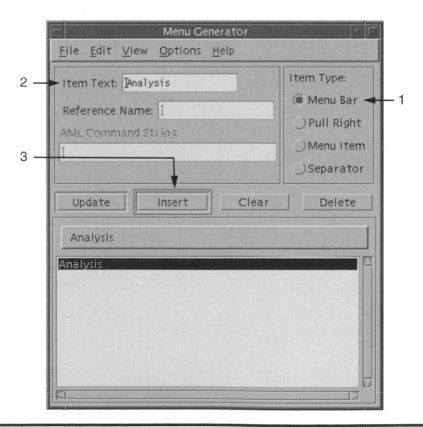

You have now created the first Menu Bar Item in your menu. Repeat steps 1–3 to add a Menu Bar Item called Draw. Now if you save and launch ACTION.MENU it will look like this:

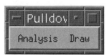

This is nice, but the Menu Bar Items don't have any choices beneath them. Next, you'll add some content to this menu with Menu Item choices. Note that this is an intermediary step in the construction of the final enhanced pulldown menu; later, you'll group the Menu Item choices under Pull Right Items.

To add Menu Item choices, follow the same procedure as above, but choose Menu Item for the Item Type.

First, in the Menu Item Display window, highlight the Menu Bar Item to which you want to add a Menu Item. The Menu Item will appear beneath it. For ACTION.MENU, the following Menu Items need to be inserted:

Under Analysis:	Under Draw:
Buffer	Draw Points
Identity	Draw Arcs
Intersect	Shade Polygons
Near	Outline Polygons
Union	

Once the Menu Item choices are inserted beneath the Menu Bar Items,
ACTION.MENU should look like this:

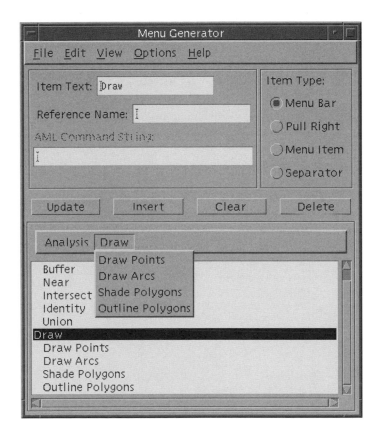

In the example, there are two Menu Bar Items, each with several Menu Item
choices. You can test how the menu will look by clicking on the Menu Bar and
pulling down the choices.

Although the menu looks good, it doesn't reflect the fact that there are really two
separate categories of analysis: overlay and proximity. To create a more
organized menu, you'll use Pull Right Items to group similar Menu Items.

To create a new Pull Right Item, follow the same procedure as for the Menu Bar Item, but select the Pull Right Item item type. First, in the Menu Item Display window, select the Menu Bar Item under which you want to add your Pull Right Item.

When you create a new Pull Right Item, by default a Menu Item called MenuItem is displayed underneath it. This item serves as a placeholder and shouldn't be deleted. (Deleting all Menu Items beneath a Pull Right Item turns the Pull Right Item itself into a Menu Item.)

For ACTION.MENU, these Pull Right Items need to be inserted:

Under Analysis:
Proximity
Overlay

After you add the new Pull Right Items, ACTION.MENU looks like this:

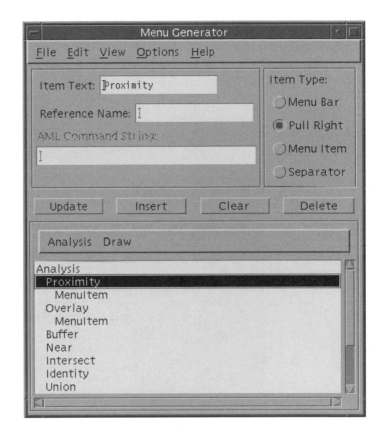

Now, using the Edit tools from the Main Menu, you can cut and paste Menu Items in the Display window to organize them under Pull Right Items:

Before you paste a Menu Item under a Pull Right Item, be sure to select an existing Menu Item underneath the Pull Right Item. Once you cut the Menu Items and paste them under the Pull Right Items, you can delete the placeholders that were created by default.

With the Menu Item choices now appropriately placed under their Pull Right Items, ACTION.MENU looks similar to this:

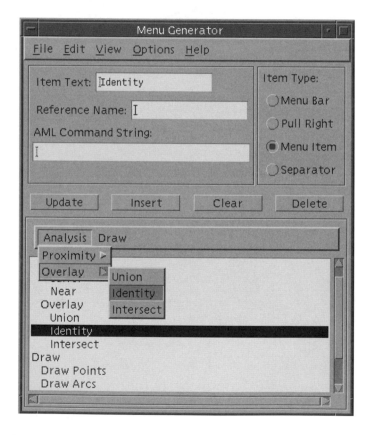

Notice that the items are automatically indented in the Menu Item Display window.

Another way to organize Menu Items is with a separator bar. Suppose, on the Draw menu, you'd like to separate the choices pertaining to polygons from the others. In the Display window, highlight the Menu Item that you want the Separator bar to follow. In the Item Type box, choose Separator, then click the Insert button. Now you can pull down the Draw menu on the Menu Bar to see the result:

Now that ACTION.MENU is formatted just the way you want, you're ready to use it in an application. Well, not quite; ACTION.MENU doesn't do anything yet. You need to incorporate return strings for your menu choices. To add return strings, follow these steps:

1. Select a Menu Item in the Menu Item Display Window.

2. Enter a return string in the AML Command String input field.

3. Click the Update button.

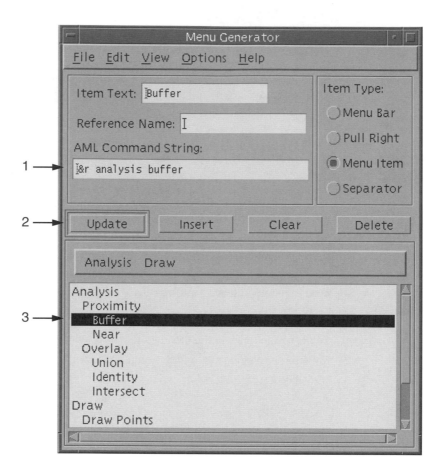

The return string in the example implies that there's an AML program called ANALYSIS.AML with many subroutines. When the user chooses BUFFER (from the Proximity submenu of the Analysis menu), the buffer routine in the AML program executes. This AML program doesn't actually exist; it simply illustrates the type of return string a Menu Item might execute.

Finally, you may want to add comments to your menu. To do this, choose View from the MenuEdit main menu and select Comments. This opens a simple text editor. Your comments will appear at the top of the menu file just like comments for any other menu.

FYI

To convert a preexisting pulldown menu to an enhanced pulldown menu, just open the old menu file in MenuEdit. When you save the menu, it's saved as an enhanced pulldown menu. To convert a sidebar, you must first change the menu number from a 2 (Sidebar) to a 1 (Pulldown).

Looking at the code

The code for enhanced pulldown menus is very different from that of other menu types. The syntax for the menu file is made up of keywords that begin with an ampersand (&). These keywords have a specific purpose in the menu file and shouldn't be confused with AML directives. The keywords define blocks of code much like the &DO &END blocks that you looked at in chapters 4 and 5.

Following is the code for ACTION.MENU:

```
[1]    8
[2]    /* action.menu
[3]    &BEGIN_MENU
[4]    &BEGIN_BLOCK "&Analysis"
[5]       &BEGIN_BLOCK "PROXIMITY"
[6]          &MENUITEM "BUFFER" &REF %all &r analysis buffer
[7]          &MENUITEM "NEAR" &REF %point  &r analysis near
[8]       &END_BLOCK
[9]       &BEGIN_BLOCK "OVERLAY"
[10]         &MENUITEM "UNION" &REF %poly     &type union
[11]         &MENUITEM "IDENTITY" &REF %all
[12]         &MENUITEM "INTERSECT" &REF %all     &type intersect
[13]      &END_BLOCK
[14]   &END_BLOCK
[15]   &BEGIN_BLOCK "&Draw"
[16]      &MENUITEM "Draw points" &REF %point     &type points
[17]      &MENUITEM "Draw arcs" &REF %line
[18]      &SEPARATOR
[19]      &MENUITEM "Shade polygons" &REF %poly     &type polygonshade
[20]      &MENUITEM "Outline polygons" &REF %poly     &type polygons
[21]   &END_BLOCK
[22]   &END_MENU
```

- The first number in the file indicates that this is Menu Type 8, or an enhanced pulldown menu. [1]

- Following the first identification number are the comments. [2]
- All enhanced pulldown menus begin and end with an &BEGIN_MENU [3] and an &END_MENU [22]. Anything outside of this block isn't evaluated as part of the menu.
- &BEGIN_BLOCK and &END_BLOCK defines a block of choices. By nesting these blocks, you can create Pull Right Items. [5], [8], [9], [13]
- To name an item, put the name on the same line as the &BEGIN_BLOCK. The name must be contained within double quotes. [4], [5], [9], [15]
- Menu Items are identified by the &MENU_ITEM keyword followed by a name and return string. [6], [7], [10–12], [16], [17], [19], [20]
- Separator bars are created with the &SEPARATOR keyword. [18]
- Reference names are identified with the &REF keyword after the name of the item. [6], [7], [10–12], [16], [17], [19], [20]

The next lesson discusses reference names in detail.

The syntax isn't too difficult, but it can involve a lot of typing. Using MenuEdit, you can create menus without needing to know the menu file syntax.

In summary

Enhanced pulldown menus are composed of three kinds of items (Menu Bar Items, Pull Right Items, and Menu Items) that can be inserted, updated, or deleted using MenuEdit. Enhanced pulldown menus should use Pull Right Items to organize the final menu. Each Menu Item should have a return string that executes a command or directive. As with any other menu or AML program, comments should be added to describe the enhanced pulldown menu.

Exercises

Exercise 14.2.1

Use MenuEdit to create an enhanced pulldown menu called UTILITIES.MENU to accomplish the following tasks:

- Build a coverage for point, line, poly, or node topology
- Clean a coverage for line or polygon topology
- Describe a coverage
- Display the log file for a coverage
- Pop up a text file

UTILITIES.MENU should use [GETCOVER] and [GETFILE] functions to perform operations like BUILD, CLEAN, DESCRIBE, and &POPUP. When complete, UTILITIES.MENU should look something like this:

Lesson 14.3—MenuEdit power tools

In this lesson

Enhanced pulldown menus allow you to do more than organize your menus into a more sophisticated format. They can also include accelerator keys and grayed out choices. In addition, enhanced pulldown menus can be embedded within form menus.

This lesson addresses ways to make your enhanced pulldown menus a powerful part of your application. You'll learn how to make your menus easier to navigate, and how to tie menus into the rest of your application.

Using and creating accelerator keys

Enhanced pulldown menus support accelerator keys, which allow you to move quickly through menu choices with the aid of arrow keys and letters. To create an accelerator key, put an ampersand (&) in front of the letter that you want to act as the accelerator key. For example, to make an accelerator key for a Menu Bar Item like Draw, you could name the item &Draw. This activates the letter "D" as the accelerator key for that item. You could just as easily make the accelerator "R" by naming the item D&raw.

The ampersand isn't displayed in the text of the Menu Bar Item. Instead, the accelerator key is underlined.

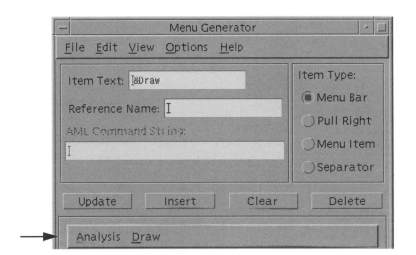

To access an accelerator key, use Alt-<accelerator letter>. In this case, Alt-A would access Analysis. Once the first accelerator key has been accessed, you no longer have to use the Alt key. The arrow keys or any additional accelerator keys can be used to navigate through the menu.

Graying out choices

Many times it's not appropriate for the user to have access to every choice in a menu. For instance, if the user chooses to work with polygons in a particular menu, it wouldn't make sense for her to have access to choices dealing with points. The &ENABLE directive allows you to gray out choices that aren't appropriate. Use &ENABLE as follows:

```
Usage: &ENABLE <&ON | &OFF> <thread name> <control>
```

The way you disable menu choices is by naming the thread and the control. (In MenuEdit, the control argument is specified in the Reference Name input field. For the rest of this lesson, therefore, the control argument will be called the Reference Name.) References can be set for every menu choice. MenuEdit allows you to add references very simply, as follows:

1. Select the item to reference in the Menu Item Display window.
2. Enter a name in the Reference Name input box.
3. Click the Update button.

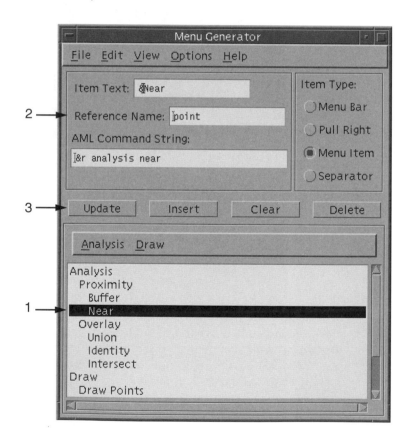

Any name can be used as a Reference Name. You can have the same Reference Name for multiple items or a single Reference Name for each item. In general, you should choose reference names on the basis of functionality. For instance, you may have a Menu Item that uses the LABELTEXT command and another Menu Item that uses the POLYGONTEXT command in ARCPLOT. You might give both of these items the same Reference Name, because they both draw labels in ARCPLOT.

If you want to disable everything on your menu that has the reference "point," you can do so with the &ENABLE directive:

```
&enable &off util point
```

The thread name in this example is util (as specified when you created the thread with the &THREAD directive).

When the menu is displayed, the Near Menu Item is grayed out.

 FYI
To have MenuEdit generate Reference Names for you when you're converting preexisting pulldown menus to enhanced pulldown menus, choose Auto References under Options. Auto References creates a Reference Name for each item in the menu. The Reference Name is the same as the name of the item.

A form and an enhanced pulldown menu can work together so that decisions made in the form affect the choices available in the enhanced pulldown menu. In the next example, a form and an enhanced pulldown menu are launched on separate threads. The form DECISION.MENU is launched on a thread named decision, while the enhanced pulldown menu ACTION.MENU is launched on a thread named action:

```
/* launch.aml
&thread &create decision &menu decision ~
   &position &ul &screen &ul ~
   &pinaction '&thread &delete &others'
&thread &create action &menu action ~
   &position &ul &thread decision &ur ~
   &pinaction '&thread &delete &others'
&thread &delete &self
&return
```

The two menus are positioned adjacent to each other and are set to dismiss each other if either one of them is closed. Depending on the feature type selected in DECISION.MENU, different items in ACTION.MENU will be grayed out.

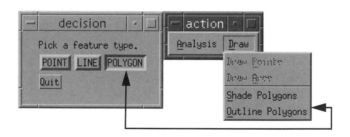

It's interesting to see how this was done. The two menus were created with FormEdit and MenuEdit, respectively. Within DECISION.MENU, there's one choice field and one button (Quit). The choice field sets a variable called TOPO that's used in enabling ACTION.MENU. The choice field for DECISION.MENU is defined as follows:

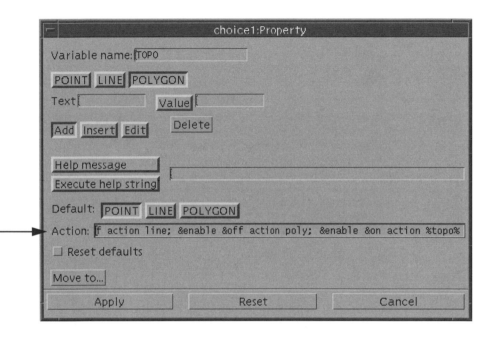

The return string in the Action: input field disables all three Reference Names in the action thread and then enables the reference for the feature type stored in the variable TOPO. Therefore, whenever a feature type is chosen in DECISION.MENU, the items in ACTION.MENU that pertain to the other two feature types are grayed out. Note that all the items in ACTION.MENU that apply to a particular feature type (e.g., polygon) must have the same Reference Name (e.g., poly).

Embedding a menu in a form

Rather than having separate menus on different threads, you can embed the enhanced pulldown menu within the form. Embedding a menu within a form menu is a function of the form menu, not the enhanced pulldown menu. The menu to be embedded is specified in the form options. The code looks like this:

```
%FORMOPT menu <menu file>
```

Within FormEdit, you pull down the Edit menu and select Preferences... to open the Preferences sheet. This is the Preferences sheet for DECISION.MENU:

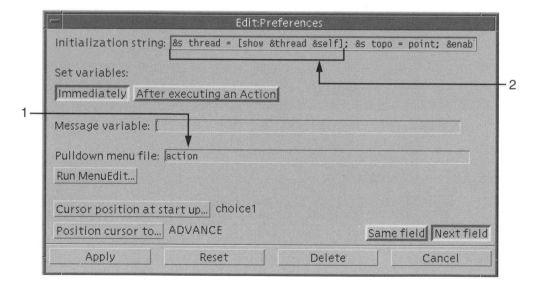

In the Pulldown menu file: input box, you specify the name of the menu to be embedded (in this case, the ACTION menu) (1). In the Initialization string: input box, you set a variable to contain the name of the thread that the form is on (2). (Here the variable is called thread.) Now click the Apply button and examine the form in FormEdit. It looks the same. The embedded menu doesn't display within FormEdit, but it does display when the form is launched.

Within FormEdit   **Final menu**

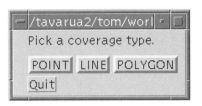   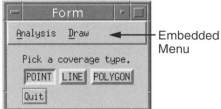

Embedded
Menu

While in FormEdit, you don't need to make space for the embedded menu. FormEdit automatically formats the form menu to accommodate the embedded menu.

When the form menu is launched, the embedded menu is on the same thread as the form menu. The name of the thread is set in the initialization string of the form, which you saw on the last page:

```
&s thread = [show &thread &self]
```

Using the variable `thread`, you can gray out choices in the embedded menu. This variable is used in the Action: input field of the FormEdit Choice Property sheet.

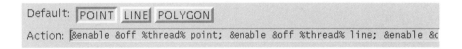

Compare the return string above with the return string set in the example on page 14-23. Notice that the thread name `action` has been replaced with the variable `%thread%`.

The embedded menu now grays out choices depending on the type of coverage selected in the form menu:

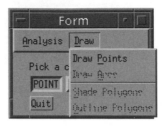

In summary

Enhanced pulldown menus give you several helpful design tools. You can add accelerator keys, you can gray out menu choices, and you can even embed pulldown menus in a form.

Accelerator keys allow the user to navigate through the menu using only the keyboard. Grayed out menu choices ensure that the user doesn't make inappropriate selections from the menu. Embedded pulldown menus contribute to a seamless interface and make it easier to group similar utilities together.

Exercises

Exercise 14.3.1

Modify UTILITIES.MENU, created in the last exercise, to include accelerator keys. Call the new menu UTILITIES2.MENU.

Exercise 14.3.2

Embed UTILITIES2.MENU into a form menu that allows you to pick the coverage or file to work with from scrolling lists. You'll have to modify UTILITIES2.MENU to use a variable set in the form instead of the [GETCOV] and [GETFILE] functions. Call the new menu UTIL_EMBED.MENU. When complete, the menu should look like this:

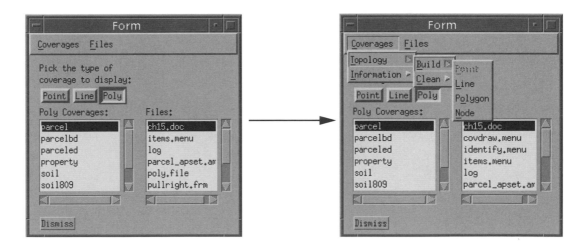

The menu should accomplish the following tasks:

- Change the scrolling list of coverages depending on the type of coverage the user selects.
- Allow the user to choose a file to view from a second scrolling list.
- If Point is selected for the coverage type, gray out the pulldown menu choices for building and cleaning a polygon coverage.
- If Poly is selected for the coverage type, gray out the pulldown menu choice for building points.

To gray out the menu choices, write an AML program called CHECKCOV.AML that determines which coverage type has been selected and disables the appropriate items. CHECKCOV.AML should look something like this:

```
/* checkcov.aml
/* USAGE: checkcov <topology_type>
/* purpose: To check a coverage for what type of topology
/*          it has.
&severity &error &routine bailout
&args topo
&s thread = [show &thread &self]
&select [keyword %topo% point line poly]
&when 1
  &do
    &enable &off %thread% poly
    &enable &on %thread% point
  &end
&when 2
  &do
    &enable &on %thread% poly
    &enable &on %thread% point
  &end
&when 3
  &do
    &enable &on %thread% poly
    &enable &off %thread% point
  &end
&otherwise
  &type unknown topology type
&end /* select
&return /*from AML
```

15 / *Building generic tools: ESRI's ArcTools*

If you diligently worked through all the previous lessons in this workbook, you now have a broad knowledge of AML programs and menus. This chapter presents ArcTools as a model for synthesizing AML programs and menu interface design.

The ArcTools menu system is a set of AML-based tools included with the ARC/INFO package. ArcTools delivers the most frequently used ARC/INFO functionality through a menu interface. In addition to acting as stand-alone tools that provide easy access to ARC/INFO functionality, ArcTools provides a set of AML macros and menus that you can incorporate into your applications. ArcTools is designed for beginners who aren't familiar with the ARC/INFO commands and casual users who understand GIS and command-line ARC/INFO, but want a menu interface for frequently performed tasks.

In this chapter, you'll build an ArcTool from start to finish. Once you complete this chapter, you'll know how to modify the functionality of an ArcTool and create your own tools that can be integrated with the ArcTools menu system.

This chapter covers the following topics:

Lesson 15.1—Introducing ArcTools

- Components of the complete tool
- The program
- The menu
- The help file

Lesson 15.2—Modifying existing tools to serve your needs

- Modifying the menu
- Modifying the AML program
- Creating the help file

Lesson 15.3—Designing your own ArcTools

- Utilities for creating new tools
- Adding custom tools to the ArcTools interface
- Accessing custom applications

Lesson 15.1—Introducing ArcTools

In this lesson

This lesson examines the ArcTools coding strategy by analyzing the construction of the MAPJOIN tool.

Components of the complete tool

The ArcTools menu interface contains five subsystems based on ARC/INFO functionality. These are Map Tools, Edit Tools, Grid Tools, Command Tools, and Land Records. Each subsystem displays a main menu that accesses the appropriate functionality.

There are two general types of ArcTools: utility tools and object tools.

Utility tools perform straightforward operations. For example, a utility tool might browse a file system and return the name of a selected coverage or issue the IDENTIFY command for a specified coverage and feature class.

Object tools are more complex. They allow the user to specify properties for an object, store these properties, and perform an action based on them. The user can edit these properties and recall them for later use. For example, an object tool for a line coverage might draw the coverage based on properties specified by the user and store the properties so the user can reuse or edit them.

The AML code for both types of tools is organized in modules used to create the overall system. The ArcTools goal is to design, implement, and revise the module independently without affecting or depending unnecessarily on other modules. This follows the concept of modular programming that was introduced in chapter 8 and applied in chapters 10 and 11.

The discussion in this chapter is limited to utility tools, but you'll find that mastering them will help you create and modify object tools as well.

In general, an ArcTool consists of a form menu, an associated AML program, and a help file. The tool's functionality is encapsulated in the menu file and associated AML program. Form menu widgets define the set of tool properties or perform an associated action. The properties for the tool are stored as global variables that follow a rigid naming convention.

The ArcTools system is made up of a number of directories that are stored under the main ArcTools directory: $ATHOME. Some of these directories and their generic contents are listed in the table below:

Directory	Contents
aelib	Contains ARCEDIT tools for editing data.
aplib	Contains ARCPLOT tools for displaying and querying data.
arclib	Contains ARC command tools that can be used in any ARC/INFO system.
customlib	This is where you can install your own tools. This directory is automatically in the AML and menu paths for ArcTools, so once a tool is installed here it's immediately accessible. It's first in the search path, so tools of the same name in this directory will override the standard tools.
gridlib	Contains ARC/INFO GRID tools for doing grid analysis.
icons	Contains all of the icons used by ArcTools. These icons are stored in xbitmap format, and can be copied and edited using any editor that accepts this format (e.g., the UNIX bitmap command). If you don't have an icon browser on your system, you can use the iconbrowse tool (&run $ATHOME/lib/iconbrowse) located in the lib directory to see what these icons look like.
lib	Contains tools that are used throughout the ArcTools system, including message tools and system file browsers.
misclib	Contains tools that aren't used by the ArcTools system but may be useful for your applications.
sysfiles	Contains miscellaneous files used by ArcTools, including AML programs that set global variables containing values to be presented to the user in a scrolling list. Refer to the README file under $ATHOME/sysfiles for a brief description of the files located under this directory.
util	Contains a few utility tools for creating ArcTools tool templates, performing a syntax check on your AML programs, and running a spell check on your help files. Refer to the README file under $ATHOME/util for further information.

For a complete list of directories, consult the online help under *Customizing ARC/INFO with AML>> Programming ArcTools >> Getting familiar with tools*.

This lesson examines the MAPJOIN tool to demonstrate the design standards of an ArcTool. The MAPJOIN tool is one of the utility tools stored in the ARCLIB directory under the main ArcTools directory.

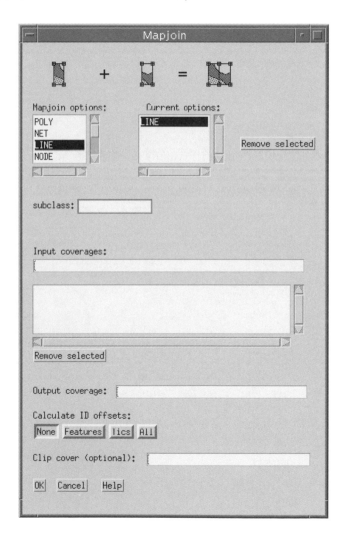

This tool can combine two or more coverages containing polygon, or arc and polygon, features in one coverage using the ARC command, MAPJOIN.

 FYI These examples discuss ArcTools as stand-alone tools (i.e., lines are commented out that access other parts of the editing application). You may notice, therefore, that the files for each tool on the disc included with this workbook are slightly different from the tools as they exist in the ArcTools system.

The program

ArcTools programs use routines to accomplish tasks as described in lesson 8.3. The program never actually runs from top to bottom; instead, routines are called as needed. The body of the program is similar for all ArcTools. Here's the body of MAPJOIN.AML:

```
&args routine arglist:rest
&severity &error &routine bailout

&if not [null %routine%] &then
  &call %routine%
&else
  &call usage
&return
```

The program is passed a routine name and possibly some other arguments, then executes the named routine. If no routine is passed, the program calls the USAGE routine. The USAGE routine is one of the five routines that should appear in every tool. These five routines are listed here and described below:

- USAGE
- BAILOUT
- EXIT
- HELP
- INIT

The USAGE routine

If the user doesn't run the program using the proper arguments, the USAGE routine is executed. Many ArcTools require arguments in addition to the routine name.

The usage routine for MAPJOIN.AML follows:

```
/* --------------
&routine USAGE
/* --------------
&type Usage: MAPJOIN INIT {'''position'''} {'''stripe'''}~
  {MODELESS | MODAL}
&type Usage: MAPJOIN <routine>
&return &inform
```

Most ArcTools have two usages. One is for a routine called INIT (discussed later in this lesson) that includes arguments such as the position of the menu and stripe text for the menu, the modality of the thread (i.e., whether the thread with the menu should be the only one that's focused), and the variables to assign. The other usage is for a routine name, and is used by the menu to execute specific routines in the program as the user makes selections from the menu.

The BAILOUT routine

If the program encounters errors, the &SEVERITY setting executes the BAILOUT routine. The standard ArcTools BAILOUT routine calls an EXIT routine and returns a message indicating the name of the routine in which the error occurred. As you can imagine, this ability to pinpoint a routine that doesn't work is useful in the debugging process. This is possible because the routine name is passed as an argument.

```
/* ----------------
&routine BAILOUT
/* ----------------
&severity &error &ignore
&call exit
&return &warning An error occurred in routine:~
  %routine% (mapjoin.aml).
```

The performance of the BAILOUT routine shouldn't be new to you. Chapter 9 presented BAILOUT routines similar to this one.

Question 15-1: Why should you set &SEVERITY to &ERROR &IGNORE in the BAILOUT routine?

(Answer on page 15-29)

The EXIT routine

The EXIT routine deletes any global variables that were defined exclusively for the tool. If there are any, they're usually named .TOOLNAME$VARNAME (e.g., .MAPJOIN$VARNAME). It also checks to see if any threads were created by the tool and deletes any that still exist.

```
/*------------
&routine EXIT
/*------------
/* Get rid of working message tool in case it is left up
/* after a problem...
&if [show &thread &exists tool$msworking] &then
  &thread &delete tool$msworking
&dv .mapjoin$*
&if [show &thread &exists tool$mapjoin] &then
  &thread &delete tool$mapjoin
&return
```

The HELP routine

The HELP routine executes a tool called DISP_HELP.AML stored in the LIB directory under the main ArcTools directory. The current tool's name is passed as an argument so DISP_HELP.AML can find the help file named with that prefix. This works because all three of the tool's components use the same prefix (e.g., MAPJOIN).

```
/* ------------
&routine HELP
/* ------------
&run disp_help mapjoin
&return
```

By default, DISP_HELP.AML launches the HyperHelp® page on UNIX and WinHelp on Windows NT for the specified tool. Before DISP_HELP.AML launches the Help page, it searches all &AMLPATH structures until it reaches the ArcTools CUSTOM directory for the ASCII file TOOLNAME.HLP (e.g., MAPJOIN.HLP). This program and an example help file are discussed later in this lesson.

The INIT routine

The INIT routine initializes the tool. INIT checks environments, assigns variables, determines menu position and stripe, launches the menu, and determines modality. Examine these properties in the INIT routine for the MAPJOIN tool.

The INIT routine for MAPJOIN.AML first sets the global variables that the menu references. Notice that the variable names all begin with the .MAPJOIN$ prefix. This naming convention (i.e., naming the variables after the tool) makes it easier for the EXIT routine to clean up the program environment.

```
/*-------------
&routine INIT   /*  {'position'} {'stripe'} {MODELESS | MODAL}
/*-------------
/* Establish initial option...
&set .mapjoin$in_subclass
&set .mapjoin$optionlist POLY NET LINE NODE ANNO. REGION. ~
ROUTE. SECTION. TEMPLATE...
&set .mapjoin$option
/* Establish initial dialog list
&set .mapjoin$dialoglist
/* Establish template cover
&set .mapjoin$template_coverage
/* Establish initial offset settings...
/* Establish initial icon settings...
&set .mapjoin$offset := NONE
&set .mapjoin$1_32.icon := mapjoin_left_32.icon
&set .mapjoin$2_32.icon := add_32.icon
&set .mapjoin$3_32.icon := mapjoin_right_32.icon
&set .mapjoin$4_32.icon := equals_32.icon
&set .mapjoin$5_32.icon := mapjoin_out_32.icon
```

Next, INIT examines the contents of the ARGLIST variable. In the MAPJOIN tool, the arguments POSITION, STRIPE, and MODALITY aren't set explicitly by the &ARGS directive. Instead, they're all stored in a variable named ARGLIST. If these values aren't passed in the &RUN statement, or if placeholders (#) are substituted, the program uses default values.

```
&set position := [extract 1 [unquote %arglist%]]
&set stripe   := [extract 2 [unquote %arglist%]]
&set modality := [extract 3 [unquote %arglist%]]
&if [null %position%] or %position%_ = #_ &then
  &set position = &cc &screen &cc
&if [null %stripe%] or %stripe%_ = #_ &then
  &set stripe = Mapjoin
&if [null %modality%] OR %modality%_ = #_ &then
  &set mode = &modal
&else
  &if [translate %modality%] = MODAL &then
    &set mode = &modal
  &else
    &set mode =
```

This section of code checks to see if the tool was launched from the command line. If the tool was launched from the command line, the AML program deletes its own thread after launching the menu. This saves the user from having to type &return on the command line to focus the tool.

```
/* Issue thread delete self if thread depth = 2 and input is
/* tty
&if [show &thread &depth] = 2 and [extract 1 [show &thread
&stack]] = tty &then
  &set launch = &thread &delete &self
&else
  &set launch
```

Next, the routine checks to see if the thread it's about to create already exists. If so, INIT deletes it so two copies of the menu won't be displayed at one time. Finally, INIT launches the menu with position and stripe arguments. If the menu is dismissed through the system's window manager, the program uses the &PINACTION option to call the EXIT routine and clean up environments.

```
&if [show &thread &exists tool$mapjoin] &then
   &thread &delete tool$mapjoin
&thread &create tool$mapjoin %mode% ~
  &menu mapjoin.menu ~
  &position [unquote %position%] ~
  &stripe [quote [unquote %stripe%]] ~
  &pinaction '&run mapjoin.aml exit'
%launch%
&return
```

The menu

The menu is the part of the tool with which the user interacts. The menu updates variables and executes specific actions by running the tool's AML program with the appropriate routine as an argument. Consider MAPJOIN.MENU. Widgets that perform an action have the following return string:

```
&run mapjoin.aml <routine>
```

Most tasks performed by the menu are accessed through the specific routines in MAPJOIN.AML. ArcTools menus generally contain a menu stripe, menu widgets (specific to each task), and standard control buttons. The upcoming pages discuss these features using MAPJOIN.MENU as an example. MAPJOIN.MENU and its widget return strings are displayed here:

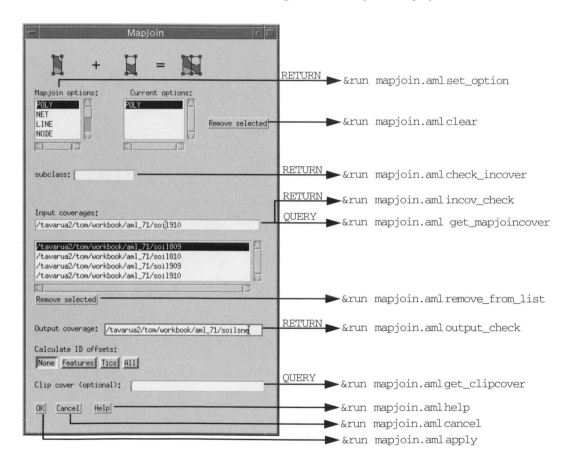

Menu stripe

A tool menu should include a menu stripe with text that describes the purpose of the tool. The INIT routine passes the stripe when it executes the program. If the stripe isn't passed to the program, or if a placeholder (#) is used, the program sets a default stripe. The stripe text is used when the new thread is created as follows:

```
&thread &create tool$mapjoin %mode% ~
  &menu mapjoin.menu ~
  &position [unquote %position%] ~
  &stripe [quote [unquote %stripe%]] ~
  &pinaction '&run mapjoin.aml exit'
```

Menu widgets

Each tool menu contains widgets to accomplish the tool's task. Examine the routines executed by each widget in MAPJOIN.MENU, beginning with the Mapjoin options: scrolling list shown here:

The Mapjoin options: scrolling list executes the following return string:

```
&run mapjoin.aml set_option
```

Here's the SET_OPTION routine in MAPJOIN.AML. Notice how other tools are called from this routine:

```
/*---------------------
&routine SET_OPTION
/*-----------------
/* setting current option
&set .mapjoin$dialoglist
&select %.mapjoin$opt%
&when NOTEST
  &do
   &if [null %.mapjoin$in_subclass%] &then
          &do
            &set .mapjoin$option NOTEST
            &set .mapjoin$template_coverage
            &return
          &end
        &else
          &do
          &set .mapjoin$msg subclass not valid for this option
            &set .mapjoin$in_subclass
            &set .mapjoin$template_coverage
            &return
            &end
   &end
&when TEMPLATE...
  &do
     &set .mapjoin$option TEMPLATE...
     &set .mapjoin$mapjoincover
     &run getcover init .mapjoin$template_coverage *
     &if ^ [null %.mapjoin$template_coverage%] &then
      &do
      &set .mapjoin$template_coverage := ~
       [pathname %.mapjoin$template_coverage%]
     &set .mapjoin$template_dispcoverage := template coverage
     [pathname %.mapjoin$template_coverage%]
     &end
     &else
      &do
        &set .mapjoin$template_coverage
        &set .mapjoin$option := TEMPLATE...
      &end
   &end

&when REGION., ANNO., ROUTE., SECTION.
    &do
       &if [null %.mapjoin$in_subclass%] &then
          &do
```

```
              &set .mapjoin$msg Please specify a subclass for ~
this option
              &return
              &end
              &else
              &set .mapjoin$opt ~
                %.mapjoin$opt%%.mapjoin$in_subclass%
              &call set_optionlist
          &end
  &otherwise
     &do
        &if [null %.mapjoin$in_subclass%] &then
           &do
            &call set_optionlist
            &end
           &else
           &do
           &set .mapjoin$msg subclass not valid for this option
           &set .mapjoin$in_subclass
           &call set_optionlist
           &end
        &end
&end
&return
```

This routine uses the data set in the scrolling list to determine the feature classes
to join. Choosing Template causes the SETUP_CLASS routine to invoke another
tool, GETCOVER. GETCOVER.AML is stored under the LIB directory under
the main ArcTools directory. This program, which sets a variable to a chosen
coverage, has the following usage:

```
Usage: getcover INIT <variable_name> {wildcard}
                     {coverage_type} {menu_position} {stripe}
Usage: getcover <routine_name>
```

Notice that there's no MODAL | MODELESS option in the usage. The GETCOVER
tool is designed to be modal so, while it's in use, the MAPJOIN tool is
unfocused. The GETCOVER tool stays on the screen until a coverage is selected
or it's dismissed by pressing the OK or Cancel button. When the GETCOVER
tool is dismissed, the MAPJOIN tool is refocused.

On the MAPJOIN tool, the names of the coverages to join are entered as shown in this example:

```
Input coverages:
/tavarua2/tom/workbook/aml_71/soil910
```

If the user doesn't know the names of the coverages to join, pressing the right mouse button executes the following statement, which runs the GET_MAPJOINCOVER routine:

```
&run mapjoin get_mapjoincover
```

GET_MAPJOINCOVER also runs the GETCOVER tool to provide a list of available coverages from which to choose.

```
/*----------------------
&routine GET_MAPJOINCOVER
/*----------------------
/* Get an input coverage...
&if [null %.mapjoin$option%] &then
     &set .mapjoin$option POLY
&if [keyword  TEMPLATE... %.mapjoin$option%] > 0 &then
     &do
     &set .mapjoin$mapjoincover
     &run getcover init .mapjoin$mapjoincover * all
     &set .mapjoin$dialoglist ~
       %.mapjoin$dialoglist% %.mapjoin$mapjoincover%
     &set .mapjoin$mapjoincover
     &return
     &end
&if [keyword NOTEST %.mapjoin$option%] > 0 &then
     &do
     &run getcover init .mapjoin$mapjoincover * all
     &set .mapjoin$dialoglist ~
   %.mapjoin$dialoglist% %.mapjoin$mapjoincover%
     &set .mapjoin$mapjoincover
     &return
     &end
&if [token -count %.mapjoin$option%] = 1 &then
   &do
     &if [before %.mapjoin$option% .] = REGION &then
     &do
     &run getcover init .mapjoin$mapjoincover * region
     &end
```

```
     &else
      &do
      &run getcover init .mapjoin$mapjoincover ~
        * %.mapjoin$option%
      &end
     &end
&else
  &do
  &run getcover init .mapjoin$mapjoincover *
  &call incov_check
  &return
&end
&if ^ [null %.mapjoin$mapjoincover%] &then
&do
  &call incov_check
  &set .mapjoin$mapjoincover
&end
&return
```

After a user enters a coverage name (whether by typing or choosing from the
GETCOVER tool's list), the INCOV_CHECK routine is called to verify the data.
This routine also builds a list of the selected coverages and stores it in the
variable named .MAPJOIN$DIALOGLIST.

```
/*--------------------
&routine INCOV_CHECK
/*--------------------
/* check if cover has all mapjoin options
&if [null %.mapjoin$option%] &then
   &set .mapjoin$option POLY
&if [null %.mapjoin$mapjoincover%] &then
   &return
&set cnum = 1
&set .mapjoin$mapjoincover := [pathname ~
%.mapjoin$mapjoincover%]
&if ^ [null %.mapjoin$mapjoincover%] &then
  &if [keyword %.mapjoin$option% NOTEST TEMPLATE] = 0 &then
/* check it out
  &do &until [null [extract %cnum% %.mapjoin$option%]]
    &if [exists %.mapjoin$mapjoincover% -[extract %cnum% ~
%.mapjoin$option%]] &then
    &do
       &set cnum = %cnum% + 1
       &end
    &else
    &do
       &set .mapjoin$msg ~
```

```
        [entryname %.mapjoin$mapjoincover%] does not contain~
          [extract %cnum% %.mapjoin$option%] features
        &set cnum = %cnum% + 1
        &set .mapjoin$mapjoincover
        &return
        &end
   &end
 /* accept any coverage - this is screened in menu
  &do
    &set .mapjoin$dialoglist ~
      %.mapjoin$dialoglist% %.mapjoin$mapjoincover%
/*    &set .mapjoin$mapjoincover
  &end
&return
```

The data stored in the variable .MAPJOIN$DIALOGLIST is presented as a scrolling list. The Remove selected button removes a selected coverage from the list (and from the MAPJOIN operation).

The Remove selected button executes the following statement in order to run the REMOVE_FROM_LIST routine:

```
&run mapjoin.aml remove_from_list
```

The REMOVE_FROM_LIST routine removes the selected coverage from the scrolling list by determining its position in the list and using this information to build a new list.

```
/*-----------------------
&routine REMOVE_FROM_LIST
/*-----------------------
/* Remove token from the working list...
/* Delete an entry from the dialog listing
&if [null %.mapjoin$dialoglistval%] &then
  &do
    &set .mapjoin$message = No coverage selected to delete
    &return
```

```
   &end
/*
/* Get position of the selected token to be removed...
&set position = [keyword [quote %.mapjoin$dialoglistval%] ~
                %.mapjoin$dialoglist%]
&set temp_list
&set i := 1
&do &until [null [extract %i% %.mapjoin$dialoglist%]]
  &if %i% ne %position% &then /* keep in list only if ^ token
                              /* to be removed
    &do  /* rebuild the list
      &set token = [extract %i% %.mapjoin$dialoglist%]
      &set temp_list = %temp_list% %token%
    &end
  &set i := %i% + 1
&end
&set .mapjoin$dialoglist = %temp_list%
&if [null %.mapjoin$dialoglist%] &then &set~
.mapjoin$dialoglist
/*&set .mapjoin$mapjoincover
&return
```

After the list of coverages to MAPJOIN is entered and checked, you can enter the name of the coverage to create as follows:

Output coverage: `/tavarua2/tom/workbook/aml_71/soilsne`

The Output coverage: field executes the following statement, which runs the OUTPUT_CHECK routine:

```
&run mapjoin.aml output_check
```

The OUTPUT_CHECK routine converts the coverage name to its full pathname and performs a number of tests. The name must contain less than thirteen characters and must not already exist (with or without its pathname). If the coverage name fails either of these tests, the routine runs MSINFORM.AML. This is another generic tool located under $ATHOME/LIB.

```
/*-------------------
&routine OUTPUT_CHECK
/*-------------------
/*  check to see if output cover exists
&set mapjoin$ok := .TRUE.
&if ^ [null %.mapjoin$out_cover%] &then
```

```
&do
   &if [LENGTH [entryname %.mapjoin$out_cover%]] gt 13 &then
      &do
         &run msinform init ~
            [quote Coverage names cannot be longer than 13 ~
               characters.] [quote Please shorten the output ~
               coverage name.]
         &set .mapjoin$out_cover
         &return
      &end
   &set .mapjoin$out_cover := [pathname %.mapjoin$out_cover%]
   &if [exists %.mapjoin$out_cover% -DIR] OR ~
   [exists %.mapjoin$out_cover% -FILE] &then
      &do
         &run msinform init ~
         [quote Output coverage [unquote ~
         [value .mapjoin$out_cover]]]~
         [quote already exists!]
         &set .mapjoin$out_cover
         &return
      &end
&end
&else
   &if ^ [exists [dir  %.mapjoin$out_cover% ] -dir ] &then
      &do
         &run msinform init ~
            [quote Output Directory does not exist! ]
         &set .mapjoin$out_cover
         &return
      &end
&return
```

The MAPJOIN tool also offers the option of specifying a clip coverage:

```
Clip cover (optional): [
```

If the user doesn't know the name of the clip coverage to use, pressing the right mouse button executes the following statement, which runs the GET_CLIPCOVER routine:

```
&run mapjoin get_clipcover
```

This routine also runs the GETCOVER tool. This is a consistent ArcTools design. Pressing the right mouse button when the cursor is in an input field launches a new tool that either gives access to the needed data or displays a string explaining the purpose of the field.

```
/*--------------------
&routine GET_CLIPCOVER
/*--------------------
/* browse for a clip coverage
&run getcover init .mapjoin$tmpclip_cover * poly ~
  '&ll &thread tool$mapjoin &ll' 'Select a Clip Coverage -
Type: POLY'
&if ^ [null %.mapjoin$tmpclip_cover%] &then
  &set .mapjoin$clip_cover = %.mapjoin$tmpclip_cover%
&return
```

The input is set to accept only coverages: if the coverage isn't valid, the input field doesn't accept it. There's no need for another routine to check the validity of the coverage.

Standard control buttons

There are standard control buttons that appear in ArcTools menus. The text for these buttons is formatted with an initial capital letter and the remaining letters are lowercase (e.g., Remove selected). The exception is the OK button, which is entirely uppercase. The following table lists the standard button names and their actions:

Button name	Action
Apply	Apply changes made in the menu and leave the menu displayed on the screen.
Cancel	Dismiss the menu without applying any changes made since the Apply button was last pressed. Paired with OK or Apply.
Dismiss	Dismiss the menu. Use when settings made in the menu can't be undone. This is usually used without OK or Apply.
Help	Display an ASCII file containing information about the tool.
OK	Apply any changes made in the menu, then dismiss the menu.
Reset	Reset the values in a menu to a previous state—the state present when Apply was last pressed, or the initial menu state.

The difference between the Apply and OK buttons is that the menu isn't dismissed when the Apply button is pressed. The Apply button allows you to see the result of the changes you've made on the display screen (e.g., color, size, location, etc.). You may want to experiment with various settings before making a final choice and dismissing the menu.

MAPJOIN.MENU includes the three standard buttons shown here:

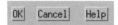

When pressed, the OK button executes the following statement to dismiss the menu and execute the APPLY routine in MAPJOIN.AML:

```
&return; &run mapjoin.aml apply
```

The APPLY routine contains the code that executes the ARC MAPJOIN command, no matter which subsystem the user is in when the menu is executed.

The APPLY routine also uses the MSWORKING tool to tell the user that the MAPJOIN is being performed. Finally, the APPLY routine calls the EXIT routine to clean up the thread and global variable environments.

```
/*--------------
&routine APPLY
/*--------------
/* Perform the mapjoin...
&run msworking init [quote Mapjoining....]
&severity &error &ignore
&data arc                /* use data block, even if in arc, to
                         /* eliminate menu flash
  &if [keyword %.mapjoin$option% TEMPLATE...] = 1 &then
    mapjoin %.mapjoin$out_cover% ~
            %.mapjoin$template_coverage%  %.mapjoin$offset% ~
            %.mapjoin$clip_cover%
  &else
    mapjoin %.mapjoin$out_cover% %.mapjoin$option% ~
            %.mapjoin$offset% %.mapjoin$clip_cover%
  &do string &list %.mapjoin$dialoglist%
    [extract 1 [unquote %string%]]
  &end
end
  QUIT
```

```
&end
&severity &error &routine bailout
&run msworking close
&type Mapjoin process complete
&call exit
&return
```

The Cancel button dismisses the menu with &RETURN and executes the CANCEL routine in the MAPJOIN.AML. The CANCEL routine calls the EXIT routine to clean up the thread and variable environments.

```
/* --------------
&routine CANCEL
/* --------------
&call exit
&return
```

The Help button executes the HELP routine in MAPJOIN.AML, which uses another ArcTool called DISP_HELP to display the help file:

```
/* -------------
&routine HELP
/* -------------
&run disp_help mapjoin
&return
```

FYI

Any button that displays a menu should include an ellipsis (. . .) as part of the button name. This is a naming convention used to indicate to the user that another menu results from pressing the button (e.g., Preferences..., Select by attribute..., etc.).

The help file

DISP_HELP.AML is run with the name of the tool passed as an argument
(e.g., MAPJOIN). This program, located in the LIB directory, is used by all
ArcTools to display their respective help files. The main help file system for
ArcTools is HyperHelp on UNIX and WinHelp on Windows NT. If you modify
an ArcTool, then you'll need to create a new help file. Following is the main body
of $ATHOME/LIB/DISP_HELP.AML:

```
/* disp_help.aml
&args topic alternate

&severity &error &routine bailout

/* Check argument
&if [null %topic%] &then &call usage
&else &call INIT

&return
```

This main body of code calls the INIT routine, which determines the help file to
be displayed. One of three different help files can be displayed:

- An ASCII text file

    This is a file that you have created for a modified tool. This file must be in, or
    before, the $ATHOME/customlib directory in your &AMLPATH. For
    instance, if your &AMLPATH looks like this:

    ```
    /tavarua2/tom/athelp
    /san1/arcexe71/arctools/customlib
    /san1/arcexe71/arctools/aelib
    ```

    then your text file will have to be in either

    ```
    /tavarua2/tom/athelp
    ```

    or

    ```
    /san1/arcexe71/arctools/customlib
    ```

    The directory /san1/arcexe71/arctools/aelib will not be searched.

- The compiled help file for the tool

    This is the HyperHelp (UNIX) or WinHelp (Windows NT) help file for the
    tool. This file can't be modified. (If you customize a tool and want to change
    the help for it, you have to create a new help file.) Later in this chapter, you'll
    learn how to export a compiled help file to an ASCII text file.

- A compiled help file that you've created

  This is a help file that you can generate in an application like WinHelp or HyperHelp. If you've created a help file like this, you must pass the help file name to DISP_HELP.AML in the argument ALTERNATE.

The INIT routine decides which type of help file is displayed. The first routine called from INIT is the ASCII routine, which checks for the existence of an ASCII text file. (This ASCII file may be a file that you've created for a modified tool.) If no ASCII file is found, then INIT launches the default help file or a compiled help file that you created.

```
/*-----------
&routine INIT
/*-----------
/*
/* First look down the amlpath as far as customlib for ascii
/* help file
&call ASCII
&if %found% &then
  &return

/* Replace invalid special characters with underscores
/*   (hyperhelp chokes on "+" and "-")
&set topic = [subst [subst %topic% - _] + _]

/* Call appropriate routine based on contents of alternate.
&if [null %alternate%] &then
    &call default
&else
    &call external_helpfile
&return
```

The DEFAULT routine launches the appropriate HyperHelp or WinHelp file.

```
/*--------------
&routine DEFAULT
/*--------------
/* Call the help topic up from atmenu.hlp, using
/* winhelp/hyperhelp.
/* NOTE: this tool does not check for the topic within the
/* helpfile
/*
/* handle case where tool is called with "tool.hlp", rather
/* than just "tool"
&set helptopic = [subst %topic% .hlp]
```

```
&select [extract 1 [ show &os]]
&when Windows_NT
    &sys start %ARCHOME%\bin\winhlp32 -i %helptopic% -W ~
        atmenu %ADOCHOME%\atmenu.hlp
&otherwise
    &sys $HHHOME/bin/hyperhelp $ADOCHOME/atmenu.hlp ~
        -m "JumpID(%helptopic%)" &
&end /* select
/*
&return
```

The EXTERNAL_HELPFILE routine is used in case you created and compiled your own help file.

```
/*-----------------------
&routine EXTERNAL_HELPFILE
/*-----------------------
/* Call the help topic up from an alternate helpfile, using
/* winhelp.
/* NOTE: this tool does not check for the existence of the
/* external helpfile or topic within that helpfile
/*
/* handle case where tool is called with "tool.hlp", rather
/* than just "tool"
&set helptopic = [subst %topic% .hlp]
&select [extract 1 [ show &os]]
&when Windows_NT
    &sys start %ARCHOME%\bin\winhlp32 -i %helptopic%
%alternate%
&otherwise
    &sys $HHHOME/bin/hyperhelp %alternate% -m
"JumpID(%helptopic%)" &
&end /* select
/*
&return
```

This is the ASCII routine that searches through the current &AMLPATH locations for an ASCII textfile with the name <toolname>.HLP (e.g., MAPJOIN.HLP). The routine stops when it reaches $ATHOME/CUSTOMLIB.

```
/*------------
&routine ASCII
/*------------
/* Look for the ascii help file in the current AMLPATH
/* directories as far as customlib.  If no helpfile is found,
```

```
/* &return.
/*
/* Check that amlpath has been set.
&if [null [show &amlpath]] &then
  &return No &amlpath has been set.
/*
/* look for the help file
&if [null [after %topic% .]] &then
  &set helpfile = %topic%.hlp
&else &set helpfile = %topic%
&set i = 1   /* keep track of the amlpaths to extract from the
list of paths
&set done = .FALSE.
&do &while not %done%
  &set path = [extract %i% [show &amlpath]]; &set i = %i% + 1
  &if [exists [joinfile %path% %helpfile% -file]] &then
    &do
       &popup [joinfile %path% %helpfile% -file]
       &set done  = .TRUE.
       &set found = .TRUE.
    &end
  &else
  &if [null %path%] or [locase [entryname %path%_]] = ~
customlib_ &then
    &do
      &set done  = .TRUE.
      &set found = .FALSE.
    &end
&end
/*
&return

/*------------
&routine USAGE
/*------------
&return &warning Usage: DISP_HELP.AML <help_topic> {helpfile}
/*--------------
&routine BAILOUT
/*--------------
&severity &error &ignore
&severity &warning &ignore
&return &warning An error has occurred in DISP_HELP.AML
```

Here's the default help file for the MAPJOIN tool:

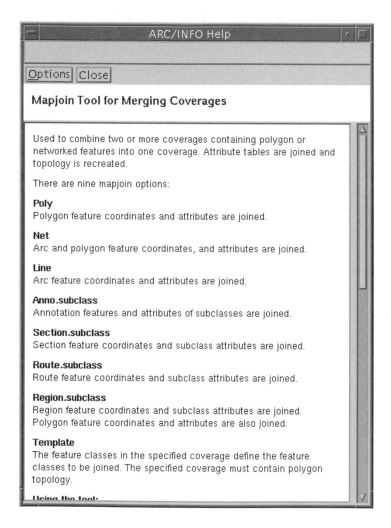

Help files should explain the tool's arguments, how and when to use the tool, and anything that the user needs to know to operate the tool correctly.

In summary

All ArcTools have three parts: a program, a menu, and a help file. The MAPJOIN tool was used to illustrate the coding method for an ArcTool as well as to show ArcTool standards for menus and AML programs.

Answer 15-1: Set &SEVERITY &ERROR &IGNORE in the BAILOUT routine because, without it, the BAILOUT routine is called repeatedly in an infinite loop if the program encounters an error in either the BAILOUT or EXIT routines.

Exercises

Exercise 15.1.1

What does an ArcTool INIT routine do if the thread it's supposed to create already exists?

Exercise 15.1.2

Does every menu widget in an ArcTool have an associated routine?

Exercise 15.1.3

The OK button used in an ArcTool performs the following action:

```
&return; &run some.aml ok
```

What would happen if the &RETURN followed the &RUN statement? Which other standard control buttons use &RETURN?

Lesson 15.2—Modifying existing tools to serve your needs

In this lesson

The modular design of ArcTools programs allows them to be easily modified. In this lesson, you'll learn to alter the functionality of an ArcTool.

Adding functionality to a tool is as easy as adding another widget to the menu and another routine to the program. To remove functionality, remove the widget. With the widget gone, the routine that stored the action can never execute.

Modifying the menu

To add functionality to an ArcTool, you must do these three things:

1. Add a widget to the menu to execute the new action.
2. Add a routine to the program that accomplishes the new action.
3. Add information to the help file that describes the tool's new functionality.

This is an easy process after you've designed the new functionality.

As an example, consider the TOOL BROWSER ArcTool. This tool runs from all the ArcTools main menus under the ArcTools pulldown. The TOOL BROWSER allows you to look though all the ArcTool libraries for the tool you want.

The menu for the TOOL BROWSER ArcTool looks like this:

By itself, the TOOL BROWSER ArcTool is useful for browsing AML, menu, and help files. However, it would be easier to know if a particular tool is the one you want if you could see its menu. The new button `View tool` will allow you to display the menu for a selected tool.

FYI

The FormEdit menus in this chapter appear as they would in the UNIX Common Desktop Environment. The appearance is different on the Windows NT platform, but the contents are the same.

Before working on the TOOL BROWSER tool, you need to copy it to a local directory. This will allow you to modify the tool without affecting the ArcTools library. In the next lesson, you'll see how to add this modified tool to ArcTools. The TOOLBROWSER tool is located in the ARCTOOLS/LIB directory. To fully modify it, you need to copy the AML file as well as the menu file:

```
Arc: &type [copy $ATHOME/lib/toolbrowser.aml toolbrowser.aml]
0
Arc: &type [copy $ATHOME/lib/toolbrowser.menu ~
toolbrowser.menu]
0
```

Once the files have been copied to your local workspace, you can begin modifying the tool. The three necessary steps are listed here and explained on the following pages.

1. Add the widget using the FormEdit menu editor.

2. Add a routine to TOOLBROWSER.AML that displays the appropriate menu.

3. Create a help file that describes the tool.

Add the button widget in FormEdit

You can use the FormEdit menu editor to add the View tool button. Specify the action &RUN TOOLBROWSER VIEW_TOOL on the button property sheet as follows:

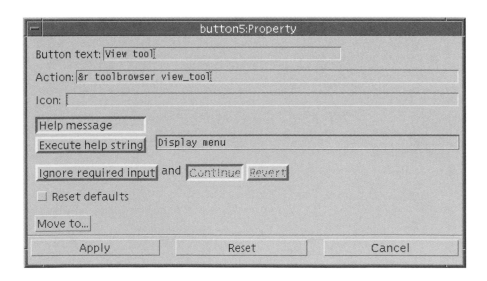

Saving the changes in FormEdit and executing the menu from the `Arc:` prompt displays the menu with the new button:

```
Arcedit: &run toolbrowser init
```

If you select the new button now, you'll receive an error message from TOOLBROWSER.AML because there's no routine associated with it yet.

Modifying the AML program

Selecting the `View tool` button should display the menu for the currently selected tool. The routine that you create should accomplish the following tasks:

- Check to make sure a tool was selected
- Manipulate the file name
- If there's no menu file, stop and inform the user
- If another menu is currently displayed, delete it
- Launch a thread that displays the selected menu
- Unfocus the displayed menu so the user is unable to interact with it

Each section of the following code completes one of the tasks. The comments explain what the routine is doing:

```
/*----------------
&routine VIEW_TOOL
/*----------------
/* Display menu file for specified tool
/*
/* .toolbrowser$tool is the variable that holds the name
/* of the selected tool.  This variable is set in
/* toolbrowser.menu.  If this variable has not been set then
/* a tool has not been selected and the AML returns to
/* toolbrowser.menu.
&if [null %.toolbrowser$tool%] &then
  &return

/* Here we establish the name of the tool without its
/* extension.  For example mapjoin.aml would be changed to
/* mapjoin.  [JOINFILE] is then used to concatenate the
/* toolname with the specific library that this tool is in.
&set toolname = [before [entryname %.toolbrowser$tool%] .]
&set toolfile = [joinfile [dir %.toolbrowser$tool%]~
  %toolname% -file]

/* Once we have the name of the file in the correct format
/* we test to see if that tool has a menu file.  If it
/* doesn't we give a helpful message to the user and return.
&if not [exists [joinfile %toolfile% menu -ext] -file] &then
  &set .toolbrowser$msg File %toolname% does not exist

/* If the menu file does exist we test to see if the thread
/* tool_temp exists.  If it does then another menu is being
/* displayed so we delete the thread.
&else
  &do
    &if [show &thread &exists tool_temp] &then
      &thread &delete tool_temp

    /* Create the thread with the selected tool menu
    /* on it.  This menu displays on the right side of the
    /* Tool browser.
    &thread &create tool_temp ~
      &menu %toolfile% ~
      &position &ul &thread tool$toolbrowser &ur ~
      &stripe [entryname %toolname%]

    /* The focus for tool_temp is turned off so that
    /* the user cannot interact with the menu.
```

```
        &thread &focus &off tool_temp
    &end
&return
```

The reason that you unfocus the thread after creating it is that you're only displaying the menu file. The AML program that usually initializes this menu was never run. If the user could interact with the menu, AML errors would result.

The EXIT routine should also be changed to delete the tool_temp thread when the tool finishes:

```
/*----------------
&routine EXIT
/*----------------
&if [show &thread &exists tool_temp] &then
  &thread &delete tool_temp
&dv .toolbrowser$*
&if [show &thread &exists tool$toolbrowser] &then
  &thread &delete tool$toolbrowser
&return
```

Now when you click on the View tool button you see this:

```
Arc: &r toolbrowser init
```

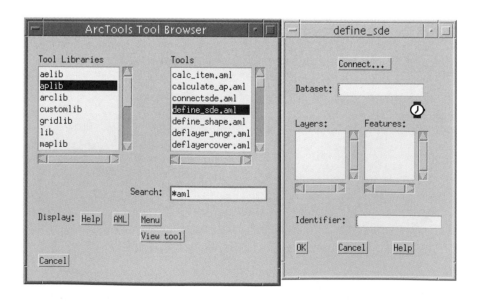

Creating the help file

Because the TOOL BROWSER tool doesn't have its own help file, you'll create one for it from scratch. This requires that you modify the tool to add a help button to its menu. As you may have noticed, however, there is already a `Help` button on the menu. This existing button displays the help file for whichever tool is selected in the scrolling list. Use FormEdit to modify the name and routine of this widget to `Tool help` to indicate that it displays the help file for the selected tool. The button property sheet should look like this:

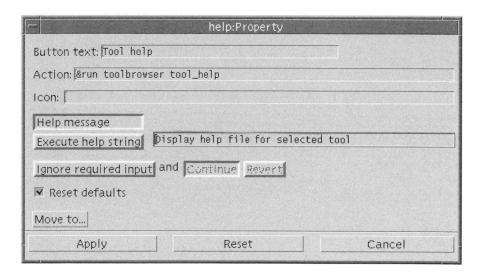

Change the current HELP routine name to TOOL_HELP in the TOOLBROWSER.AML.

Now you can add a new widget to call a help routine for the TOOL BROWSER tool itself.

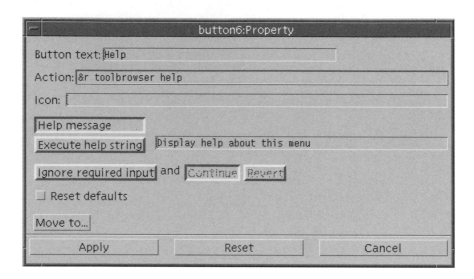

All you need to do to TOOLBROWSER.AML is add the following help routine:

```
/*-------------------
&routine HELP
/*-------------------
/* Popup help about this tool
&run disp_help toolbrowser
&return
```

This routine is running DISP_HELP.AML in the ARCTOOLS/LIB directory. The AML program searches the current &AMLPATH structures until it reaches ARCTOOLS/CUSTOMLIB. For now, you can put the help file in the current directory and execute it from there. Notice that the full pathname to DISP_HELP isn't provided. Instead of setting individual &AMLPATH structures to access other ArcTools utilities, you can run $ATHOME/LIB/SETPATHS.AML before running TOOLBROWSER.AML. SETPATHS.AML sets the paths for all ArcTools applications. In the next lesson, you'll see how to access this help file while in ArcTools.

The tool should now look like this:

Now that the tool is set up, you can create an ASCII help file. Following is a list of elements that a help file should include:

- Tool name (required)
- Brief description (required)
- How to use the tool (required)
- List of operations (optional)
- Additional topics (optional)

Following these guidelines, the help file should look something like this:

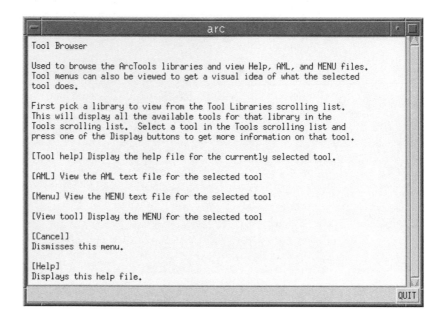

 FYI On the Windows NT platform you can create your own help file in WinHelp (or any software that conforms to the WinHelp 32 help file structure). You can then access this help file by specifying the topic and alternate file where the topic can be found when you run DISP_HELP. This is also possible on the UNIX platform, but requires purchasing HyperHelp software.

If the menu file for a tool already exists, you can't modify the compiled help file. You can, however, copy the text from the default file. At the top of all the help pages is an Options pulldown menu. Under this pulldown menu is a Copy option. By selecting the Copy option, you can copy and paste the text into a text editor. (On Windows NT, the Copy option is under the Edit pulldown menu.)

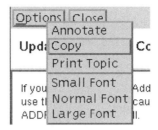

For a more thorough explanation of how to create a help file, consult the online help under *Programming ArcTools >> Customizing tools >> Customizing ARC/INFO with AML.*

In summary

You can modify an ArcTool by adding and deleting widgets, or modifying the actions they perform. You can use FormEdit to manipulate the widgets and their properties. With any modifications of the menu file, you should also edit the routines in the tool's AML program. Finally, you should document all changes in the tool's help files.

In the next lesson, you'll learn how to integrate your tools into the ArcTools system.

Exercises

Exercise 15.2.1

Add a widget to the TOOL BROWSER tool in the AMLWB/CH15/LS2 directory that will allow you to copy the AML program and menu for a selected tool from the $ATHOME directory. At present, the tool looks like this:

The widget should access the SAVE AS tool in ArcTools and copy the selected tool.

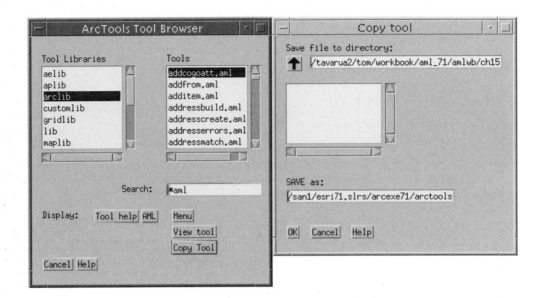

Follow these steps to create the COPY_TOOL routine:

- Extract the tool name from the file extension in `%.toolbrowser$tool%`.
- Set the names of the AML and menu files to variables.
- Check to see if the AML and menu files exist.
- If both files exist, then &RUN SAVEFILEAS.AML. The usage for SAVEFILEAS.AML is as follows:

```
Usage: savefileas INIT <in_file_name> <variable_name>
                  {ASCII | COVER| INFO} {'position'} {'stripe'}
Usage: savefileas <routine_name>
in_file_name: The name of the file to be saved to a new location.
variable_name: The full name of the new file.
ASCII | COVER| INFO: The type of file to be copied.
```

- SAVEFILEAS.AML doesn't actually save anything, it simply sets the new location for the file to the variable passed to it in the VARIABLE_NAME argument. This means that once SAVEFILEAS.AML sets the location, the COPY_TOOL routine must do the copying. Before copying the tool, the routine should also delete any existing files that have the same name in the specified location. (The Save As menu prompts the user if existing files are going to be overwritten. If the user chooses not to overwrite them, then SAVEFILEAS.AML returns a null value for VARIABLE_NAME.)

- If the VARIABLE_NAME is null, return without saving the files.

- Otherwise, delete the files and save the new files.

- If both AML and MENU files were copied, then return to the menu.

- If only one of the tool files exists, execute the copying process for that file only. This is necessary because there are AML files in the ArcTools libraries that don't have an associated MENU file.

- Tell the user which files were copied and to which directory they were copied.

Lesson 15.3—Designing your own ArcTools

In this lesson

Now that you know how to modify a stand-alone tool, it's time to assimilate the tool into ArcTools. This lesson introduces programs that help create new tools and shows you how to incorporate your custom tools into ArcTools.

Utilities for creating new tools

The ARCTOOLS/UTIL directory contains several scripts that you can use to create template AML programs and menus for both utility and object tools. These scripts are called make util-aml, make-util-menu, make-obj-aml, make-obj-menu, and make-hlp. Each of these scripts takes the name of the tool it creates as an argument. Using these scripts provides a foundation for creating a tool. You can also copy the following files from the UTIL directory and rename them appropriately: tmplt_util.aml, tmplt_util.menu, tmplt_obj.aml, tmplt_obj.menu, and tmplt_hlp.

If you execute the following statement,

```
% $ATHOME/util/make-util-aml freq
freq.aml
```

the following file is created:

```
/*----------------------------------------------------------
/*           Environmental Systems Research Institute, Inc.
/*----------------------------------------------------------
/*    Program: FREQ.AML
/*    Purpose:
/*
/*----------------------------------------------------------
/*      Usage: FREQ INIT {'position'} {'stripe'}
/*                       {MODELESS | MODAL}
/*      Usage: FREQ <routine> {args}
/*
/* Arguments: routine  - name of the routine to be called.
```

```
/*          position - (quoted string) opening menu position.
/*          stripe  - (quoted string) menu stripe displayed.
/*          MODELESS | MODAL    - keyword for creating modal
/*                          thread.
/*
/*   Routines: BAILOUT - error handling.
/*            EXIT - cleanup and exit tool.
/*            HELP - display tool help file.
/*            INIT - initialize tool and invoke menu.
/*            USAGE - return tool usage.
/*
/*   Globals:
/*------------------------------------------------------------
/*      Calls: freq.menu disp_help.aml
/*------------------------------------------------------------
/*      Notes:
/*------------------------------------------------------------
/*   History: tom -  02/25/97 - Original coding
/*============================================================
/*
&args routine arglist:rest
/*
&severity &error &routine bailout
/*
/* Check arguments
&if ^ [null %routine%] &then
  &call %routine%
&else
  &call usage
&return


/*-----------
&routine INIT  /*  {'position'} {'stripe'} {MODELESS | MODAL}
/*-----------
/* Initialize tool interface
/*
&set position = [extract 1 [unquote %arglist%]]
&set stripe  = [extract 2 [unquote %arglist%]]
&set modality = [extract 3 [unquote %arglist%]]
&if [null %position%] or %position%_ = #_ &then
  &set position = &cc &screen &cc
&if [null %stripe%] or %stripe%_ = #_ &then
  &set stripe = Put Your Menu Stripe Here
&if [null %modality%] or %modality%_ = #_ &then
  &set mode =
&else
  &if [translate %modality%] = MODAL &then
    &set mode = &modal
```

```
    &else
      &set mode =
/*
/* Issue thread delete self if thread depth = 2
/* and input is tty
&if [show &thread &depth] = ~
2 and [extract 1 [show &thread &stack]] = tty &then
  &set launch = &thread &delete &self
&else
  &set launch
/*
&if [show &thread &exists tool$freq] &then
   &thread &delete tool$freq
&thread &create tool$freq %mode% ~
  &menu freq ~
  &position [unquote %position%] ~
  &stripe [quote [unquote %stripe%]] ~
  &pinaction '&run freq exit'
%launch%
/*
&return

/*-------------------
&routine USER_ROUTINE
/*-------------------
/*
/* >>>>>>   INSERT USER ROUTINE   <<<<<<
/*
&return

/*-----------
&routine HELP
/*-----------
/* Display help for this tool
&run disp_help freq
&return



/*-----------
&routine USAGE
/*-----------
/* Display usage for this tool
&type Usage: FREQ INIT {'''position'''} {'''stripe'''} ~
{MODELESS | MODAL}
&return &inform

/*-----------
```

```
&routine EXIT
/*-----------
/* Clean up and exit menu
&dv .freq$*
&if [show &thread &exists tool$freq] &then
  &thread &delete tool$freq
&return

/*--------------
&routine BAILOUT
/*--------------
&severity &error &ignore
/* &call exit
&return &warning An error has occurred in routine: %routine%
(FREQ.AML).
```

Notice that the dirty work is already done. The only routines you need to add are the ones that actually perform the operation. These routines should replace &routine USER_ROUTINE in the code. The comment in front of &call exit in the BAILOUT routine should also be removed.

A menu file can be created just as easily by typing the following command:

```
% $ATHOME/util/make-util-menu freq

7
/*------------------------------------------------------------
/*          Environmental Systems Research Institute, Inc.
/*------------------------------------------------------------
/*    Menu: FREQ.MENU
/* Purpose: freq tool interface.
/*------------------------------------------------------------
/*    Calls: freq
/*------------------------------------------------------------
/* Globals: .freq$*
/*------------------------------------------------------------
/*    Notes:
/*------------------------------------------------------------
/* History: tom -  02/25/97 - Original coding
/*============================================================

 %ok  %cancel  %help
%ok BUTTON ~
    HELP 'Apply settings and quit menu' ~
    OK &run freq ok
%cancel BUTTON CANCEL ~
    HELP 'Quit from this menu' ~
    'Cancel' &run freq exit
```

```
%help BUTTON RETURN KEEP ~
    HELP 'Display help about this menu' ~
    'Help' &run freq help
%FORMOPT SETVARIABLES IMMEDIATE MESSAGEVARIABLE .freq$msg
```

By opening this menu in FormEdit, you can add and arrange widgets until you achieve the look and feel that you want.

Creating the basic format for the help file is just as easy using make-hlp.

```
% $ATHOME/util/make-hlp freq

Note: Please refer to $ATHOME/util/help_guidelines for more
detailed information on creating help files

freq
    *Tool Name - (REQUIRED)


This tool is used to...
    *Brief description - (REQUIRED) - one to three sentences


Tool operations:
    *List of operations the tool does - (OPTIONAL)


Using the tool:
    *Brief list of how to use the tool - (REQUIRED)
    step-by-step guide on how to perform each operation
    with the tool
[OK] executes the selections on this menu.

[Cancel] dismisses this menu.

[Help] displays this help file.


Additional topics: (OPTIONAL)
- topic 1
- topic 2
```

Using these three tools, you can easily create the framework for a new tool. From this framework, you can concentrate on creating the functionality of the tool.

FYI

There are also three more tools in the $ATHOME/UTIL directory that are available on the UNIX platform: `amlcheck`, `about`, and `spellcheck`. Of the three, only `amlcheck` is available for Windows NT. For a complete plot of all the icons in ArcTools, look for `iconplot.eps` in the $ATHOME/UTIL directory. For a more detailed discussion of all these tools, open $ATHOME/UTIL/READ.ME in a text editor.

Adding custom tools to the ArcTools interface

If you want to add a new tool to the default ArcTools system, you need to create a way for ArcTools to access it. This may involve adding a choice to a scrolling list, adding a button to a form, or adding a choice to a pulldown menu. If you modified an existing tool, you simply want it to be displayed instead of the original tool. The best way to accomplish either of these goals is to modify the ATPREFS.AML in the $ATHOME/SYSFILES directory.

When ArcTools is initialized, $ATHOME/DRIVERS/ATPREFS_DRIVER.AML is executed. This AML program attempts to execute three copies of ATPREFS.AML in this order: $ATHOME/SYSFILES/ATPREFS.AML, $HOME/ATPREFS.AML, and an ATPREFS.AML in your current directory. ATPREFS.AML sets up the default preferences for that ArcTools session. Each copy, if it exists, of ATPREFS.AML overwrites the last settings. This allows you to copy $ATHOME/SYSFILES/ATPREFS.AML to your $HOME directory and customize it as you like. Your customized ATPREFS.AML will take precedence over the ATPREFS.AML in the SYSFILES directory.

Using ATPREFS.AML, you can establish which directories will be in the &AMLPATH and &MENUPATH before the standard paths are set. Any tools that you place in these preset directories will be executed before the tools in the ArcTools libraries. Looking at ATPREFS.AML, you can see that the change is quite simple:

```
/*******************************************************
/*   USER SPECIFIED AML AND MENU PATHS   *
/************************************
/*
/* Sets user specified AML and MENU paths.  These paths will
/* be searched prior to the ArcTools paths.  By setting the
```

```
/* following two lines, ArcTools can be customized with
/* system, user, or project specific paths for AMLs and
/* menus.  Multiple paths can be set using spaces between
/* the paths.  Either full paths or local paths can be used.
/* For example, /tmp/projects/tools could be specified, or,
/* if ArcTools is started from /tmp/projects, then just
/* "tools" could be used as the AMLPATH and MENUPATH.  These
/* paths will be appended with the ArcTools paths, which are
/* set by $ATHOME/lib/setpaths.aml.

/* Uncomment these lines to set:
/* &amlpath <set the path here>
/* &menupath <set the path here>
```

All you have to do is change these two lines to include the directories of your custom libraries.

```
&amlpath /tavarua2/tom/atcustom /tavarua2/tom/athelp
&menupath /tavarua2/tom/atcustom
```

Now, if you put the TOOL BROWSER tool created in the last lesson into the ATCUSTOM directory, and the TOOLBROWSER.HLP file into the ATHELP directory, the tool will become a part of ArcTools:

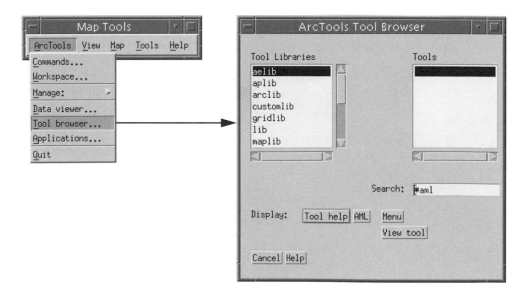

FYI

Placing customized tools in the ARCTOOLS/CUSTOM directory also allows your tools to be incorporated into ArcTools. This is an option if you want to make the tool available to everyone in your organization. In order for this to work, you must have write permission to the CUSTOM directory.

Accessing custom applications

You may want to bundle a group of tools together into a separate application and make this application accessible from ArcTools. To launch a small application, you may simply want to modify an ArcTool. For a large application, you may want to launch it directly from the ArcTools main menu, which looks like this:

The scrolling list is created by accessing the file TOOLSETS.AML in the SYSFILES directory. This program sets the variable that's used by the scrolling list to display the sets of tools. When the user selects a tool set (e.g., Command Tools), the variable is set to the name of the driver AML program with its arguments. The scrolling list uses the CHOICE -PAIRS option within the main menu.

```
&set .arctools$tool_sets = ~
 'Map Tools'          'map_tools INIT'     ~
 'Edit Tools'         'edit_tools INIT'    ~
 'Grid Tools'         'grid_modeler INIT'  ~
 'Command Tools'      'command_tools INIT' ~
 'Land Records'       'property_tools INIT'

&return
```

Selecting the OK button runs the AML program indicated by the subsystem
selected by the user.

To modify the main menu:

• Copy ARCTOOLS/SYSFILES/TOOLSET.AML into your personal custom
  directory.

• Modify TOOLSET.AML to add your application.

• Add a line to your ATPREFS.AML (after setting $AMLPATHS) to run
  TOOLSETS.AML.

With the new application added to ArcTools, the main menu looks like this:

If you want to change the main menu for everyone in your organization, you can
change the TOOLSETS.AML in the SYSFILES directory. Any changes (outside
the custom directory) made to the ARCTOOLS directory may be overwritten if
the software is reinstalled.

There are several other tool set files like TOOLSETS.AML for other scrolling lists. For more information, see the online help topic under *Programming ArcTools >> Customizing ARC/INFO with AML* or examine the READ.ME file in the ARCTOOLS/SYSFILES directory.

Applications picker

Another way to make your application accessible from ArcTools is to put it in the Applications picker. This tool is located under every main ArcTools pulldown menu:

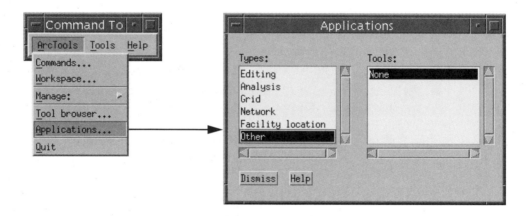

The applications available in the Applications picker are set in the ATPREFS.AML:

```
/* TOOLS:

    /* EDITING APPLICATIONS:
    &set .atprefs$edit_tools = ~
    None              'applpicker NOT_INSTALLED'

    /* ANALYSIS APPLICATIONS:
    &set .atprefs$analysis_tools = ~
    None              'applpicker NOT_INSTALLED'

    /* GRID APPLICATIONS:
    &set .atprefs$grid_tools = ~
    None              'applpicker NOT_INSTALLED'
```

```
/* NETWORK APPLICATIONS:
&set .atprefs$vr_tools = ~
None                    'applpicker NOT_INSTALLED'

/* FACILITY LOCATION APPLICATIONS:
&set .atprefs$la_tools = ~
None                    'applpicker NOT_INSTALLED'

/* OTHER APPLICATIONS:
&set .atprefs$other_tools = ~
None                    'applpicker NOT_INSTALLED'
```

Right now, none of the choices has an application associated with it. To make the
TOOL BROWSER tool show up in the Applications picker under the Other
choice, you simply add one line to ATPREFS.AML:

```
/* OTHER APPLICATIONS:
&set .atprefs$other_tools = ~
'Tool browser'        'toolbrowser init'
```

Now when you launch ArcTools, the tool is available in the Applications picker:

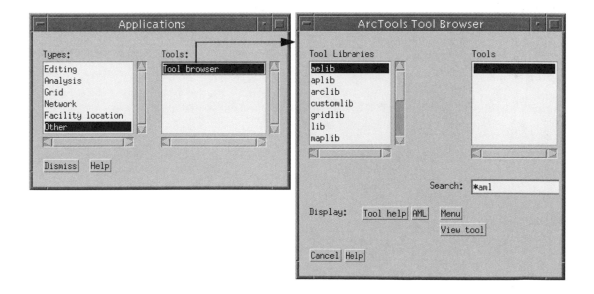

FYI

One of the most common things AML programmers do with ArcTools is extract tools to use in their applications. The TOOL BROWSER can help you find useful tools, and the about script can tell you which tools and icons a particular tool uses. This tool can then be copied into your application. When you work with copied tools, follow the format for modifying them that was presented in this chapter.

In summary

This process of creating your own tools can be simplified with the scripts provided in the ARCTOOLS/UTIL directory. You can integrate a new tool with existing tools by modifying ATPREFS.AML.

Exercises

Exercise 15.3.1

Use ATPREFS.AML to add the TOOL BROWSER tool to your ArcTools interface.

16 Communicating with other software applications

In previous chapters, you've learned how to take advantage of AML in developing ARC/INFO applications. So far, with the exception of the [TASK] function, these applications have been limited to the tools ARC/INFO can directly provide.

Through the process of Inter-Application Communication (IAC), ARC/INFO can communicate with other software packages. IAC allows ARC/INFO either to pass or receive commands to and from other applications. This makes it possible for an AML application to exploit the capabilities of other applications.

In general, computers communicate through a Remote Procedure Call (RPC) protocol. On most machines, RPC is a standard mechanism. Throughout the software industry, two RPC protocols exist: the Open Networking Consortium (ONC) RPC protocol, and the Distributed Computing Environment (DCE) RPC protocol. IAC in ARC/INFO utilizes ONC RPC. This is significant because it means that any software package that supports ONC RPC can communicate directly with ARC/INFO. Although it's more difficult, some packages using DCE RPC can be programmed through low-level code to communicate with ONC RPC.

RPC isn't standard on Microsoft Windows NT. In order to have some Windows applications communicate directly with ARC/INFO, you may have to purchase commercial RPC packages (for instance, NobleNet's EZRPC or ONC RPC for Windows). This chapter teaches you how to use IAC with ARC/INFO and how to package RPC commands through AML.

This chapter covers the following topics:

Lesson 16.1—Introducing IAC

- IAC basics: The client/server architecture
- IAC basics: Asynchronous processing
- When to use IAC

Lesson 16.2—ARC/INFO as the IAC server

- Overview: The IAC ARC/INFO server
- Opening the ARC/INFO server process
- Processing client requests: An ARC/INFO server and ARC/INFO client
- Closing the ARC/INFO server process
- Processing client requests: An ARC/INFO server and ArcView® GIS client
- Processing client requests: An ARC/INFO server and C client
- Processing client requests: An ARC/INFO server and Visual Basic client
- Using multiple ARC/INFO servers

Lesson 16.3—ARC/INFO as the IAC client

- Overview: Establishing ARC/INFO as a client to another software's server
- Using an ARC/INFO client with an ARC/INFO server
- Using an ARC/INFO client with an ArcView GIS server
- Using an ARC/INFO client with a C server

Lesson 16.1—Introducing IAC

In this lesson

ARC/INFO software's IAC allows software applications on local or remotely networked machines to communicate with each other. Using a communication based on a client/server architecture, IAC makes it possible to develop AML applications that seamlessly integrate the capabilities of other applications.

The ability to communicate with other software applications is an immensely powerful tool. In this lesson, you'll learn the underlying concepts of how IAC operates and how it can help your ARC/INFO application become more versatile and robust.

IAC basics: The client/server architecture

As far as IAC is concerned, an ARC/INFO session (or process) behaves in one of two ways: as a client to another package, or as a server for another package. A *client* makes requests of a server; the *server* processes the requests. Optionally, the client can ask for results from the server. The client and server are separate processes that usually communicate through a network (though they can be on the same machine).

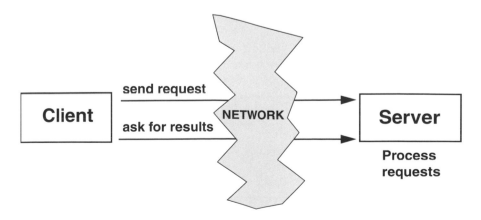

You can have ARC/INFO clients to another ARC/INFO session. For example, multiple clients can simultaneously pass display information to one ARC/INFO server that's operating in ARCPLOT. You can have ArcView GIS as a client to ARC/INFO. The ArcView client would then have access to ARC/INFO's geoprocessing tools. You can also have Microsoft Visual Basic or C program clients access an ARC/INFO server.

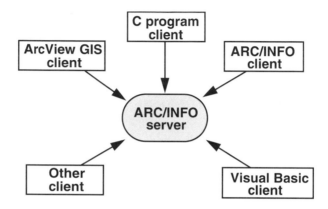

ARC/INFO doesn't have to be the server. You can have it as a client to any other software package that supports ONC RPC. It's important to note that an ARC/INFO session can be both a client and a server at the same time.

FYI

If you're running an ARC/INFO server on one machine and an ARC/INFO client on another machine, you need two ARC/INFO licenses. If you're running an ARC/INFO server and ARC/INFO client on the same machine, you only need one license.

On Windows NT, if you have a Windows NT ARC/INFO server with networked PC machines, you can use IAC to communicate between the satellites without purchasing additional software licenses.

Q

Question 16-1: Do an IAC client and server have to be on networked machines or can they operate on the same machine?

(Answer on page 16-7)

IAC basics: Asynchronous processing

Remember that IAC uses RPC to communicate between machines. By nature, RPC is synchronous. This means that after you send a request, you can't do any work until the request is finished. (This is similar to launching a nonthreaded AML program from a menu: the menu is busy until the AML program completes processing.) IAC in ARC/INFO, however, appears asynchronous to the clients. *Asynchronous processing* means that commands can be executed simultaneously on both server and client. In other words, the IAC client sends requests to the ARC/INFO server, then continues processing. The server process receives client requests and can execute commands from the server keyboard or requests received from the client.

The ARC/INFO server's ability to appear asynchronous is managed through a dispatcher. This server dispatcher acts as a manager of all the requests that clients pass to the server. You'll learn about the dispatcher in more detail in the next lesson.

Asynchronous processing allows the client and server to act independently. This means that in some cases the client may have to query the server for results. If the client asks the server to display a coverage in ARCPLOT, the server simply displays the coverage. The client doesn't have to do anything else. If, on the other hand, the client asks the server to do a mathematical calculation, the client must then query the server to get the results of the calculation.

When to use IAC

IAC allows a program like ARC/INFO to react to real-time events without user intervention. Take the example of an Emergency 911 Dispatch Operation. First, assume that many police precincts within a city have networked machines tied into a central dispatch office. These satellite precincts collect crime address data and geocode the data using their local machine. At the same time, they could send the coordinate information for each crime to the central dispatch office. The central office would receive all the data from the precincts and perform emergency vehicle routing. In this example, the information flow is from the clients to the server.

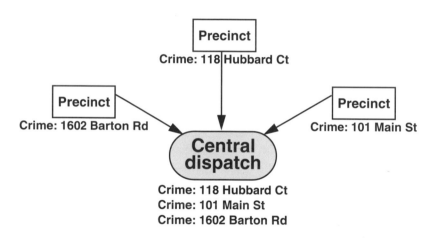

Another example is an Automated Vehicle Location (AVL) application. AVL systems track the position and movement of vehicles in real time and are used in a wide variety of transportation industries to make decisions about routing and dispatch. Multiple ARC/INFO users could simultaneously receive information regarding current vehicle location and update their individual map displays with everyone else's data. In this example, the information flow is from the client to the server and then back to the client.

IAC can also be used to speed up applications that involve intensive computer processing. For example, ARC/INFO can handle some statistical processing, but isn't designed for complex statistical computations. While it's true that you can write a sophisticated AML program to work through some complicated statistics, these AML programs tend to slow down your machine dramatically. In cases like this, it's more efficient to have a smaller AML program use IAC to pass the computation to a C program.

In summary

Inter-Application Communication uses a client/server method to communicate between software applications. IAC is a powerful tool that allows ARC/INFO to be either a server or a client to other programs. The ARC/INFO client and server can behave asynchronously by executing commands independently and simultaneously. AML can be used to seamlessly integrate the capabilities of other software packages into an ARC/INFO application. IAC has improved Emergency 911 Dispatching and Automated Vehicle Location systems that update information in real time.

Answer 16-1: Client and server processes are usually on separate networked machines, but they can be on the same machine.

Exercises

Exercise 16.1.1

What is meant by IAC's client/server architecture?

Exercise 16.1.2

Why is it important to know what type of RPC protocol ARC/INFO supports?

Exercise 16.1.3

What type of RPC package comes with Microsoft Windows NT?

Lesson 16.2—ARC/INFO as the IAC server

In this lesson

The basic premise of IAC is that client and server pass information to one another. This lesson teaches you how to set up ARC/INFO as a server to a number of clients, including ARC/INFO, ArcView GIS, Visual Basic, and C.

AML requests and directives drive IAC. In this lesson, you'll also learn how AML programs can be used to automate the client/server interaction process. Some client connections to an ARC/INFO server are briefly outlined. First, it will be helpful to have an overview of the IAC ARC/INFO server.

Overview: The IAC ARC/INFO server

An ARC/INFO server can react to requests from different clients while at the same time supporting user interaction from the server keyboard. To accomplish this, the ARC/INFO server processes client requests whenever it's waiting for the user to finish typing a command. Incoming client requests are queued while the ARC/INFO server processes another client request or a command executed directly at the server terminal. ARC/INFO can be an IAC server to multiple clients at the same time. For instance, an ARC/INFO server can receive requests from both ARC/INFO and ArcView GIS clients.

FYI

There's no interrupt capability with IAC. If the ARC/INFO server is busy executing a command or an AML program, all client requests continue to queue until the first command is completed. Even commands executed from the server must occasionally wait.

Remember, there can be only one input source to the AML processor at a time. Note that the following commands wait for user interaction. In these instances, no other requests are processed until the current request is completed.

```
ARC:     &setvar name [RESPONSE 'Enter your name']

ARCPLOT: mapextent *
```

After an ARC/INFO IAC server process is initiated, it returns control to the user. In other words, the ARC/INFO server is running in the background. This means that you can still issue commands at the server keyboard, run AML programs, or interact with menus while also processing requests from clients. Again, remember that client requests may be queued while the user on the server keyboard completes a command.

Although it's possible to set up multiple servers, it's simpler to consider a single server application first. Clients of an ARC/INFO server don't actually communicate directly with the server, but rather with an intermediary called a *dispatcher*. The dispatcher is an active program that runs in memory (there's no separate window for it).

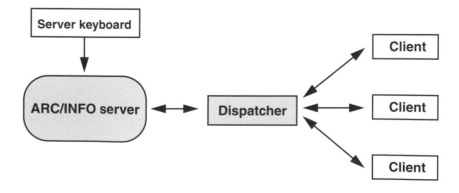

Requests from the clients are passed to the dispatcher. The dispatcher checks to see if the server is busy before giving any instructions. If the server isn't busy, the dispatcher gives the request to the server and the server executes the command. If the server is busy, the dispatcher queues all client requests until the server is free. When the server is free, the dispatcher gets client requests from the queue until the queue is empty.

There are three main parts to the running of an IAC ARC/INFO server: opening the server process, processing client requests, and closing the server process.

Opening the ARC/INFO server process

One of the primary uses of IAC is with an ARC/INFO server and another application as a client. To put an ARC/INFO process in server mode, start ARC and use the [IACOPEN] function.

```
Usage: [IACOPEN {connect_file}]

&type [IACOPEN]
Host: carmel Program no: 40000000 Version no:1
0
```

Remember that AML functions return information. The first returned line (starting Host:) is the connect information returned by the server. This information includes the server hostname (carmel), program number (40000000), and version number (1).

The 0 on the second line shows that the server has successfully started. Any number other than zero indicates some type of error. See the chart below for a list of errors you may encounter.

Returned value from [IACOPEN]	Meaning
0	Server opened successfully
5	Bad number of arguments
100	Cannot open connect file
101	Server port already open
102	Bad program number
103	Bad version number
105	Cannot create connect file
106	Cannot start dispatcher

You can be at any ARC/INFO prompt to start the server. For example, the following commands do the same thing:

```
Arc: &type [IACOPEN]
Arcplot: &type [IACOPEN]
Grid: &type [IACOPEN]
Arcedit: &type [IACOPEN]
...
```

FYI

If you're using IAC on Microsoft Windows NT, you first need to run a program called PORTMAP. The UNIX environment automatically supports a networked environment. PCs, on the other hand, have to be told to support a TCP/IP networked environment. The PORTMAP command instructs your PC to build a port mapping environment so it can utilize RPC connections. PORTMAP is located in %ARCHOME%\BIN. The commands to open an IAC ARC/INFO server on Windows NT are:

```
/* In one ARC window, type the following:
%ARCHOME%\bin\portmap.exe
/* PORTMAP will lock the window, so you must open another
/* ARC window and type the following:
&type [IACOPEN]
Host:carmel Program no: 40000001 Version no:1
0
```

Opening the server using &TYPE [IACOPEN] may seem easy, but it actually results in more work later because you'll have to remember the information that was &TYPEd to the screen. To make life easier, use [IACOPEN] with the {connect_file} option to name an output text file that contains the connection information. You can later use the information in this file when connecting the client to the server. Note that even if you specify a {connect_file}, the information is still printed to the screen.

```
Arc: &type [IACOPEN connectfile]
Host: carmel Program no: 40000001 Version no:1
0
```

The file `connectfile` contains the host name, the program number, and the version number. To view the contents of the file, type `&popup connectfile`.

Note that the first time the server was opened (page 16-11), the program number was 40000000. The next time it was 40000001. When you first open the server, the program number starts at 40000000; it then increments by 1 for each dispatcher. So another way of looking at the program number is to see it as the address of the dispatcher.

FYI

> Although the ARC/INFO server ID (e.g., 40000000) appears to be a decimal number, you should remember that it's actually a hexadecimal number. This point becomes significant if you're connecting an ARC/INFO server to another software client, like ArcView GIS. For example, if the server ID is 40000000, the ArcView client references it as 0x40000000.

ARC/INFO doesn't automatically give you an error message if you're unable to open a server process. Be sure to set [IACOPEN] to a variable so you can test the value of the variable to make sure the server opened successfully.

```
&setvar openstatus = [IACOPEN connectfile]
&type %openstatus%
0
```

Instead of typing these commands each time, you can have an AML program
open the ARC/INFO server, then test the value of the status variable:

```
/* Some aml that opens the IAC ARC/INFO server
/* and tests to make sure it opened successfully.
&setvar openstatus = [IACOPEN connectfile]
&if %openstatus% ne 0 &then
  &do
    &return Unable to open IAC server, error %openstatus%
  &end
&else &type ARC/INFO server opened successfully!
```

FYI

You'll save time in the long run by using [IACOPEN] with the {connect_file}
option. Since clients need this connection information to access the server,
having a connect file simplifies the connection process. This will be discussed
again later in the context of [IACCONNECT].

Once you use [IACOPEN] successfully, a server port and dispatcher are opened
for the current ARC/INFO session. Clients will then be able to connect and pass
requests. If you look at the server window, you'll see the usual `Arc:` prompt.
Remember that you're still able to execute any command and/or menu from this
server window.

Q

Question 16-2: Fill in the missing command lines to check for different
types of errors that may occur when the ARC/INFO server opens.

```
/* Some aml that opens the IAC ARC/INFO server
/* and tests to make sure it opened successfully.
&setvar openstatus = [IACOPEN connectfile]
&select %openstatus%
  &when 0 &type ARC/INFO server opened successfully!
  &when _____ &return &error Bad number of arguments
  &when _____ &return &error Cannot open connect file
  &when _____ &return &error Server port already open
  &when _____ &return &error _____
  &when _____ &return &error _____
  &when _____ &return &error _____
  &when _____ &return &error _____
&end
```

(Answer on page 16-24)

FYI

[IACOPEN] with the -AUTH option enables you to control who's permitted to use the server. Although anyone can connect to any network server, only clients listed in the user list can send requests. Consult the online help under [IACOPEN] for more information.

Processing client requests: An ARC/INFO server and ARC/INFO client

You can use your ARC/INFO session as a server to another ARC/INFO client. Follow the instructions below to see how IAC works in this situation.

- First, use the same computer terminal to open two ARC/INFO windows pointing to the same workspace.
- Second, designate one of the windows to be the server window. Open the IAC server channel. If you're using Windows NT, see the FYI box on page 16-12.

In the server window, type:

```
Arc: &type [IACOPEN connectfile]
Host: carmel Program no: 40000000 Version no: 1
0
```

- Third, from the client window, use [IACCONNECT] to connect to the server. Note that because the connect file has the connection parameters, the ARC/INFO client knows which ARC/INFO server to communicate with. View the connection file that the server created. Now connect to the server. Try the following:

In the client window, type:

```
&popup connectfile
&sv server_id = [IACCONNECT connectfile connectstatus]
```

> It's very important to keep track of which directory the server and client point to. If the server located at `/carmel3/ridland/iac` creates a connect file, then the client must either be at that location or specify a path to the connect file. For example, if the client is located at `/tavarua2/tom`, it must access the connect file by typing:
>
> ```
> &sv server_id = [IACCONNECT ~
> /carmel3/ridland/iac/connectfile  connectstatus]
> ```

- Fourth, from the client window, use [IACREQUEST] to issue commands you want to execute in the server window.

```
&sv  job_id =  [IACREQUEST %server_id% 1 '&type hello' ~
                reqstat]
```

If everything worked, you should see `hello` in the server window. Although the example is a simple one, it shows you how an ARC/INFO server can be used with an ARC/INFO client. In the IAC client process, [IACCONNECT] and [IACREQUEST] will be discussed in much greater detail. The next step would be to close the ARC/INFO server process.

Closing the ARC/INFO server process

Use the [IACCLOSE] function to disable your ARC/INFO process as a server. After completing this command, your machine will still accept `arc` commands from your server's keyboard. You're only closing down the server to client machines—you're not closing your own `arc` process.

```
Usage: [IACCLOSE]
```

```
Arc: &type [IACCLOSE]
0
```

> [IACCLOSE] must be executed in the same ARC/INFO session (window) in which [IACOPEN] is issued.

The returned integer is 0 when the server port closes successfully. Any number other than zero indicates some type of error. See the chart below for a list of errors you may encounter.

Returned value from [IACCLOSE]	Meaning
0	Closed successfully
100	No server
101	Other error

Like [IACOPEN], [IACCLOSE] doesn't automatically notify you if your server successfully closed. You can write AML code that sets [IACCLOSE] to a variable, then tests the value of the variable. For example:

```
&setvar closestat = [IACCLOSE]
&if %closestat% ne 0 &then
  &do
    &return Trouble closing server, error %closestat%
  &end
```

Or you can eliminate the need for a local variable called closestat by using the following code:

```
&if [IACCLOSE] ne 0 &then
  &do
    &return Trouble closing server, error %closestat%
  &end
```

Or you can test each value and report to the user the name of the error:

```
&select [IACCLOSE]
  &when 0 &type Server closed successfully!
  &when 100 &return No server detected.
  &when 101 &return Undocumented error.
&end
```

Processing client requests: An ARC/INFO server and ArcView GIS client

If you have ArcView GIS users in your organization, you may want them to be able to access the functionality of ARC/INFO. The following AML code and Avenue™ script allows users to enter and execute ARC/INFO commands from within their ArcView project. The example assumes that you have a copy of ArcView GIS Version 2.1 (or later version) and know how to create, compile, and execute an Avenue script. (Avenue is the ArcView scripting language.)

Follow the instructions below for a quick self-run demonstration of IAC with an ARC/INFO server and ArcView client.

- First, open an ARC/INFO window and, from another window, launch the ArcView application.
- Second, open the ARC/INFO IAC server channel. Note that this example doesn't use a {connect_file}.

In the ARC/INFO server window, type:

```
Arc: &type [IACOPEN]
Host: carmel Program no: 40000000 Version no: 1
0
```

- Third, within your ArcView application, create a new project and a new script with the following lines in it:

```
                          iac script
'iac avenue script for using ArcView as a
'client to an ARC/INFO server
client = RPCClient.Make("carmel",0x40000000,1)
aiCommand = msgBox.input
            ("Enter the ARC/INFO command",
             "iactest","")
jobID = client.Execute(1,aiCommand,String)
client.close
```

Note that the host, program, and version number from the ARC/INFO window are duplicated in the RPCClient.Make request. Note also that ArcView GIS needs to have the program number written in hexadecimal format (0x40000000) instead of 40000000. (You could give the equivalent decimal value if you knew it.) Duplicate these connection parameters so that ArcView knows which ARC/INFO dispatcher to communicate with. The Client.Execute request passes whatever the aiCommand is to the ARC/INFO server.

Compile and run the Avenue script. You should see an input box appear in your ArcView project that allows you to type in ARC/INFO commands.

These ARC/INFO commands are then passed to the ARC/INFO server. ArcView is still active while ARC/INFO processes the command. In other words, after the ARC/INFO command is passed to the server, control is returned to the ArcView user.

A sample command you can use is MAPJOIN. Since ArcView can't combine (merge) coverages, have ArcView pass the MAPJOIN command to the ARC/INFO server. You can also query the ARC/INFO server to find out if it's finished merging the coverages. (Querying the server is discussed in lesson 16.3.) After the server completes the MAPJOIN, you can add the new coverage as a theme to your ArcView view.

Again, this is an example. For further information on implementing IAC in ArcView GIS, refer to the ArcView online help.

FYI

There are sample IAC Avenue scripts that you can download from the ESRI home page (www.esri.com).

Processing client requests: An ARC/INFO server and C client

Any program using ONC RPC can access the functionality of ARC/INFO. The C programming language, for example, doesn't have an inherent georelational data model. In other words, a C program can have a file of x,y coordinates, but can't easily make a coverage from it. It can, however, take advantage of the ARC/INFO data model. For instance, you can use a C client to pass raw coordinates to ARC/INFO and use ARC/INFO commands to create and display a coverage from them.

The next example assumes that you know how to create, compile, and execute a C program.

FYI

There's a sample C program that comes with ARC/INFO software. This program (located at $ARCHOME/AICLIENT) allows users to enter ARC/INFO commands from within their C program. Inside the directory is a README file for more information on how to use this C program.

Follow the instructions below for a quick, self-run demonstration of IAC with an ARC/INFO server and C program client.

• First, open an ARC/INFO window, then open a UNIX or DOS window.

If you're using Microsoft Windows NT, remember that RPC doesn't come with the standard Windows package. You must obtain an additional software program, such as NobleNet EZRPC, or download another package (even ONC RPC) from the Internet before you can use an ARC/INFO server and C client.

• Second, from the ARC/INFO window, initiate the ARC/INFO server. (If you don't remember how, refer back to page 16-11.) Make sure you use the {connect_file} option.

- Third, copy the needed C program from where ARC/INFO is located. If the [COPY] function returns a zero, the file was successfully copied.

```
Arc: &type [COPY $ARCHOME/aiclient/arcclient arcclient -FILE]
0
```

- Fourth, from the client window, launch the sample C program by typing arcclient. You'll be prompted for the name of the connection file.

In the client C window, run the ARCCLIENT program.

```
arcclient
Give the connection file name: connectfile
RESULT hello from arc
Give request string (or type exit): <enter ARC/INFO command>
```

From the C program prompt, you can issue individual ARC/INFO commands or run an AML program. These commands are passed to, then executed on, the ARC/INFO server. While the server is executing your command, the C client waits for you to enter another ARC/INFO command. In other words, the C application is paused while ARC/INFO processes your command. When the user types exit, the C program client closes the connection to ARC/INFO. Remember that this is only one example of how a C program client can interact with ARC/INFO.

FYI

> Writing a C client to ARC/INFO requires a thorough knowledge of the C language and RPC programming techniques. As C is considered a low-level programming language (low-level languages give programmers more control at the expense of simplicity), beginners often find it easier to write IAC clients in a high-level scripting language, such as AML or Avenue.

Processing client requests: An ARC/INFO server and Visual Basic client

If you're proficient with C, you can construct a Dynamic Link Library (DLL) from the client source code. This allows communication with ARC/INFO, via RPC, from application builders such as Visual Basic, Power Builder, and Delphi. Remember that ONC RPC-compliant libraries aren't standard on Windows NT, and you must link against these libraries when building your client.

Using multiple ARC/INFO servers

If you're using IAC with many clients but only one server, the demand may be too much for your server. As a result, the server may be unacceptably slow. IAC allows you to use multiple ARC/INFO servers to process client requests.

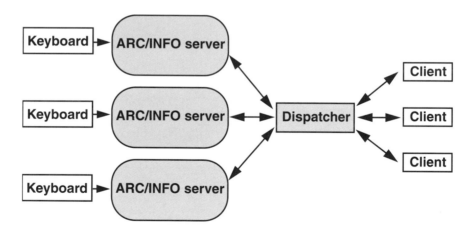

The [IACOPEN] function with the -DISPATCH option allows multiple servers to be connected to the same dispatcher. The dispatcher then sends requests to the first available server. If no servers are available for processing, a queue is formed. When a server becomes available, the dispatcher retrieves a request from the queue.

If you want all requests to go through the same dispatcher, you must specify that dispatcher for each new server. For example, start a server with its dispatcher:

```
&sv openstatus [iacopen connectfile]
```

One server process and one dispatcher have been initiated. Now, to have the next server use the same dispatcher, type:

```
&sv openstatus2 [iacopen -dispatch connectfile]
```

Note that all you need to do is use the same connection parameters (stored in the connect file). To add a third server on another machine and have it use the same dispatcher, type:

```
&sv openstatus3 [iacopen -dispatch connectfile]
```

All the client sees is one server (available through the single connect file called `connectfile`).

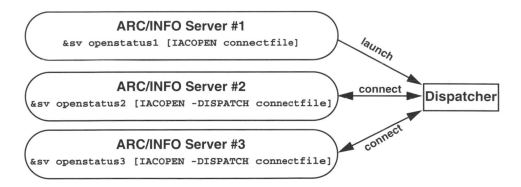

Multiple servers are extremely useful when you have a large volume of IAC traffic. The same dispatcher (accessed through the {connect_file} argument) can be used to simplify client interaction.

In summary

When ARC/INFO is a server, up to 100 clients can simultaneously connect. The server can react to requests from these clients while supporting user interaction from the server machine itself.

Other programs can use ARC/INFO by connecting to it as a client. ArcView GIS and other programs like C or Visual Basic can execute ARC/INFO commands, thus making their own applications more powerful.

Opening the ARC/INFO server involves [IACOPEN]. There are optional arguments for naming a system file to contain the server connection information. You can control who has access to the server with security measures. If you experience a large volume of RPC traffic, you have the ability to use multiple servers, thus increasing your system's performance. The server stays open regardless of which ARC/INFO subsystem you're in. For example, if you start the server while in ARC, then move to ARCPLOT, the server remains active. To close the server, use [IACCLOSE].

Answer 16-2: Fill in the missing command lines to check for different types of errors when the ARC/INFO server opens.

```
/* Some aml that opens the IAC ARC/INFO server
/* and tests to make sure it opened successfully.
&setvar openstatus = [IACOPEN connectfile]
&select %openstatus%
  &when 0 &type ARC/INFO server opened successfully!
  &when 5 &return &error Bad number of arguments
  &when 100 &return &error Cannot open connect file
  &when 101 &return &error Server port already open
  &when 102 &return &error Bad program number
  &when 103 &return &error Bad version number
  &when 105 &return &error Cannot create connect file
  &when 106 &return &error Cannot start dispatcher
&end
```

Exercises

Exercise 16.2.1

Why is there no interrupt capability with IAC?

Exercise 16.2.2

Why is it a good idea to use a {connect_file} with the [IACOPEN] function?

Exercise 16.2.3

What does the following line of AML code indicate?

```
&type [IACOPEN myfile]
105
```

Exercise 16.2.4

Why is it necessary to use the PORTMAP command if you're running IAC on Windows NT?

Exercise 16.2.5

Can ARC/INFO be an IAC server to both an ArcView GIS and ARC/INFO client at the same time? If so, which client has priority when sending commands?

Exercise 16.2.6

What's wrong with the following code?

```
&type [IACCLOSE myfile]
```

Lesson 16.3—ARC/INFO as the IAC client

In this lesson

While the previous lesson focused on ARC/INFO as a server, this lesson deals with the IAC ARC/INFO client. An ARC/INFO client can simultaneously connect to up to 100 servers, such as statistics packages, image-processing software, or GIS, like ARC/INFO or ArcView. Thus IAC makes it possible for an AML application to exploit the capabilities of other applications. Connections to other servers are maintained when the ARC/INFO client moves from one environment to another (e.g., from ARC to ARCPLOT).

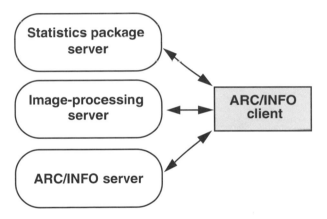

In this lesson, you'll learn how to set up and use an ARC/INFO client with an ARC/INFO server, and an ARC/INFO client with an ArcView server. Connecting an ARC/INFO client to another application, like a C program, will also be discussed briefly.

Overview: Establishing ARC/INFO as a client to another software's server

To connect your ARC/INFO session client to a server, you need to know the host name of the machine the server is on, the program number, and the version number of the server. If you're accessing an ARC/INFO server, this information is available in the optional {connect_file} argument used with [IACOPEN]. If you're accessing another software's server, you need to obtain the server information some other way, perhaps by asking the system administrator.

Using an ARC/INFO client with an ARC/INFO server

There are four general steps to using ARC/INFO as an IAC client to an ARC/INFO server. These steps are (1) connecting to the server, (2) sending requests to the server, (3) optionally getting results from the server, and (4) disconnecting from the server.

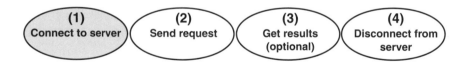

Step 1: Connecting to the server

To connect to the server, use [IACCONNECT]. Recall that when an IAC server is opened, the <hostname>, <program_number>, and <version_number> are automatically generated by that server. When the connection is made to an ARC/INFO server, these three pieces of information can be stored in a {connect_file} generated by [IACOPEN]. Note that before the client can connect to the server, the server port should be opened.

Follow the steps below for a self-run demonstration of IAC with an ARC/INFO client and an ARC/INFO server.

- Use the same computer terminal to open two ARC/INFO windows.
- Designate one of the windows to be the server window, and open the IAC server channel. For instructions, see page 16-11.
- From the client window, use [IACCONNECT] to connect to the server.

```
usage: [IACCONNECT <connect_file> <status_variable>]
```

[IACCONNECT] returns an integer that represents the unique ID of the server. You use this integer when you pass requests to the server (so the ARC/INFO client remembers which server dispatcher it's connected to). It's a good idea to set [IACCONNECT] to a variable. That way, it's easier to have AML keep track of the server ID for you.

The parameters for the server can be specified using a `<connect_ file>`. Use the `<status_variable>` to test whether the connection is successful. For example, the following command would connect to the IAC server:

```
&setvar server_id = [IACCONNECT connectfile connectstatus]
```

FYI

You can also type in the connection parameters with an alternate usage of [IACCONNECT]. With the following usage, you state explicitly the host name, program number, and version number:

```
usage: [IACCONNECT <hostname><program_number> ~
                    <version_number> <status_variable>]
&setvar server_id = [IACCONNECT carmel 40000000 ~
                    1 connectstatus]
```

You should create an AML program that tests the value of the `<status_variable>`. Any number other than zero indicates some type of error. The chart below lists of the types of errors you may encounter.

Value of <status_variable> from [IACCONNECT]	Meaning
0	Connection successful
5	Bad number of arguments
100	Cannot open connect file
101	Cannot connect to server
102	No more server ports
103	Bad version number
109	Server program not registered

The following example demonstrates error checking with the `connectstatus` variable:

```
/* AML code that connects to the server then tests the
/* <status_variable> called connectstatus.
&setvar server_id = [IACCONNECT connectfile connectstatus]
&if %connectstatus% ne 0 &then
 &do
    &return IAC connection error, number %connectstatus%
 &end
```

You can give a message to the user specifying the error. For example:

```
/* AML code that connects to the server then tests the
/* <status_variable> called connectstatus.
&setvar server_id = [IACCONNECT connectfile connectstatus]
&select %connectstatus%
 &when 0 &type Connection to server %server_id% was successful
 &when 5 &return &error Bad number of IAC arguments
 &when 100 &return &error Cannot open connect file
 &when 101 &return &error Cannot connect to server
 &when 102 &return &error No more server ports
 &when 103 &return &error Bad version number
 &when 109 &return &error Server program not registered
&end
```

FYI

If you're connecting from a Windows NT ARC/INFO client to a UNIX ARC/INFO server, you may have to perform an additional setup task. For example, if your PC can't see the UNIX server, you may need your system administrator to modify your PC's *lmhosts* file. This file is located at `<drive_letter>:\winnt\system32\drivers\etc\lmhosts`. You may need to add the proper information to identify the UNIX machine you want to connect to.

After successfully connecting to the server, you're ready to send requests.

Step 2: Sending requests to the server

To send requests to a server, use [IACREQUEST]. A request can be a single ARC/INFO command or an AML program run on the server. Think of a request as an instruction or task for the server to carry out. Each task you send should have a unique job number. When you give the server tasks, you shouldn't just say "do a job," then "do another job." Instead, you should be specific and tell the server to "do job1," then "do job2," then "do job3," and so on.

To help you manage your jobs, the [IACREQUEST] function returns a unique job number. You want to set this function to a variable so you can later query the status of the job that you sent to the server.

```
usage: &setvar job_id = [IACREQUEST <server_id> ~
                         <procedure_number> <request_string> ~
                         <status_variable> {timeout_secs}]
```

If you're using an ARC/INFO server, control is returned to you after you issue the [IACREQUEST] command. Recall that IAC in ARC/INFO is asynchronous. This means that the server obtains the request, returns a job number, and allows you to continue while the server processes your job. To find out whether the job is finished, however, you have to query the server. If the server isn't an ARC/INFO server, control may not be returned to you until after the request is finished. (Note that if you're experiencing heavy network traffic or can't afford to be locked up for any length of time, you can use the {timeout_secs} argument to abort the request after a given time period.) Try the following simple request from the ARC/INFO client window:

```
&sv job1 = [IACREQUEST %server_id% 1 '&type hello' reqstat]
```

Notice that the <procedure_number> is 1 for job1. When you execute commands on the ARC/INFO server, always use procedure number 1, regardless of what the request is.

For some requests, you may not be as concerned about retrieving results from the server. Take the example of job1. The server is asked to perform a simple operation. Other simple tasks that the server might perform include:

```
&sv job2 = [IACREQUEST %server_id% 1 'arcplot' reqstat]
&sv job3 = [IACREQUEST %server_id% 1 'lineset color' reqstat]
&sv job4 = [IACREQUEST %server_id% 1 'mapextent zone' reqstat]
&sv job5 = [IACREQUEST %server_id% 1 'arcs zone' reqstat]
```

If you issue these types of requests, one after the next, it's usually safe to assume the server has completed the request. Possible complications that may arise will be discussed shortly.

Internally, the system returns a unique value that you set to a job number. This way, the system can keep track of your jobs. For example, if you listed the values of the job variables, you might see:

```
Arc: &listlocal
Local:
      job1                    1
      job2                    2
      job3                    3
      job4                    4
      job5                    5
```

Note that each job can use the same <status_variable>. You can reuse `reqstat` if you make a looping AML program that tests for the value of the status variable each time a command is sent. A <status_variable> value of zero (0) indicates that the request was successfully sent to the server. See below for a list of other values.

Value of <status_variable> from [IACREQUEST]	Meaning
0	Request successful
100	Invalid server-ID
101	Bad procedure number
110	Request timeout
199	Unknown error

The following sample code does error checking with the `reqstat` variable:

```
&if %reqstat% ne 0 &then
 &do
    &return IAC request error, number %reqstat%
 &end
```

Be careful with commands that invoke menus or prompt you to type in information. These menus or prompts are displayed on the server machine's terminal instead of the client window. For example, the following requests wouldn't be advisable:

```
&sv job2 = [IACREQUEST %server_id% 1 'mapextent ~
           [getcover]' reqstat]
&sv job2 = [IACREQUEST %server_id% 1 ~
           [QUOTE &sv qagain = [query 'Do you want to draw ~
           another coverage?']] reqstat]
```

As you've seen, connecting an ARC/INFO client to an ARC/INFO server involves some fairly convoluted commands. You can simplify the process and build your own AML program that takes care of all the functions and job numbers behind the scenes. Consider the following example:

```
/* iacloop.aml
/* Allow client to easily type commands to send to server
[1]   &sv server_id = [IACCONNECT connectfile connectstat]
[2]   &if %connectstat% eq 0 &then &type Connect successful...
[3]     &else &return Connection was not successful, error ~
              number %connectstat%
[4]   &sv cmd = [RESPONSE ~
        [QUOTE Enter command for A/I server ("<cr>" to end)] ~
        .FALSE.]
[5]   &do &while [quote %cmd%] nc [quote .FALSE.]
[6]     &sv job = [IACREQUEST %server_id% 1 %cmd% reqstat]
[7]     &if %reqstat% ne 0 &then &return Client request error,~
              number %reqstat%
[8]     &sv cmd = [RESPONSE ~
          [quote Enter command for A/I server ("<cr>" to end)]~
            .FALSE.]
[9]   &end
[10] &sv disconnectstat = [IACDISCONNECT %server_id%]
[11] &if %disconnectstat% eq 0 &then &type Disconnect was ~
              successful...
[12]    &else &return Disconnect was not successful, error ~
              number %disconnectstat%
[13] &return
```

In IACLOOP.AML, lines [1], [2], and [3] connect to the server and perform error checking to make sure the connection is successful. Note that [IACCONNECT] is using a connect file. If you are in a different workspace from the server, you have to specify a path to the connect file. You could incorporate a scrolling list of files (using the [GETFILE] function) that allow the user to choose a proper connect file.

Line [4] uses the [RESPONSE] function to ask the user for the first command to send to the server. Whichever command the user types is a quoted string. If the user simply types the carriage return, the cmd variable will default to .FALSE.

Lines [5] and [6] start a looping process and send the first command to the server. The loop continues until the user types <cr> and the cmd variable is set to .FALSE.

Line [7] does some error checking to make sure the request was successfully sent to the server.

Line [8] asks the user to type another command to send to the server.

Line [9] ends the &do &while loop.

Lines [10], [11], and [12] disconnect from the server and check for errors to make sure the disconnect is successful.

When you run the AML program from the client window, you see the following prompt:

```
Arc: &run iacloop
Enter command for A/I server ("<cr>" to end):
```

You can keep entering commands as long as you don't enter a blank line. When you type <cr> by itself, the client stops sending commands and disconnects from the server. Remember that the server can execute commands while you, as a client, are also executing a command. Although client commands may be queued, the fact that you have more than one person inputting commands may cause problems. For example:

```
Server command: ARCPLOT
Client command: &sv job5 = [IACREQUEST %server_id% 1 ~
                'MAPJOIN...' reqstat]
```

The client's MAPJOIN request would fail because the server process is actually in ARCPLOT instead of ARC. As a client, it would be wise to test the program the server is in before sending the command. For example:

```
Client command: &sv job5 = [IACREQUEST %server_id% 1 ~
                '&RUN MJ' reqstat]

/*mj.aml
&if [show program] eq ARC &then MAPJOIN ...
&if [show program] eq ARCPLOT &then ARC MAPJOIN ...
(...)
```

Now, when the server executes MJ.AML, the AML program is prepared for whatever subsystem the server is in.

So far, these examples have illustrated how you can send requests to the server. To obtain results from the server, an additional step is required.

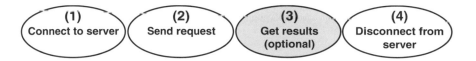

Step 3: Getting results from the server (optional)

The ARC/INFO server allows the client to specify a job ID number and query it for results. If you're doing numerical calculations or other operations on the server for which you need a result, use [IACREQUEST] with a procedure number of 1 in conjunction with the &IACRETURN directive. Try the following simple request from the client window:

```
&sv job2 = [IACREQUEST %server_id% 1 '&sv num = [calc 2 + 2]; ~
            &iacreturn %num%' reqstat]
```

The server sets a variable called num equal to the result of the [CALC] function (num = 4). The &iacreturn directive tells the server to prepare to return the value of num.

To get results from the server (i.e., to obtain results from the &IACRETURN directive), use [IACREQUEST] with a procedure number of 2 and specify the job ID of the request you want information about.

```
&type [IACREQUEST %server_id% 2 %job2% reqstat]
RESULT 4
```

Remember that when the server performed the [CALC] function it was part of job2. The command set the [CALC] results (4) in a variable called num. The usage &iacreturn %num% tells the system to return the value of %num% when [IACRETURN] is used with a procedure number of 2 and the proper job ID. In other words, to get the results you just need to query the job ID.

See the following chart for a summary of what the procedure numbers for [IACREQUEST] represent:

When you're connected to an ARC/INFO server	
<procedure_number> in [IACREQUEST]	What it does
1	Sends a job for the server to execute. To get results, use &IACRETURN with the name of the variable that contains the information to return.
2	Queries the job using the job number to retrieve the variable value from the server.

You can also have the server run an AML program. If you simply want the server to execute some commands, you don't need to use &IACRETURN. For example, the following AML program doesn't require the client to obtain results from the server:

```
/* symdump.aml
/* If user is in ARCPLOT, do a symboldump with 10 shades
&if [SHOW PROGRAM] ne ARCPLOT &then &return
clear
symboldump shade screen 1 10
&return
```

To have the server execute SYMDUMP.AML, use the following command from the client window:

```
&sv job5 [IACREQUEST %server_id% 1 '&r symdump' reqstat]
```

The server executes SYMDUMP.AML. The client doesn't need a response from the server. Note that the AML program must be stored wherever the server is pointing. If the AML program is located in another directory, you can specify a path or use &AMLPATH and &MENUPATH to give alternate directories for the files.

The next example uses a mathematical expression. Run the following AML program from the client window, then query the server for the results of the job:

```
/* calctest.aml
&sv a = 1024
&sv b = 16
&sv c = [calc %a% / %b%]
&iacreturn %c%
&return


&sv job6 [IACREQUEST %server_id% 1 '&r calctest' reqstat]
&type [IACREQUEST %server_id% 2 %job6% reqstat]
RESULT 64
```

Your client process has executed a mathematical expression on the server (this time through an AML program) and received the results.

Now look at a more sophisticated AML program that will be executed by the server. Make sure that the server is pointing to a workspace that has coverages in it before you run it.

```
/* polytst1.aml
/* Sequentially works through each coverage in a workspace
/* and reports if the coverage has polygon topology.
/* Each polygon coverage name is then added to the previous
/* coverage name and set in a variable called "covers."
/* &iacreturn returns "mess" variable string to client.
/* The "mess" variable will contain a comma separated list of
/* coverages that have polygon topology.
 &sv covers = [unquote '']  /* initiate variable to nil
 &do cov &list [UNQUOTE [LISTFILE * -cover]]
    &describe %cov%
    &if %DSC$Qtopology% &then
       &do
          &sv covers = %cov%,%covers%
       &end
 &end
 &sv mess = coverages: %covers% have polygons
 &iacreturn %mess%
 &return
```

Now that you've reviewed POLYTST1.AML, have the client execute it on the server machine. Then use [IACREQUEST] with a procedure number of 2 to retrieve the results from the server. Depending on your current workspace, you may find different names for the coverages that have polygons.

```
&sv job7 [IACREQUEST %server_id% 1 '&run polytst1' reqstat]
&ty [IACREQUEST %server_id% 2 %job7% reqstat]
RESULT coverages: counties, landuse have polygons
```

It's important to realize that POLYTST1.AML has no error checking. A prudent AML programmer could add some error checking besides the typical "Cannot connect to IAC server" or "Server request error" messages often sent to the user.

In the next example, POLYTST1.AML is modified so that if the server encounters a problem with the command, &SEVERITY executes either the warnmsg or errmsg routine. These routines capture the message string, &CALL the exit routine, and use &IACRETURN to send the last message to the user.

```
/* polytst2.aml
/* ...(same comments as polytst1.aml)...
/* Added severity to handle errors.  Also added exit routine.
 &severity &warning &routine warnmsg
 &severity &error &routine errmsg
 &sv covers = [unquote '']  /* initiate variable to nil
 &do cov &list [UNQUOTE [LISTFILE * -cover]]
    &describe  %cov%
    &if %DSC$Qtopology% &then
      &do
        &sv covers = %cov%,%covers%
      &end
&end
&sv mess = coverages: %covers% have polygons
&call exit
&return

/***************
&routine exit
/***************
 &severity &error &ignore
 &severity &warn &ignore
 &iacreturn %mess%
 &return
/***************
&routine warnmsg
/***************
```

```
&severity &warn &ignore
&sv mess = [quote %aml$message%]
&call exit
&return &error

/*******************
&routine errmsg
/*******************
 &severity &error &ignore
 &sv mess [quote %aml$message%]
 &call exit
 &return &error
```

If your workspace has no INFO subdirectory and therefore no coverages, you should see:

```
&sv job8 [IACREQUEST %sid% 1 '&run polytst3 junk' ~
           reqstat]
&ty [IACREQUEST %sid% 2 %job8% reqstat]
RESULT warning: new location is not a workspace
```

If there are no polygon coverages in the workspace, you should see:

```
&sv job9 [IACREQUEST %server_id% 1 '&run polytst2' reqstat]
&ty [IACREQUEST %server_id% 2 %job9% reqstat]
RESULT coverages:   have polygons
```

Remember that you won't receive an error or warning message if you don't have coverages present in a workspace.

In the next example, an &ARGS DATAPATH statement is added that allows you to enter a workspace name and/or path to check for polygon coverages. If the user enters a workspace name, POLYTST3.AML moves there and begins processing. In the exit routine, there's an additional line of code that moves the user back to the previous workspace.

```
/* polytst3.aml
/* ...same comments as polytst2.aml...
/* Now added &args directive to allow user to specify path to
/* where coverages are located.
 &severity &warning &routine warnmsg
 &severity &error &routine errmsg
 &args datapath
   &if not [null %datapath%] &then
     &do
       &sv oldworkspace = [show workspace]
```

```
        workspace %datapath%
      &end
  &sv covers = [unquote '']  /* initiate variable to nil
  &do cov &list [UNQUOTE [LISTFILE * -cover]]
    &describe  %cov%
    &if %DSC$Qtopology% &then
       &do
          &sv covers = %cov%,%covers%
       &end
  &end
  &sv mess = coverages: %covers% have polygons
  &call exit
  &return

/***************
&routine exit
/***************
 &severity &error &ignore
 &severity &warn &ignore
 &iacreturn %mess%
 &if [variable oldworkspace] &then workspace %oldworkspace%
 &return


/**************
&routine warnmsg
/**************
 &severity &warn &ignore
 &sv mess = [quote %aml$message%]
 &call exit
 &return &error


/******************
&routine errmsg
/******************
 &severity &error &ignore
 &sv mess [quote %aml$message%]
 &call exit
 &return &error
```

You can now enter a path to a workspace that has coverages you want
information about. For example:

```
&sv job10 [IACREQUEST %server_id% 1 ~
        '&run polytst3 /carmel3/ridland/testcovs' reqstat]
&ty [IACREQUEST %server_id% 2 %job10% reqstat]
RESULT coverages:
world_dd,wcounties,utmgrd_dd,ustates,usacounty,street,stbuf,
soils,atlanta, have polygons
```

If you enter a workspace name that doesn't exist, you should see:

```
&sv job9 [IACREQUEST %sid% 1 '&run polytst3 mydata' reqstat]
&ty [IACREQUEST %sid% 2 %job9% reqstat]
RESULT Unable to change workspace.
```

Remember that as a client you don't see the process run on the server. The advantage of the code above is that if there's a problem with the AML program as it runs on the server, the error message can be retrieved by the client.

As stated before, when you use an ARC/INFO client with an ARC/INFO server, the processes are asynchronous. This means that you can have a client give a server a command and then have the client itself execute another command. Take the buffer command as an example. The client sends a buffer command to the server. The server may take a few minutes to complete the task. While the server is executing the buffer command, the client is free to execute another command. This can be awkward, because the client AML program may need the results from the server before it can continue. For example:

```
/* Client has aml that sends buffer command to the server.
/* Server does buffer as an asynchronous process.
    &sv job11 [IACREQUEST %server_id% 1 'buffer roads ~
    roadbuff # # 100 # line' reqstat]
/* Client's aml then does some other geoprocessing.
...
...
...
/* Client's aml then wants to do identity overlay between
/* roadbuff and another coverage called floods.
/* Note that this aml will fail if server has not completed
/* the buffer command by this time.
    identity floods roadbuff fldrdbuf poly
    &return
```

Note that the example assumes that the server and client know about the same data (i.e., that they point to the same workspace). If either one points to another workspace, then you must move to where the data is or specify pathnames to the data.

It's very difficult to time asynchronous processes so as to ensure that the buffer command finishes before the identity command needs the output coverage. There are two ways to avoid this type of problem: (1) build error checking into your AML program, and (2) use the AIREQUEST ATOOL. The next example uses error checking in the AML program.

In the following example, error checking is added to prevent the AML program from failing if the server doesn't complete the buffer command in time:

```
/* Client has aml that sends buffer command to the server.
/* Server does buffer as an asynchronous process.
    &sv job12 [IACREQUEST %server_id% 1 'buffer roads roadbuff~
            # # 100 # line' reqstat]
/* client aml then does some other geoprocessing.
 ...
 ...
 ...
/* client's aml then wants to do identity overlay between
/* roadbuff and another coverage called floods.
/* Now added error checking to make sure new buffered coverage
/* exists before issuing identity command.
&if [exists roadbuff -cover] &then
  &do
    identity floods roadbuff fldrdbuf poly
  &end
&return
```

One problem with simply adding error checking like the [EXISTS] function is that if the coverage doesn't exist, or the buffer command isn't finished, the loop is skipped. On the one hand, the AML didn't fail, but on the other hand, it didn't perform the desired identity overlay command, either.

The second way to avoid having your AML program fail as a consequence of the inconsistencies of asynchronous processing is to use the AIREQUEST ATOOL. This ATOOL forces the client to wait for the server to respond with the results. It suspends asynchronous processing for whatever request you send.

```
Usage: airequest <server-id> <'request string'> ~
               <.return_variable> <.status_variable> ~
               {timeout_secs}
```

```
/* Client has AML that sends buffer command to the server.
/* Server does buffer as an asynchronous process.
/* Use airequest atool now.  Client's AML is on hold until
/* server responds with a return variable.
/* If client has not connected to dispatcher within 10
/* seconds, abort request.
    airequest %server_id% 'buffer roads roadbuff # # 100 # ~
    line'  .returnvar .reqstat 10
/* client AML then does some other geoprocessing.
 ...
```

```
. . .
. . .
/* client's AML then wants to do identity overlay between
/* roadbuff and another coverage called floods.
    identity floods roadbuff fldrdbuf poly
&return
```

If you list the value of the <.return_variable> and <.status_variable>, you'll see the following:

```
&listglobal
Global:
   .RETURNVAR                             RESULT
   .REQSTAT                               0
```

In the example, the AIREQUEST ATOOL suspends the client's AML program until the server has completed the request.

When you query the server, there are four possible values for the job status: QUEUE, indicating that the request is waiting to be processed; PROCESSING, indicating that the request is currently being executed; RESULT or DONE, indicating that the request has been completed. When the client first asks the server if it has finished a job, the server responds with RESULT. Each subsequent time the client asks for the status of that same job, the server responds with DONE. The following table summarizes the four job status values:

Result of client query to ARC/INFO server	What it means	Example of server message
QUEUE	Server is processing another job. Client's job is waiting in the queue to be processed. Server will tell client which job is currently being processed.	QUEUE PROCESSING 1
PROCESSING	Server is executing client's job.	PROCESSING
RESULT	Server has completed job. Server gives answer to client (if client used &IACRETURN directive).	RESULT RESULT 64
DONE	Server has completed job.	DONE

The following example shows you how you can query the results of your jobs.
After this is a chart that illustrates the client/server interaction.

```
[1] &sv job1 [IACREQUEST %server_id% 1 'buffer river riverbuf ~
              # # 200 # line' reqstat]
[2] &type [IACREQUEST %server_id% 2 %job1% reqstat]
    PROCESSING
[3] &sv job2 [IACREQUEST %server_id% 1 'identity zoning ~
              riverbuf composite1' reqstat]
[4] &type [IACREQUEST %server_id% 2 %job1% reqstat]
    PROCESSING
[5] &type [IACREQUEST %server_id% 2 %job2% reqstat]
    QUEUE PROCESSING 1
[6] &type [IACREQUEST %server_id% 2 %job1% reqstat]
    RESULT
[7] &type [IACREQUEST %server_id% 2 %job2% reqstat]
    QUEUE PROCESSING 1
[8] &type [IACREQUEST %server_id% 2 %job1% reqstat]
    DONE
[9] &type [IACREQUEST %server_id% 2 %job2% reqstat]
    RESULT
[10] &type [IACREQUEST %server_id% 2 %job1% reqstat]
     DONE
[11] &type [IACREQUEST %server_id% 2 %job2% reqstat]
     DONE
```

	ARC/INFO client commands	ARC/INFO server actions
[1]	Send Job #1 = buffer river coverage 200 meters	Buffering begins...
[2]	Query: What is status of Job #1?	Response: PROCESSING
[3]	Send Job #2 = identity zoning and riverbuf coverages	Server is still processing job #1
[4]	Query: What is the status of Job #1?	Response: PROCESSING
[5]	Query: What is the status of Job #2?	Response: QUEUE PROCESSING 1
[6]	Query: What is the status of Job #1?	Response: RESULT
[7]	Query: What is the status of Job #2?	Response: PROCESSING
[8]	Query: What is the status of Job #1?	Response: DONE
[9]	Query: What is the status of Job #2?	Response: RESULT
[10]	Query: What is the status of Job #1?	Response: DONE
[11]	Query: What is the status of Job #2?	Response: DONE

As you can see, when you want to query the server on the status of a job, use [IACREQUEST] with a procedure of 2 in conjunction with the job ID.

The final stage of interaction is to disconnect from the server.

Step 4: Disconnecting from the server

To disconnect from the server, use [IACDISCONNECT].

```
Usage: [IACDISCONNECT <server_id>]

&type [IACDISCONNECT %server_id%]
0
```

There are a few possible values that may result from the [IACDISCONNECT] function. Any value other than zero indicates a problem.

Possible return value from [IACDISCONNECT]	Meaning
0	Disconnect successful
100	Invalid server-ID
199	Unknown error

You can put the value in a variable and then test the variable to make sure the disconnect worked properly. For example:

```
&if [IACDISCONNECT %server_id%] ne 0 &then
  &do
    &return Error closing connection to server...
  &end
```

Using an ARC/INFO client with an ArcView GIS server

You may want to use an ARC/INFO client with an ArcView GIS server. This would allow you, for example, to use the geoprocessing tools in ARC/INFO to create coverages or grids and then display them in ArcView.

It's important to note that there's no dispatcher when you use an ArcView server. Instead, the client (ARC/INFO in this case) interacts directly with the server. This also means that the client is on hold until the server finishes processing the client's request. When the server has completed the request, control returns to the client. Users on the server machine may be in competition with client processes. Whichever process is sent to the server first is initiated, whether the request originates from the server machine or the client machine. Other requests may have to wait until the server is ready to respond.

To use ARC/INFO as a client to an ArcView server, first make sure the ArcView application is running either in a separate window from ARC/INFO or on another machine. Before you can connect to the server, you must also launch the server process. The following examples assume that you know how to create, compile, and execute an Avenue script.

In an ArcView project, initiate the ArcView server. Create an Avenue script like the one below, then compile and run it.

When initiating the ArcView server, you must enter the program and version number. For more information on starting the ArcView server, consult the ArcView online help. There's a topic that describes how to choose the proper numbers for the ONC RPC protocol. Note that you don't specify a host name. The host name automatically defaults to whatever machine name you're using.

The general process for using ARC/INFO as an IAC client to an ArcView server consists of three steps: (1) connecting to the ArcView server, (2) sending Avenue requests to the server and asking for results from the server, and (3) disconnecting from the ArcView server.

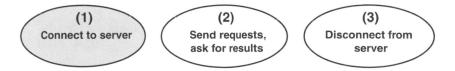

Step 1: Connecting to the ArcView server

When you connect to the ArcView server from the ARC/INFO side, the ArcView server doesn't automatically create a connect file. You must obtain the program and version number from whomever is running the server, and find out the name of the machine ArcView is running on. In the next example, the machine name is Carmel.

```
&sv server_id = [IACCONNECT carmel 40000000 1 connectstatus]
```

You should include the usual AML error checking to make sure the connection is successful. (Refer to page 16-30.) After connecting to the server, the next step is to send requests.

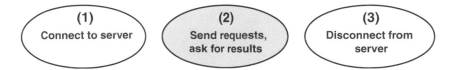

Step 2: Sending Avenue requests to the server and asking for results

It's important to remember that there's a fundamental difference in the ways the ArcView and ARC/INFO servers behave. The ARC/INFO server allows you to have two types of requests: a send job request and a query job request. The ArcView server has only one request type—the send job request. This is because ArcView doesn't use a dispatcher. The commands the client sends ArcView are therefore not processed asynchronously. When the client sends a request, that action locks the client up. There's no dispatcher to accept the request, give the client a job number, and allow the client to return to work.

FYI

> Since the client may be locked up, you can use the {timeout_secs} argument in [IACREQUEST] to control how long you want to wait. See the ARC/INFO command usage for more information.

When you're connected to an ArcView server	
<procedure_number> in [IACREQUEST]	What it does
1	Sends job for server to execute. To get the results, query the name of the variable associated with the [IACREQUEST] function.

Other than the fact that you can only have a procedure number of 1, there's no difference in the syntax of the [IACREQUEST] function on the ARC/INFO side. Of course, you must remember to send native commands to the server package. When the server is ArcView, this means that you must send Avenue statements.

```
&sv result = [iacrequest %server_id% 1 ~
              'av.getproject.getname' reqstat]
```

In the statement above, ARC/INFO asked the ArcView server for the name of the open project. The next step is to return the result to the client.

To obtain the results from the ArcView job, simply check the value of the variable returned from the [IACREQUEST] function.

```
&listlocal
Local:
  RESULT                                    iac.apr
  REQSTAT                                   0
```

Now that you know the project name, you can use other Avenue statements to carry out operations within the project.

Although there are sometimes reasons to enter Avenue statements one by one on the command line, it normally makes more sense to have the ARC/INFO client run an Avenue script from an AML program. The following example assumes you have already created and compiled an Avenue script called ShowData in your ArcView project.

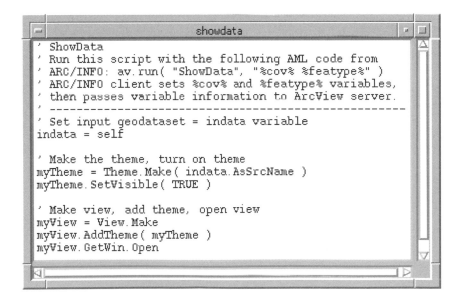

```
' ShowData
' Run this script with the following AML code from
' ARC/INFO: av.run( "ShowData", "%cov% %featype%" )
' ARC/INFO client sets %cov% and %featype% variables,
' then passes variable information to ArcView server.
' --------------------------------------------------
' Set input geodataset = indata variable
indata = self

' Make the theme, turn on theme
myTheme = Theme.Make( indata.AsSrcName )
myTheme.SetVisible( TRUE )

' Make view, add theme, open view
myView = View.Make
myView.AddTheme( myTheme )
myView.GetWin.Open
```

After compiling the script in ArcView, you're ready to execute it from ARC/INFO. The following AML program (ARC2AV.AML) connects to the ArcView server, allows the ARC/INFO user to pick a feature type and coverage from a menu choice, then passes the coverage name and feature type to the ShowData script. The script then displays the selected coverage as a theme in a view in ArcView.

```
/* arc2av.aml
/* Connect to an ArcView server.  Choose an ARC/INFO
/* coverage to display.  Then open a new ArcView view,
/* add the coverage as a theme, and make the theme visible.
[1] &sv server_id = [IACCONNECT carmel 40000000 1 ~
                         connectstatus]
[2] &sv featype [GETCHOICE poly line point -prompt 'Choose a ~
                   feature type to view']
[3] &sv cov [GETCOVER * -%featype% 'Choose a coverage to ~
                   view']
[4] &sv cmd [QUOTE av.run( "ShowData", "%cov% %featype%" )]
[5] &sv result = [IACREQUEST %server_id% 1 %cmd% reqstat]
[6] &sv closestat = [IACDISCONNECT %server_id%]
```

ARC2AV.AML first uses [IACCONNECT] to connect the ARC/INFO client to the ArcView server [1].

In line [2], from ARC/INFO, the user is asked (with the [GETCHOICE] function) to choose a feature type. The user chooses either POLY, LINE, or POINT.

The ARC/INFO user is then asked to choose a coverage from a scrolling list generated by the [GETCOVER] function [3].

Line [4] sets a variable called cmd that runs the Avenue script ShowData. It then passes this script the name of the selected coverage (%cov%) and feature type (%featype%).

Line [5] uses [IACREQUEST] to pass the cmd variable to the ArcView server. The cmd variable contains the string av.run("ShowData", "<your_coverage> <your_feature_type>"). The <your_coverage> and <your_feature_type> represent input gathered from [GETCOVER] and [GETCHOICE]. ArcView then executes the ShowData Avenue script.

Finally, line [6] disconnects from the ArcView server. Note that you should modify the AML program as needed to include error checking.

Suppose the user chooses a line coverage called World_dd. The cmd variable is set to av.run("ShowData", "world_dd line"). The ShowData script then creates the following ArcView view:

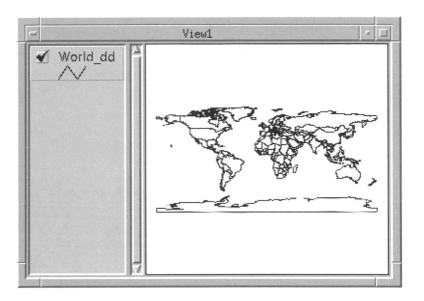

When you're finished sending requests to the ArcView server, disconnect from it.

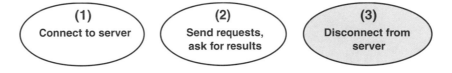

Step 3: Disconnecting from the ArcView server

To disconnect from the server, use the [IACDISCONNECT] function from the ARC/INFO client window. For example:

```
&sv disconnectstat [IACDISCONNECT %server_id%]
```

You should include the usual AML error checking to make sure the disconnection is successful. For more information, see page 16-45.

Question 16-3: Choose the answer that correctly completes the following sentence:

When an ARC/INFO client sends requests to an ArcView server, the ARC/INFO client sends (ARC/INFO / Avenue) statements.

(Answer on page 16-53)

Using an ARC/INFO client with a C server

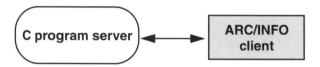

You may want to use your ARC/INFO session as a client to a C program server. The main reason to connect to a C program server is to take advantage of the C programming language's mathematical processing speed. You can use AML functions and directives for simple tasks like determining the mean, exponentiation, logarithms, and so on. Use C to calculate statistics from input data and to do other complex computations.

FYI

There are sample C programs located in `$ARCHOME/arcserver`. Be sure to review the README file. You must be an accomplished C programmer to modify the program templates that are included with the ARC/INFO software.

In summary

As an ARC/INFO client, you can access up to 100 simultaneous software packages to help with statistics, image processing, and other GIS needs.

Use [IACCONNECT] to establish the connection to the server. You can issue commands one at a time with [IACREQUEST]. If you're connecting to an ARC/INFO server, these commands can be packaged in an AML program to simplify your work. If you're connecting to an ArcView server, you must remember to use Avenue statements inside the [IACREQUEST] function. You can also run Avenue scripts from the ARC/INFO client. When commands are sent to a non-ARC/INFO server, the format and language within [IACREQUEST] depends on the server software.

You can have the server simply execute a client's commands or have the client query the server for results of the command. Depending on the server, you receive results in different formats. You can have AML process these results to maintain a seamless integration with the server package.

Answer 16-3: When an ARC/INFO client sends requests to an ArcView server, the ARC/INFO client sends Avenue statements to the server. Remember that the ARC/INFO client must send requests in whatever language the server package recognizes. For example, if ARC/INFO is a client to a Visual Basic server, you need to use [IACREQUEST] with Visual Basic commands, or use a command that runs a Visual Basic script.

Exercises

Exercise 16.3.1

Explain the hazards of sending the following requests to an ARC/INFO server:

```
&sv job1 = [IACREQUEST %server_id% 1 'polys [getcover]' reqstat]

airequest %server_id% 'marker *'  .returnvar .reqstat 10
```

Exercise 16.3.2

Fix the following AML code:

```
&sv job2 = [IACREQUEST %server_id% 1 '&sv num = [calc 2 + 2]; ~
            &iacreturn %num%' reqstat]

&type [IACREQUEST %server_id% 1 %job2% reqstat]
```

Exercise 16.3.3

What are some complications that may arise from the following AML code?

```
/* resbuf.aml

&sv job9 [IACREQUEST %server_id% 1 'reselect roads majroad ~
            line majrd.res' reqstat]

buffer majroad majrdbf # # 200 # line
```

Exercise 16.3.4

What are two programming techniques you can use to avoid problems with ARC/INFO's asynchronous processing?

Exercise 16.3.5

There are five errors in this AML program. Can you find them?

```
/* iacloop.aml
/* Allow client to easily type commands to send to server
&sv server_id = [IACCONNECT connectfile connectstatus]
&if %connectionstat% eq 101 &then &type Connect successful...
    &else &return Connection not successful, error number ~
          %connectionstat%
&sv cmd = [RESPONSE [QUOTE Enter command for A/I server ~
          ("<cr>" to end)] .TRUE.]
&do &while [quote %cmd%] nc [quote .FALSE.]
    &sv job = [IACREQUEST %server_id% 2 %cmd% reqstat]
    &if %reqstat% ne 0 &then &return Client request error, ~
          number %reqstat%
    &sv cmd = [RESPONSE [quote Enter command for A/I server ~
          ("<cr>" to end)] .TRUE.]
&end
&sv disconnectstat = [IACCLOSE %server_id%]
&if %disconnectstat% eq 0 &then &type Disconnect was ~
          successful...
    &else &return Disconnect not successful, error number ~
          %disconnectstat%
&return
```

Alphabetical list of AML directives and functions

AML directives

&ABBREVIATIONS controls whether to enable any AML abbreviations.

&AMLPATH directs the AML processor to search for AML files in the named directories if they can't be found in the current working directory.

&ARGS allows one AML file to receive arguments passed by the &RUN directive that invoked the program.

&ATOOL directs the AML processor to search for user-defined commands in the named directories.

&CALL transfers control to the specified routine block.

&CODEPAGE sets the current native code page.

&COMMANDS lists AML directives and functions at the terminal.

&CONV_WATCH_TO_AML converts user input in a watch file to an AML file.

&DALINES specifies the number of dialog lines for ANSI terminals.

&DATA submits a command to the operating system along with input to the command.

&DATE_FORMAT specifies the input and display format for date strings.

&DELVAR deletes a specified list of variables.

&DESCRIBE stores information about a specified geographic data set in AML-reserved variables.

&DO	delimits a block of statements or directives to be executed one or more times. Variants include &DO &LIST, &DO &UNTIL, &DO &WHILE.
&DO &LIST	constitutes a variant of an &DO loop in which the block is repeated once for each token in a list.
&DO &REPEAT	constitutes a variant of an &DO loop in which the block is repeated until an exit condition is met.
&DO &TO &BY	constitutes a variant of an &DO loop in which the block is repeated until its index variable falls outside a range of values.
&DO &UNTIL	constitutes a variant of an &DO loop in which the block is repeated until an expression is true.
&DO &WHILE	constitutes a variant of an &DO loop in which the block is repeated while an expression is true.
&ECHO	causes all input to be redisplayed at the terminal and in an enabled watch file.
&ENABLE	enables or disables a form control or menu choice.
&ENCODE	creates an encoded file from an AML file or a menu file.
&FLUSHPOINTS	removes any points that have been pushed on the point buffer.
&FORMAT	sets the number of decimal places for display of real numbers.
&FULLSCREEN	toggles full-screen output of program-specific listings.
&GETLASTPOINT	retrieves the last point input and places the coordinate and key values in AML-reserved variables.
&GETPOINT	gets a point from the user and stores coordinate and key values in AML-reserved variables.
&GOTO	causes control to be passed to the statement following the specified label.

&IACRETURN	returns request information from an ARC/INFO server.
&IF &THEN &ELSE	allows statements to be executed conditionally.
&LABEL	marks the location in an AML program referenced by &GOTO.
&LISTCHAR	lists the set of characters that indicates functions, AML intrinsic features, or AML variables.
&LISTFILES	lists all AML user files that are currently open.
&LISTGLOBAL	lists global variables and their values.
&LISTLOCAL	lists local variables and their values.
&LISTPROGRAM	lists program variables and their values.
&LISTVAR	lists local, global, and program variables and their values.
&MENU	specifies the menu to be activated and displayed for user input.
&MENUPATH	directs the AML processor to search for AML files in the named directories if they can't be found in the current working directory.
&MESSAGES	toggles the output of informational messages.
&PAUSE	halts program flow temporarily.
&POPUP	displays a scrollable text file at the terminal.
&PT	enables the performance timer.
&PUSHPOINT	places a point and key value in a point buffer that can be read by the next program command that needs point data.
&RETURN	terminates an AML file or the current input source.
&ROUTINE	starts a routine block.
&RUN	executes the specified AML file.

&RUNWATCH	causes input to be read from a watch file.
&SELECT &WHEN &OTHERWISE	allows a number of different statements to be conditionally executed.
&SETCHAR	sets the characters that indicate functions, AML intrinsic features, or AML variables.
&SETVAR	creates or sets the value of AML variables.
&SEVERITY	specifies what is to be done if a severity condition occurs.
&SHOW	returns the state or other values of the specified command.
&STATION	defines the workstation environment.
&STOP	terminates all files, menus, and threads.
&SYSTEM	initiates a dialog with the operating system or executes a specified operating system command.
&TB	displays a list of the current input sources and AML files that are open.
&TERMINAL	sets the terminal type and input device for menu input.
&TEST	enables and disables AML TEST mode.
&THREAD	controls multiple input sources in the same session.
&TRANSLATE	enables message translation for ARC/INFO and AML.
&TTY	causes input to be read from the terminal.
&TYPE	sends the specified message to the terminal.
&USAGE	provides the usage of an AML directive or function.
&WATCH	enables and disables a watch file.
&WORKSPACE	changes the current workspace.

AML functions

[ABS]   returns the absolute value of the number.

[ACCESS]   verifies specified access privileges for a system file.

[ACOS]   returns the angle in radians of which the given number is the cosine.

[AFTER]   returns a specified substring of a string.

[ANGRAD]   converts an angle in a valid ARC/INFO format to radians.

[ASIN]   returns the angle in radians of which the given number is the sine.

[ATAN]   returns the angle in radians of which the given number is the tangent.

[ATAN2]   returns the arc tangent of y/x in the range of $-\pi$ to π.

[BEFORE]   returns a specified substring of a string.

[CALC]   returns the result of the calculation of an ARC expression.

[CLOSE]   closes the file opened on the AML file unit.

[COPY]   copies a system file, directory, workspace, or INFO data file.

[COS]   returns the cosine of the given angle.

[CVTDISTANCE]   converts metric units of distance to meters and imperial units of distance to feet.

[DATE]   returns the current date and/or time in a number of different formats.

[DELETE]   deletes a system file, directory, or INFO data file.

[DIGNUM]   returns a key number from the digitizer cursor.

[DIR]   returns the directory part of the given file specification.

[ENTRYNAME]	returns the file name part of the given file specification.
[EXISTS]	determines whether the given object (file, coverage, workspace, etc.) exists.
[EXP]	returns *e* raised to the given number.
[EXTRACT]	extracts an element from a list of elements.
[FILELIST]	creates a file containing a list of objects of the specified type.
[FORMAT]	formats a string.
[FORMATDATE]	converts a locale-specific date/time string to a formatted string.
[GETCHAR]	displays a prompt and gets a single keystroke as a response.
[GETCHOICE]	displays a menu of a list of choices from which a selection can be made.
[GETCOVER]	displays a menu of coverages from which one can be selected.
[GETDATABASE]	displays a menu of ArcStorm databases, connections, tables, layers, libraries, or historical views, from which a selection can be made.
[GETDATALAYER]	displays a list of layers from an SDE data set.
[GETDEFLAYERS]	display a list of defined layers.
[GETFILE]	displays a menu of files, directories, or INFO data files from which a selection can be made.
[GETGRID]	displays a menu of grids from which one can be selected.
[GETIMAGE]	displays a menu of images from which one can be selected.
[GETITEM]	displays a menu of the items of a coverage or an INFO file from which one name can be selected.
[GETLAYERCOLS]	displays a list of columns from a defined layer.

[GETLIBRARY]	displays a menu of map libraries or the tiles or layers of a specified library from which a selection can be made.
[GETSTACK]	displays a menu of grid stacks from which one can be selected.
[GETSYMBOL]	displays a menu of the specified symbol type from which one can be selected.
[GETTIN]	displays a menu of TINs from which a selection can be made.
[GETUNIQUE]	displays a menu of unique item values from an INFO data file from which a selection can be made.
[IACCLOSE]	closes an ARC/INFO server.
[IACCONNECT]	connects an ARC/INFO client to a server.
[IACDISCONNECT]	disconnects an ARC/INFO client from a server.
[IACOPEN]	starts an ARC/INFO server.
[IACREQUEST]	sends a request to a server.
[INDEX]	returns the position of the leftmost occurrence of a specified string in a target string.
[INVANGLE]	calculates the polar angle between two points.
[INVDISTANCE]	calculates the distance between two points.
[ITEMINFO]	returns the description of an item in an INFO data file.
[JOINFILE]	provides system-independent handling of pathnames to files and directories.
[KEYWORD]	returns the position of a keyword within a list of keywords.
[LENGTH]	returns the number of characters in a string.
[LISTFILE]	writes to the screen, or creates a file containing a list of objects of the specified type.

[LISTITEM]	returns a list of the item names of a coverage or INFO data file.
[LISTUNIQUE]	displays a list of unique INFO item values, or writes them to a new file.
[LOCASE]	converts a string to lowercase.
[LOG]	returns the natural logarithm of a number.
[LOG10]	returns the logarithm base ten of a number.
[MAX]	returns the greater of two numbers.
[MENU]	returns a menu selection string after the display of a menu.
[MIN]	returns the lesser of two numbers.
[MOD]	returns the remainder when one integer is divided by another.
[NULL]	indicates whether a string is all blanks or null, or if it contains any characters.
[OKANGLE]	indicates whether an angle is in a valid ARC/INFO angle format.
[OKDISTANCE]	indicates whether a distance is in a valid ARC/INFO distance format.
[OPEN]	opens a system file for reading or writing.
[PATHNAME]	returns a fully expanded file specification.
[QUERY]	displays a prompt and returns a value of true or false based on a case-insensitive response of YES, NO, OKAY, QUIT, or <CR>.
[QUOTE]	places quotation marks around specified quoted or unquoted strings.
[QUOTEEXISTS]	determines if the argument is quoted, has quoted strings, or has quoted characters.

[RADANG]	converts an angle measured in radians to an angle in a valid ARC/INFO format.
[RANDOM]	returns a random number between 0 and 2,147,483,697.
[READ]	reads a record from the file opened on the specified AML file unit.
[RENAME]	renames a system file, directory, workspace, or INFO file.
[RESPONSE]	displays a prompt and accepts a response.
[ROUND]	rounds a real to an integer value.
[SCRATCHNAME]	generates a unique name for the specified type of object in the current workspace.
[SEARCH]	returns the position of the first character of a search string in a target string.
[SHOW]	returns the state or other values of the specified command.
[SIN]	returns the sine of the given angle.
[SORT]	sorts a list of elements.
[SQRT]	returns the square root of the given variable.
[SUBST]	substitutes one specified string for another in a target string.
[SUBSTR]	extracts a substring from a string starting at a specified character position.
[TAN]	returns the tangent of the given angle.
[TASK]	runs an executable program and receives a value back from the program.
[TOKEN]	allows tokens in a list to be manipulated.
[TRANSLATE]	translates one specified string into another in a target string.

[TRIM]	removes any occurrences of a specified character from the ends of a target string.
[TRUNCATE]	truncates a real to an integer value.
[TYPE]	returns a code indicating the type specification of a string.
[UNQUOTE]	removes quotation marks from each end of a quoted string.
[UPCASE]	converts a string to uppercase.
[USERNAME]	returns the current user name.
[VALUE]	returns the contents of the given variable.
[VARIABLE]	indicates if the given AML variable exists.
[VERIFY]	returns the position of the first character in a target string that doesn't occur in a search string.
[WRITE]	writes a record to the file opened on the specified AML file unit.

B Functional list of AML directives and functions

User environment directives

&ABBREVIATIONS	controls whether to enable any AML abbreviations.
&AMLPATH	directs the AML processor to search for AML files in the named directories if they can't be found in the current working directory.
&ATOOL	directs the AML processor to search for user-defined commands in the named directories.
&CODEPAGE	sets the current native code page.
&DATE_FORMAT	specifies the input and display format for date strings.
&DESCRIBE	stores information about a specified geographic data set in AML-reserved variables.
&ENCODE	creates an encoded file from an AML file or a menu file.
&FORMAT	sets the number of decimal places for display of real numbers.
&LISTCHAR	lists the set of characters that indicates functions, AML intrinsic features, or AML variables.
&LISTFILES	lists all AML uscr files that are currently open.
&MENUPATH	directs the AML processor to search for AML files in the named directories if they can't be found in the current working directory.
&SETCHAR	sets the characters that indicate functions, AML intrinsic features, or AML variables.

&STATION   defines the workstation environment.

&TERMINAL   sets the terminal type and input device for menu input.

&WORKSPACE   changes the current workspace.

Input source directives

&MENU   specifies the menu to be activated and displayed for user input.

&PAUSE   halts program flow temporarily.

&RETURN   terminates an AML file or the current input source.

&RUN   executes the specified AML file.

&RUNWATCH   causes input to be read from a watch file.

&STOP   terminates all files, menus, and threads.

&THREAD   controls multiple input sources in the same session.

&TTY   causes input to be read from the terminal.

Statement execution control directives

&CALL   transfers control to the specified routine block.

&DATA   submits a command to the operating system along with input to the command.

&DO   delimits a block of statements or directives to be executed one or more times. Variants include &DO &LIST, &DO &UNTIL, &DO &WHILE.

&DO &LIST   constitutes a variant of an &DO loop in which the block is repeated once for each token in a list.

&DO &REPEAT	constitutes a variant of an &DO loop in which the block is repeated until an exit condition is met.
&DO &TO &BY	constitutes a variant of an &DO loop in which the block is repeated until its index variable falls outside a range of values.
&DO &UNTIL	constitutes a variant of an &DO loop in which the block is repeated until an expression is true.
&DO &WHILE	constitutes a variant of an &DO loop in which the block is repeated while an expression is true.
&GOTO	causes control to be passed to the statement following the specified label.
&IF &THEN &ELSE	allows statements to be executed conditionally.
&LABEL	marks the location in an AML program referenced by &GOTO.
&ROUTINE	starts a routine block.
&SELECT &WHEN &OTHERWISE	allows a number of different statements to be conditionally executed.
&SYSTEM	initiates a dialog with the operating system or executes a specified operating system command.

Variable manipulation directives

&ARGS	allows one AML file to receive arguments passed by the &RUN directive that invoked the program.
&DELVAR	deletes a specified list of variables.
&LISTGLOBAL	lists global variables and their values.
&LISTLOCAL	lists local variables and their values.
&LISTPROGRAM	lists program variables and their values.

&LISTVAR   lists local, global, and program variables and their values.

&SETVAR   creates or sets the value of AML variables.

Help directives

&COMMANDS   lists AML directives and functions at the terminal.

&USAGE   provides the usage of an AML directive or function.

Program testing and monitoring directives

&CONV_WATCH_TO_AML   converts user input in a watch file to an AML file.

&ECHO   causes all input to be redisplayed at the terminal and in an enabled watch file.

&PT   enables the performance timer.

&SEVERITY   specifies what is to be done if a severity condition occurs.

&SHOW   returns the state or other values of the specified command.

&TB   displays a list of the current input sources and AML files that are open.

&TEST   enables and disables AML TEST mode.

&WATCH   enables and disables a watch file.

Inter-Application Communication (IAC) directives

&IACRETURN   returns request information from an ARC/INFO server.

Dialog management directives

&DALINES   specifies the number of dialog lines for ANSI terminals.

&FULLSCREEN   toggles full-screen output of program-specific listings.

&MESSAGES   toggles the output of informational messages.

&POPUP   displays a scrollable text file at the terminal.

&TRANSLATE   enables message translation for ARC/INFO and AML.

&TYPE   sends the specified message to the terminal.

Coordinate input directives

&FLUSHPOINTS   removes any points that have been pushed on the point buffer.

&GETLASTPOINT   retrieves the last point input and places the coordinate and key values in AML-reserved variables.

&GETPOINT   gets a point from the user and stores coordinate and key values in AML-reserved variables.

&PUSHPOINT   places a point and key value in a point buffer that can be read by the next program command that needs point data.

Mathematical and trigonometric functions

[ABS]   returns the absolute value of the number.

[ACOS]   returns the angle in radians of which the given number is the cosine.

[ANGRAD]   converts an angle in a valid ARC/INFO format to radians.

[ASIN]	returns the angle in radians of which the given number is the sine.
[ATAN]	returns the angle in radians of which the given number is the tangent.
[ATAN2]	returns the arc tangent of y/x in the range of $-\pi$ to π.
[CALC]	returns the result of the calculation of an ARC expression.
[COS]	returns the cosine of the given angle.
[CVTDISTANCE]	converts metric units of distance to meters and imperial units of distance to feet.
[EXP]	returns *e* raised to the given number.
[INVANGLE]	calculates the polar angle between two points.
[INVDISTANCE]	calculates the distance between two points.
[LOG]	returns the natural logarithm of a number.
[LOG10]	returns the logarithm base ten of a number.
[MAX]	returns the greater of two numbers.
[MIN]	returns the lesser of two numbers.
[MOD]	returns the remainder when one integer is divided by another.
[OKANGLE]	indicates whether an angle is in a valid ARC/INFO angle format.
[OKDISTANCE]	indicates whether a distance is in a valid ARC/INFO distance format.
[RADANG]	converts an angle measured in radians to an angle in a valid ARC/INFO format.
[RANDOM]	returns a random number between 0 and 2,147,483,697.

[ROUND]	rounds a real to an integer value.
[SIN]	returns the sine of the given angle.
[SQRT]	returns the square root of the given variable.
[TAN]	returns the tangent of the given angle.
[TRUNCATE]	truncates a real to an integer value.

String manipulation functions

[AFTER]	returns a specified substring of a string.
[BEFORE]	returns a specified substring of a string.
[EXTRACT]	extracts an element from a list of elements.
[FORMAT]	formats a string.
[FORMATDATE]	converts a locale-specific date/time string to a formatted string.
[INDEX]	returns the position of the leftmost occurrence of a specified string in a target string.
[KEYWORD]	returns the position of a keyword within a list of keywords.
[LENGTH]	returns the number of characters in a string.
[LOCASE]	converts a string to lowercase.
[NULL]	indicates whether a string is all blanks or null, or if it contains any characters.
[QUOTE]	places quotation marks around specified quoted or unquoted strings.
[QUOTEEXISTS]	determines if the argument is quoted, has quoted strings, or has quoted characters.

[SEARCH]	returns the position of the first character of a search string in a target string.
[SORT]	sorts a list of elements.
[SUBST]	substitutes one specified string for another in a target string.
[SUBSTR]	extracts a substring from a string starting at a specified character position.
[TOKEN]	allows tokens in a list to be manipulated.
[TRANSLATE]	translates one specified string into another in a target string.
[TRIM]	removes any occurrences of a specified character from the ends of a target string.
[UNQUOTE]	removes quotation marks from each end of a quoted string.
[UPCASE]	converts a string to uppercase.
[VERIFY]	returns the position of the first character in a target string that doesn't occur in a search string.

Reporting functions

[DATE]	returns the current date and/or time in a number of different formats.
[DIGNUM]	returns a key number from the digitizer cursor.
[ITEMINFO]	returns the description of an item in an INFO data file.
[LISTITEM]	returns a list of the item names of a coverage or INFO data file.
[LISTUNIQUE]	displays a list of unique INFO item values, or writes them to a new file.
[SHOW]	returns the state or other values of the specified command.

[TASK]	runs an executable program and receives a value back from the program.
[TYPE]	returns a code indicating the type specification of a string.
[USERNAME]	returns the current user name.
[VALUE]	returns the contents of the given variable.
[VARIABLE]	indicates if the given AML variable exists.

User-file input/output functions

[CLOSE]	closes the file opened on the AML file unit.
[DELETE]	deletes a system file, directory or INFO data file.
[OPEN]	opens a system file for reading or writing.
[READ]	reads a record from the file opened on the specified AML file unit.
[WRITE]	writes a record to the file opened on the specified AML file unit.

File management functions

[ACCESS]	verifies specified access privileges for a system file.
[COPY]	copies a system file, directory, workspace, or INFO data file.
[DELETE]	deletes a system file, directory, or INFO data file.
[DIR]	returns the directory part of the given file specification.
[ENTRYNAME]	returns the file name part of the given file specification.

[EXISTS]	determines whether the given object (file, coverage, workspace, etc.) exists.
[FILELIST]	creates a file containing a list of objects of the specified type.
[JOINFILE]	provides system-independent handling of pathnames to files and directories.
[LISTFILE]	writes to the screen, or creates a file containing a list of objects of the specified type.
[PATHNAME]	returns a fully expanded file specification.
[RENAME]	renames a system file, directory, workspace, or INFO file.
[SCRATCHNAME]	generates a unique name for the specified type of object in the current workspace.

User input functions

[GETCHAR]	displays a prompt and gets a single keystroke as a response.
[GETCHOICE]	displays a menu of a list of choices from which a selection can be made.
[GETCOVER]	displays a menu of coverages from which one can be selected.
[GETDATABASE]	displays a menu of ArcStorm databases, connections, tables, layers, libraries, or historical views, from which a selection can be made.
[GETDATALAYER]	displays a list of layers from a SDE data set.
[GETDEFLAYERS]	displays a list of defined layers.
[GETFILE]	displays a menu of files, directories, or INFO data files from which a selection can be made.
[GETGRID]	displays a menu of grids from which one can be selected.

[GETIMAGE]	displays a menu of images from which one can be selected.
[GETITEM]	displays a menu of the items of a coverage or an INFO file from which one name can be selected.
[GETLAYERCOLS]	displays a list of columns from a defined layer.
[GETLIBRARY]	displays a menu of map libraries or the tiles or layers of a specified library from which a selection can be made.
[GETSTACK]	displays a menu of grid stacks from which one can be selected.
[GETSYMBOL]	displays a menu of the specified symbol type from which one can be selected.
[GETTIN]	displays a menu of TINs from which a selection can be made.
[GETUNIQUE]	displays a menu of unique item values from an INFO data file from which a selection can be made.
[MENU]	returns a menu selection string after the display of a menu.
[QUERY]	displays a prompt and returns a value of true or false based on a case-insensitive response of YES, NO, OKAY, QUIT, or <CR>.
[RESPONSE]	displays a prompt and accepts a response.

Inter-Application Communication (IAC) functions

[IACCLOSE]	closes an ARC/INFO server.
[IACCONNECT]	connects an ARC/INFO client to a server.
[IACDISCONNECT]	disconnects an ARC/INFO client from a server.
[IACOPEN]	starts an ARC/INFO server.
[IACREQUEST]	sends a request to a server.

Form-building controls and options

BUTTON   creates a button control and returns a string to the AML
                            processor.

CHECKBOX   creates a check box control that can be used to toggle the
                            setting of a condition.

CHOICE   creates an exclusive setting control that sets the value of an
                            AML variable and returns a string for evaluation.

DISPLAY   creates a field that displays the value of an AML variable.

&ENABLE   enables or disables a form control or menu choice.

FORMINIT   returns an initialization string to the AML processor when the
                            form is first displayed.

FORMOPT   positions the cursor in the form and specifies the manner in
                            which variables are assigned.

INPUT   creates a text field and/or a scrolling list control that accepts a
                            value for an AML variable and displays it.

SLIDER   creates a slider bar control that is used to set the value of an
                            AML variable.

C ESRI license agreement

ESRI license agreement

Read the following license agreement carefully and make sure you are willing to accept all of its terms and conditions before opening the sealed media package containing the CD–ROM.

ESRI LICENSE AGREEMENT

This is a license agreement and not an agreement for sale. This ERSI License Agreement (Agreement) is between the end user (Licensee) and Environmental Systems Research Institute, Inc. (ESRI), and gives Licensee certain limited rights to use the proprietary AML Workbook software, data, and related materials (Software, Data, and Related Materials). All rights not specifically granted in this Agreement are reserved to ESRI.

Reservation of Ownership and Grant of License: ESRI retains exclusive rights, title, and ownership of any copy of the Software, Data, and Related Materials licensed under this Agreement and, hereby, grants to Licensee a personal, nonexclusive, nontransferable license to use the Software, Data, and Related Materials pursuant to the terms and conditions of this Agreement. From the date of receipt, Licensee agrees to use reasonable effort to protect the Software, Data, and Related Materials from unauthorized use, reproduction, distribution, or publication.

Copyright: The Software, Data, and Related Materials are owned by ESRI and are protected by United States copyright laws and applicable international treaties and/or conventions.

Permitted Uses:

Licensee may install the Software, Data, and Related Materials onto one computer system.

Licensee may make only one (1) copy of the Software, Data, and Related Materials for archival purposes unless the right to make additional copies is granted to Licensee in writing by ESRI.

Licensee may use, copy, alter, modify, merge, reproduce, and/or create derivative works of the Software, Data, and Related Materials for Licensee's own internal use. The portions of the Software, Data, or Related Materials merged with other software, data, or related materials shall continue to be subject to the terms and conditions of this Agreement and shall provide the following copyright attribution notice acknowledging ESRI's proprietary rights in the Software, Data, and/or Related Materials: "Portions of this work include intellectual property of ESRI and are used herein by permission. Copyright © 1993–1997 Environmental Systems Research Institute, Inc. All rights reserved."

Uses Not Permitted:

Licensee may not sell, rent, lease, sublicense, lend, time-share, or transfer, in whole or in part, or provide unlicensed Third Parties access to prior or present versions of the Software, Data, and Related Materials, any updates, or Licensee's rights under this Agreement.

Licensee may not remove or obscure any ESRI copyright or trademark notices.

Term: The license granted by this Agreement is coterminous with Licensee's ESRI Software License Agreement for ARC/INFO software subject to ESRI's then-current licensing policies.

Limited Warranty: ESRI warrants that any media upon which the Software, Data, and Related Materials are provided will be free from defects in materials and workmanship under normal use and service for a period of ninety (90) days from the date of receipt. EXCEPT FOR THE LIMITED WARRANTIES SET FORTH ABOVE, THE SOFTWARE, DATA, AND RELATED

MATERIALS CONTAINED THEREIN ARE PROVIDED "AS-IS," WITHOUT WARRANTY OF ANY KIND, EITHER EXPRESS OR IMPLIED, INCLUDING, BUT NOT LIMITED TO, THE IMPLIED WARRANTIES OF MERCHANTABILITY AND FITNESS FOR A PARTICULAR PURPOSE.

ESRI does not warrant the accuracy of the map data contained herein. ESRI is not inviting reliance on the data, and the Licensee should always check actual data. The map data and other information contained in this document are subject to change without notice.

Exclusive Remedy and Limitation of Liability: During the warranty period, ESRI's entire liability and Licensee's exclusive remedy shall be to return the license fees paid upon the Licensee returning the Software, Data, and Related Materials to ESRI or its Distributors with a copy of Licensee's receipt. ESRI shall not be liable for indirect, special, incidental, or consequential damages related to Licensee's use of the Software, Data, and Related Materials, even if ESRI is advised of the possibility of such damage.

U.S. Government Restricted Rights: Use, duplication, and disclosure by the U.S. Government are subject to restrictions as set forth in FAR §52.227-14 Alternate III (g)(3) (JUN 1987), FAR §52.227-19 (JUN 1987) and/or FAR §12.211/12.212 (Commercial Technical Data/Computer Software), DFARS §252.227-7015 (NOV 1995) (Technical Data), and/or DFARS §227.7202 (Computer Software), as applicable. Contractor/Manufacturer is Environmental Systems Research Institute, Inc., 380 New York Street, Redlands, California 92373-8100, USA.

Governing Law: This Agreement shall be governed by the laws of the United States of America and the State of California without reference to conflict of laws principles.

Entire Agreement and Amendments: The parties agree that this constitutes the sole and entire agreement of the parties as to the matter set forth herein and supersedes any previous agreements, understandings, and arrangements between the parties relating hereto. Any Amendments to this Agreement must be in writing and signed by an authorized representative of each party.

Exercise solutions

Exercise 1.2.1

```
ARCEDIT
EDITCOVER VEGETATION
DRAWENVIRONMENT ARC NODE ERROR
EDITFEATURE ARC
NODESIZE DANGLE .2
NODECOLOR DANGLE 2
NODESIZE PSEUDO .15
NODECOLOR PSEUDO 5
DRAW
```

Exercise 1.2.2

(a) When you run an AML program with &RUN, control passes from the <u>keyboard</u> to the <u>AML program</u>.

(b) Control returns to where it originated when &RETURN is encountered.

Exercise 1.2.3

You can type directives at the keyboard or write them in AML programs. ARC/INFO accepts input from many different sources including the keyboard, menus, and AML programs.

Exercise 1.2.4

(a) All AML programs should include comments and the &RETURN directive.

(b) You may want to edit an AML program that was converted from a watch file to remove the &WATCH &OFF. In addition, you may need to edit the program to fix typing errors, change the subsystem from which the AML program runs, remove commands, or modify command arguments.

Exercise 1.2.5

Delete the ARCPLOT command, the mistyped LINESYMBOL command (line 5), and the &WATCH directive. Add &RETURN on the last line, and also add comments to provide information about the program.

Exercise 1.2.6

Use &WATCH with the &COORDINATES option to capture it, then use &CWTA with the &COORDINATES option to copy it to the AML program.

Exercise 1.3.1

`mike` is printed on the screen. Variable names aren't case sensitive, so %name% and %NAME% are the same variable. The second &SV overwrites the first.

Exercise 1.3.2

(a) %SNAPTOL% is a local variable and can only be used in the program where it was assigned.

(b) You can make %SNAPTOL% global by preceding its name with a period (i.e., .SNAPTOL). Global variables are usable in all programs run during an ARC/INFO session.

Exercise 1.3.3

DSC$ variables provide information about <u>coverages</u>. PNT$ variables provide information about <u>coordinates</u>.

Exercise 1.3.4

(a) Directives   <u>c</u>   perform an operation and return a value

(b) Variables   <u>a</u>   execute an operation

(c) Functions   <u>b</u>   store and return values

Exercise 1.3.5

The AML code &workspace /gis/project5/[username] changes the current user workspace to a directory named the same as the user's login name under the /gis/project5 directory.

Exercise 1.3.6

If used directly in a command line, the value of a function can't be used later in the program. Assign the value of a function to a variable if it needs to be referenced many times in a program.

Exercise 1.3.7

```
POLYGONSHADES LANDUSE [calc [response 'Enter a shade symbol number'] + 100]
```

Exercise 1.3.8

```
&sv letter = [upcase b]
&sv letter = B
```

Exercise 1.3.9

```
KILL SOILS_OLD ALL
KILL PARCELS_OLD ALL
KILL FIREDIST_OLD ALL
```

Exercise 1.3.10

```
ARCS (!ZONE LANDUSE ROAD!)
```

Exercise 1.3.11

```
CALCULATE BLD_SPACE = [calc %side% - [response 'Enter front offset']]
CALCULATE BLD_SPACE = [calc 250 - [response 'Enter front offset']] CALCULATE
BLD_SPACE = [calc 250 - 35]
CALCULATE BLD_SPACE = 215
```

Exercise 2.1.1

```
&sv cover = [getcover * -poly 'Select a coverage to shade']
&sv item = [getitem %cover% -poly -integer ~
  'Select an item to shade with'] POLYGONSHADES %cover% %item%
```

Exercise 2.1.2

```
TEXTSET [getfile '/gis/symbolsets/*.txt' 'Select a TEXTSET']
```

Exercise 2.1.3

```
&sv soiltype = [getunique soil -poly soil_type ~
  'Select a Soil type' -none]
```

Exercise 2.1.4

```
EDITFEATURE [getchoice ARC NODE LABEL TIC LINK -prompt ~
  'Select an EDITFEATURE']
```

Exercise 2.1.5

```
EDITCOVERAGE [getcover * -all 'Select a coverage to edit']
```

Exercise 2.2.1

Only tablet and digitizer menus are interchangeable like pulldown and sidebar menus. All other menu types have different formats and can't be dynamically changed by the options on the &MENU directive (&PULLDOWN, &SIDEBAR, &DIGITIZER, etc.).

Exercise 2.2.2

```
3
0 0

'Change Workspace'
'Choose a coverage to edit'
'Choose an EDITFEATURE'
'Quit from menu'
```

Exercise 2.2.3

```
1 pulldown
'Page size'
  '8.5 X 11'       PAGESIZE 8.5 11
  '11 X 17'        PAGESIZE 11 17
  '17 X 32'        PAGESIZE 17 32
  '22 X 34'        PAGESIZE 22 34
  '34 X 46'        PAGESIZE 34 46
'Page units'
  Inches           PAGEUNITS INCHES
  Centimeters      PAGEUNITS CM
'Map Limits'       MAPLIMITS *
```

Exercise 2.2.4

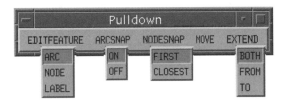

Exercise 2.2.5

```
SHADESET LANDUSE.SHD
POLYGONSHADES LANDUSE LU_CODE
POLYGONS LANDUSE
```

Exercise 2.3.1

Change the menu choice to the following:

```
'Draw a coverage'    &run drawcov.aml
```

and write the following AML:

```
/* drawcov.aml
&sv cov = [getcover]
MAPE %cov%
&sv feat = [getchoice ARC POLY POINT NODE -prompt ~
'Select a featureclass to draw']
%feat%s %cov%
&return
```

Exercise 2.3.2

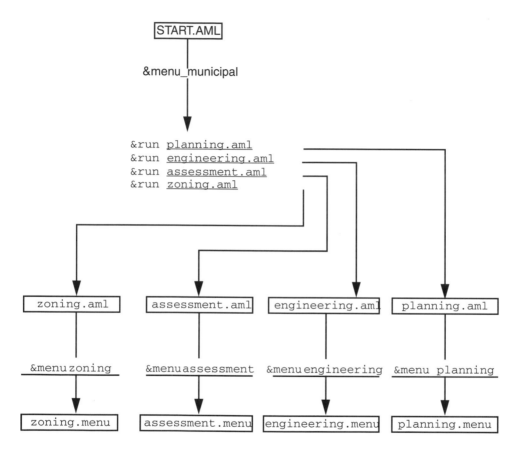

Exercise 3.1.1

```
&amlpath /usr/urban/transportation /usr/urban/utilities
&menupath [show &amlpath]
```

Exercise 3.1.2

```
&amlpath /usr/urban/transportation /usr/urban/utilities ~
   /usr/natural/soils /usr/natural/hydrology
```

or

```
&amlpath [show &amlpath] /usr/natural/soils /usr/natural/hydrology

&menupath /usr/urban/transportation /usr/urban/utilities ~
   /usr/natural/soils /usr/natural/hydrology
```

or

```
&menupath [show &menupath] /usr/natural/soils /usr/natural/hydrology
```

Exercise 3.1.3

```
/usr/urban/utilities
```

Exercise 3.1.4

c $ARCHOME/startup/arcplot.aml

a /usr/.arcplot

b /usr/natural/soils/arcplot.aml

Exercise 3.2.1

```
Arcplot: &abbreviations &on
Arcplot: &sv al = arclines street street_code
```

Exercise 3.2.2

Unwanted results occurred because an AML variable is set with the same name as a command when AML abbreviations are activated. AML evaluates the variable edit because it's the first character string on the line. The ARCEDIT command EDIT won't execute until either the variable is deleted or AML abbreviations are deactivated.

Exercise 3.2.3

The directories storing AML programs aren't named correctly. ATOOL commands must be stored in program-specific directories that are named the same as the ARC subsystem for which the ATOOLs apply (e.g., ARC, ARCEDIT, ARCPLOT).

Exercise 3.2.4

```
&atool /usr/gis/hydrology  /usr/gis/transportation
```

Exercise 3.2.5

ARCEDIT doesn't have a FORMS command (it's really an ATOOL). AML hasn't been directed to search any user-specified directories, so AML searches the system ATOOL directory ($ARCHOME/atool/arcedit) for an AML program with the same name as the command. In this case, AML would find FORMS.AML in the ARCEDIT program-specific directory and execute it as an ATOOL command.

Exercise 3.3.1

(a) ```
    &amlpath /usr/gis/planning/transportation /usr/gis/planning/utilities
    &menupath /usr/gis/planning/transportation /usr/gis/planning/utilities
    ```

(b) ```
    &atool /usr/gis/atool
    ```

Exercise 3.3.2

```
&encode &encrypt update.aml
```

```
&encode &encrypt plan.menu
```

Exercise 3.3.3

```
update.eaf
```

```
plan.emf
```

Exercise 3.3.4

Move UPDATE.AML and PLAN.MENU to a location to which only selected personnel have access. Never delete the original code—the encoding process isn't reversible.

Exercise 4.1.1

```
&sv incover = [getcover * -all 'Select a coverage to BUILD']
&sv topology = [getchoice POLY LINE POINT NODE -prompt -
  'Select a topology type']
&if [query 'Do you wish to crate a new outcoverage? (Y/N)'] &then &do
  &setvar outcover = [response 'Please enter the name of the outcover']
  &if [null %outcover%] &then
    &return You must enter a outcoverage name
  copy %incover% %outcover%
  &sv incover = %outcover%
&end
build %incover% %topology%
&return
```

Exercise 4.1.2

```
&if [locase [username]] = jorge &then &do
  &type Good morning jorge, have a good day.
  &workspace /project23/jorge
  ARCPLOT
&end
&else &do
  &type You may only work in the guest account on this machine.
  &workspace /guest
&end
&return
```

Exercise 4.1.3

```
&if [locase [username]] = jorge &then &do
   &sv timeofday = [before [date -time] .]
   &if %timeofday% le 12 &then
      &type Good morning jorge, have a good day.
   &else
      &type Good afternoon
   &workspace /project23/jorge
   ARCPLOT
&end
&else &do
   &type You may only work in the guest account on this machine.
   &workspace /guest
&end
&return
```

Exercise 4.2.1

```
&select [upcase [username]]
   &when jorge; &do
      &workspace /instructors/training
      &type you are now in the training directory
   &end /*jorge
   &when tom; &do
       &workspace /courses/project/hydro
       &type you are now in the hydrology directory
   &end /*tom
   &otherwise; &do
       &workspace /guest
       &type you may only work in the guest directory
   &end
&end /*select
&return
```

Exercise 4.2.2

```
&sv covername = [getcover * -all 'Select a coverage to draw']
MAPEX %covername%
&sv featuretype = [getchoice POINT LINE POLY NODE -prompt ~
   'Select a feature to display']
&if %featuretype% = POINT  &then
   POINTS %covername%
&else &if %featuretype% = LINE  &then
   ARCS %covername%
&else &if %featuretype% = POLY  &then
```

```
  POLYS %covername%
&else &if %featuretype% = NODE  &then
  NODES %covername%
&return
```

Exercise 4.2.3

```
&setvar covername = [getcover * -all 'Select a coverage to draw']
MAPEX %covername%
/*
&setvar features
&if [exists %covername% -point] &then
  &sv features = %features% POINT
&if [exists %covername% -line] &then
  &sv features = %features% LINE NODE
&if [exists %covername% -poly] &then
  &sv features = %features% POLY
/*
&if [null %features%] &then
  &return The coverage %covername% has no point, line, or polygon ~
  features.
/*
&if [null [extract 2 %features%]] &then &do
  &sv featuretype = %features%
  &type %features% is the only feature type contained in the coverage ~
%covername%
&end
&else
&sv featuretype = [getchoice %features%]
/*
&if %featuretype% = POINT &then
  POINTS %covername%
&else &if %featuretype% = LINE &then
  ARCS %covername%
&else &if %featuretype% = POLY &then
  POLYS %covername%
&else &if %featuretype% = NODE &then
  NODES %covername%
&return
```

Exercise 4.2.4

```
&setvar covername = [getcover * -all 'Select a coverage to draw']
MAPEX %covername%
&setvar featuretype = [getchoice POINT LINE POLY NODE ~
  -prompt 'Select a feature to display']
/*
```

```
&if [exists %covername% -%featuretype%] &then &do
  &if %featuretype% = POINT &then
    POINTS %covername%
  &else &if %featuretype% = LINE &then
    ARCS %covername%
  &else &if %featuretype% = POLY &then
    POLYS %covername%
  &else &if %featuretype% = NODE &then
    NODES %covername%
&end
&else
  &return A feature type of %featuretype% does not exist~
 for the coverage %covername%
&return
```

Exercise 4.2.5

```
&setvar covername = [getcover * -all 'Select a coverage to draw']
MAPEX %covername%
&setvar featuretype = [getchoice POINT LINE POLY NODE -prompt ~
  'Select a feature to display']
&select %featuretype%
&when POINT
  POINTS %covername%
&when LINE
  ARCS %covername%
&when POLY
  POLYS %covername%
&when NODE
  NODES %covername%
&end
&return
```

Exercise 4.2.6

Generally, [GETCHOICE] is better because it ensures that the user enters valid values. [RESPONSE], on the other hand, allows the user to enter any value, so extensive error checking may be needed to ensure that the input is valid.

Exercise 4.2.7

For the nesting structure to work correctly, the project number must be known before entering the first &IF statement. [GETCHOICE] needs to be issued for *each* level of the nested structure. This requires repeated entry by the user, adding significant opportunities for error.

Exercise 4.3.1

```
&if [locase [username]] = jorge &then &do
  &setvar timeofday [before [date -time] .]
  &if %timeofday% le 6 &then
    &type Here a little early!
  &else &if %timeofday% gt 6 and %timeofday% le 12 &then
    &type Good morning jorge, have a good day.
  &else &if %timeofday% gt 12 and %timeofday% lt 17 &then
    &type Good afternoon.
  &else
    &type Good evening, time to go home.
  &workspace /project23/jorge
  ARCPLOT
  &end
&else &do
  &type You may only work in the guest account on this machine.
  &workspace /guest
  &end
&return
```

Exercise 4.3.2

```
&sv covername = [getcover * -all 'Select a coverage to draw']
MAPEX %covername%
/*
&sv features
&if [exists %covername% -point] &then
  &sv features = %features% POINT
&if [exists %covername% -line] &then
  &sv features = %features% LINE NODE
&if [exists %covername% -poly] &then
  &sv features = %features% POLY
/*
&if [null %features%] &then
  &return The coverage %covername% has no point, line, or polygon ~
features.
/*
&if [quote %features%] cn POINT &then
  POINTS %covername%
&if [quote %features%] cn LINE &then
  ARCS %covername%
&if [quote %features%] cn POLY &then
  POLYS %covername%
&if [quote %features%] cn NODE &then
  NODES %covername%
&return
```

Exercise 4.3.3

```
&sv cov = [getcover * -line]
&describe %cov%
  &if %dsc$qedit% &then &do
    &type You must reconstruct coverage topology.
    &sv choice = [getchoice BUILD CLEAN -prompt ~
     'Select a command to rebuild topology']
    &type Your original coverage will be renamed to %cov%old
RENAME %cov% [entryname %cov%old]
    &if %choice% = CLEAN &then
      CLEAN %COV%OLD %COV% # # line
    &else
       BUILD %COV% LINE
  &end
&return
```

Exercise 5.1.1

The focus of this exercise is the counted &DO loop. There's no error checking in this solution. To include error checking you'd need to include an &IF statement to verify that the user entered an appropriate base coverage name and number, and optionally to verify whether the user wants to continue deleting. For more information about error checking using &IF statements, see chapter 4.

```
LISTCOVERAGES
&sv covername = [response 'Enter base coverage name']
&sv covernum = [response 'Enter the highest coverage number']
copy %covername%%covernum% %covername%final
&do deletecov = 1 &to %covernum% &by 1
  kill %covername%%deletecov% all
&end
LISTCOVERAGES
&return
```

Exercise 5.1.2

Change 20 to -20 for the numbers to progress in descending order.

Exercise 5.1.3

```
&do projfile &list mercator.prj lambert.prj utm.prj perspective.prj
  MAPPROJECTION AUTOMATIC %projfile%
  MAPEXTENT STATES
```

```
   POLYGONS STATES
   &pause Press return to continue
   CLEAR
&end
&return
```

Exercise 5.2.1

```
&sv cover
&do &while [null %cover%]
   &sv cover = [getcover * -all 'Select a coverage to continue' -none]
&end
```

Exercise 5.2.2

```
&do &until %qfinish%
  &sv grafile = [getfile *.gra -file]
  &if not [null %grafile%] &then &do
    &if [delete %grafile%] = 0 &then &type %grafile% deleted
    &else &type %grafile% not deleted successfully
  &end
  &sv qfinish = [query 'Are you finished deleting graphics files' ~
    .false.]
&end
&return
```

Exercise 5.2.3

```
&sv qfinish = .true.
&do &while %qfinish%
  &sv infofile = [getfile temp* -info]
  &if [delete %infofile% -info] = 0 &then &type %infofile% deleted
    &else &type %infofile% not deleted successfully
  &sv qfinish = [query 'Continue deleting INFO files' .true.]
&end
&return
```

Exercise 5.3.1

SOILS1, SOILS2, SOILS3, STREETS1, STREETS2, STREETS3, ZONING1, ZONING2, ZONING3.

Exercise 5.3.2

As soon as the task is complete, the loop stops. Without this clause, the loop repeats until the count variable falls out of range, even if the task was completed much sooner. This wastes computer resources and time.

Exercise 5.3.3

```
/*  indexvar2.aml
&sv number = 1
&do owner &list HILL, DAVIS, MINAMI
   &sv own%number% = %owner%
   &type Value of own%number% = [value own%number%]
   &sv number = %number% + 1
&end
&return
```

Exercise 5.3.4

There are four workspaces: TRACT1 through TRACT4; each contains three coverages: ZONING, SOILS, and PARCELS. The AML program moves to each workspace and creates plot files at two sizes for each of the three coverages; therefore, twenty-four plot files are created (twelve at 8.5 x 11 and twelve at 14 x 26).

Exercise 5.3.5

```
/* Answer for exercise 5.3.5
/* Streamintersect2.aml
/* Finds the node where two streams intersect.
/*
/* User picks stream names from a menu
CLEAR
ASELECT STREAM ARC
MAPEXTENT STREAM
&do strpick = 1 &to 2
  &sv str%strpick% = [getunique stream -arc strm_name 'Pick a stream']
  RESELECT STREAM LINE STRM_NAME = [value str%strpick%]
  &type Drawing [value str%strpick%]...
  ARCLINES STREAM %strpick%
  &sv numstr%strpick% = [extract 1 [show select stream line] ]
  /*  create arrays
    &do count = 1 &to [value numstr%strpick%]
       &sv str%strpick%node%count% = [show select stream line ~
         %count% item tnode#]
```

```
   &end   /* do count
  ASELECT STREAM ARC
&end       /* do strpick
/* search for matching nodes in arrays str1node and str2node
&sv intersection = 0
&sv qdone = .false.
&type /&Searching for intersection...
&do index1 = 1 &to %numstr1% &until %qdone%
   &do index2 = 1 &to %numstr2% &until %qdone%
      &if [value str1node%index1%] = [value str2node%index2%] &then
         &do
            &sv intersection = [value str2node%index2%]
            &sv qdone = .true.
         &end   /* if matched
   &end   /* &do index2
&end     /* &do index1
&if %intersection% = 0 &then
   &type /&No intersection between %str1% and %str2% was found.
   &else &type /&Intersection located at node %intersection%
&return
```

Exercise 6.1.1

```
/* 0.5 km radius reselection with chosen point at center of circle
/* Reset selected sets for the coverage
CLEAR
MAPEX ZONING
ASELECT STREETS ARCS
ASELECT SITES POINTS
ASELECT ZONING POLYS
LINECOLOR WHITE
ARCS STREETS
LINECOLOR GREEN
POLYGONS ZONING
MARKERCOLOR CYAN
POINTS SITES
/*
/* Ensure that all operations are in map units
/*
UNITS MAP
&getpoint &map
/*
/* Do the reselections
/*
RESELECT STREETS ARCS   CIRCLE %pnt$x% %pnt$y% 500 PASSTHRU
RESELECT SITES   POINTS CIRCLE %pnt$x% %pnt$y% 500 PASSTHRU
RESELECT ZONING  POLYS  CIRCLE %pnt$x% %pnt$y% 500 PASSTHRU
```

```
/*
/* Mapextent into the extent of the circle
/*
&sv xlow =  %pnt$x% - 500
&sv xhigh = %pnt$x% + 500
&sv ylow =  %pnt$y% - 500
&sv yhigh = %pnt$y% + 500
MAPEXTENT %xlow% %ylow% %xhigh% %yhigh%
/*
/* Redraw screen
/*
CLEAR
LINESET COLOR
ARCLINES STREETS 14
POINTMARKERS SITES 2
POLYGONLINES ZONING 3
LINESYMBOL 6
CIRCLE %pnt$x% %pnt$y% 500
&return
```

Exercise 6.1.2

```
/* One square km reselection with chosen point at center of square
/* Reset selected sets for the coverage
CLEAR
MAPEXTENT ZONING
ASELECT STREETS ARCS
ASELECT SITES POINTS
ASELECT ZONING POLYS
ARCS STREETS
POINTS SITES
POLYGONS ZONING
/*
/* Ensure that all operations are in map units
/*
units map
&getpoint &map
&sv xlow  = %pnt$x% - 500
&sv xhigh = %pnt$x% + 500
&sv ylow  = %pnt$y% - 500
&sv yhigh = %pnt$y% + 500
/*
/* Do the reselections
/*
RESELECT STREETS ARCS   BOX %xlow% %ylow% %xhigh% %yhigh%
RESELECT SITES   POINTS BOX %xlow% %ylow% %xhigh% %yhigh%
RESELECT ZONING  POLYS  BOX %xlow% %ylow% %xhigh% %yhigh%
```

```
/*
/* Mapextent into the extent of the square
/*
MAPEXTENT %xlow% %ylow% %xhigh% %yhigh%
/*
/* Redraw screen
/*
CLEAR
/*
/* Set the lineset to color
/*
LINESET COLOR
ARCLINES STREETS 14
POINTMARKERS SITES 2
POLYGONLINES ZONING 3
LINESYMBOL 6
BOX %xlow% %ylow% %xhigh% %yhigh%
&return
```

Exercise 6.1.3

```
/* The first part is the colorref.aml
/* colorref.aml
/*
&sv ystart = 0
&sv xstart = 0
&sv ysize = .5
&sv xsize = .5
&sv xtext = %xsize% * .5
&sv ytext = %ysize% * .5
/*  set text characteristics
TEXTSET FONT.TXT
TEXTSYMBOL 2
TEXTSIZE .12
TEXTCOLOR BLACK
TEXTJUSTIFICATION CC
SHADESET COLOR.SHD
/* start creating adjacent solid color boxes for each
/* shade with the shade symbol number centered in the box
&do colorbox = 1 &to 15
  SHADESYM %colorbox%
  PATCH %xstart% %ystart% [calc %xstart% + %xsize%] ~
    %ysize%
  /*  now label each box with its shade symbol number
  MOVE [calc %xstart% + %xtext%] %ytext%
  TEXT %colorbox%
  /*  move the start point of the next box
```

```
  &sv xstart = %xstart% + %xsize%
&end
/*
/* ex6_1_3.aml
/*
&getpoint &page
&if %pnt$y% gt 0.5 &then &do
  &sv .color$choice = 0
  &type Please select color from row of color boxes
  &type    Color set to 0 due to improper cursor selection.
  &return
&end
&if %pnt$x% ge 7.5 &then &do
  &sv .color$choice = 0
  &type Please select color from row of color boxes
  &type    Color set to 0 due to improper cursor selection.
  &return
&end
&sv mult100 =         [truncate [calc %pnt$x% * 100] ]
&sv .color$choice = [truncate [calc %mult100% / 50 ] ] + 1
MAPEX ZONING
POLYGONSHADE ZONING %.color$choice%
&return
```

Exercise 6.2.1

```
/* Polygon reselection from three different coverages
/* Polygon specified once by cursor
/*
MAPEX STREETS SITES ZONING
LINESET COLOR
UNITS MAP
/*
/* Flush the point input buffer
/*
&flushpoints
&type Use 1 key to outline temporary area, 9 to close poly
/*
/*   Use magenta to draw the outline while selecting vertices of the
/*   polygon.  Because you are using &GETPOINT, the polygon boundaries
/*   aren't drawn automatically as they are when you specify a polygon
/*   boundary with RESELECT.
/*
LINESYMBOL 6
/*
/* Choose a point in map units with the cursor
/*
```

```
&getpoint &map &cursor
/*
/* Initialize counter for array storage of points
/*
&sv count 1
/*
/* Place the selected point into the first element of x- and y-arrays
/* by assigning "array" variables to the key, x-, and y-coordinates
/* chosen.
/*
&sv key%count% = %pnt$key%
&sv x%count%   = %pnt$x%
&sv y%count%   = %pnt$y%
/*
/* Make the last point equal to the first point for polygon closure
/* (XLAST and YLAST won't be loaded into buffer).
/*
&sv keylast =   %pnt$key%
&sv xlast   =   %pnt$x%
&sv ylast   =   %pnt$y%
/*
/* Set the variables CURRENT$X and CURRENT$Y equal to the first
/* point.
/*
&sv current$x  = %pnt$x%
&sv current$y  = %pnt$y%
/*
/* Keep accepting points until the user presses the 9 key.
/*
&do &while %pnt$key% ne 9
  /*
  /* Choose a point in map units with the cursor.
  /*
  &getpoint &map &cursor
  &sv count = %count% + 1
  /*
  /* Place the selected point into the current element of the x- and
  /* y-arrays by assigning "array" variables to the key, x-, and
  /* y-coordinates chosen.
  /*
  &sv key%count% = %pnt$key%
  &sv x%count%   = %pnt$x%
  &sv y%count%   = %pnt$y%
  /*
  /* If the user hasn't pressed a 9, draw a line from the last
  /* x- y-coordinate (CURRENT$X,CURRENT$Y) to the coordinate just
  /* chosen.  Then update the variables CURRENT$X and CURRENT$Y to
  /* contain the coordinate just chosen.
```

```
/*
&if %pnt$key% ne 9 &then &do
   LINE %current$x% %current$y% %pnt$x% %pnt$y%
   &sv current$x = %pnt$x%
   &sv current$y = %pnt$y%
&end
&else &do
   LINE %current$x% %current$y% %xlast% %ylast%
&end
&end

&if %count% gt 3 &then &do
   /*
   /*   If the count is less than or equal to 3, not enough points were
   /*   chosen to make a polygon.  If the count is 3, the last one is
   /*   the the 9 key, leaving only two points for the polygon.  Two
   /*   points doesn't define a polygon.  CLear the selected sets.
   /*
   ASELECT STREETS ARCS
   ASELECT SITES    POINTS
   ASELECT ZONING   POLYS
   /*
   /*   Push the selected points into the point buffer for
   /*   reselection of the first coverage.
   /*
   &do i = 1 &to %count%
      &pushpoint [value key%i%]   [value x%i%]   [value y%i%]
   &end
   RESELECT STREETS ARC POLYGON * PASSTHRU
   /*
   /* The reselect emptied the buffer, but clear it just to be safe.
   /*
   &flushpoints
   /*
   /* Repeat the same procedure for the other two coverages.
   /*
   &do i = 1 &to %count%
      &pushpoint [value key%i%]   [value x%i%]   [value y%i%]
   &end
   RESELECT SITES POINT POLYGON * PASSTHRU
   &flushpoints
   /*
   /*
   &do i = 1 &to %count%
      &pushpoint [value key%i%]   [value x%i%]   [value y%i%]
   &end
   RESELECT ZONING POLYS POLYGON * PASSTHRU
   /*
```

```
/* Now, using the same points in the array, find the MAPEXTENT
/* that completely contains the user-specified polygon.
/* NOTE:  The last point (i.e., the 9 key) doesn't count.
/*
&sv mape_countlimit  %count% - 1
/*
/* Set some ridiculously low and high numbers--make the initial
/* lows really high and the initial highs really low.
&sv xlow  = 9999999999.9
&sv ylow  = 9999999999.9
&sv xhigh = -9999999999.9
&sv yhigh = -9999999999.9
/*
/* Loop through the points in the x- and y-arrays looking for
/* minimums and maximums.
/*
&do i = 1 &to %mape_countlimit%
  &sv xlow  = [min %xlow%  [value x%i%] ]
  &sv ylow  = [min %ylow%  [value y%i%] ]
  &sv xhigh = [max %xhigh% [value x%i%] ]
  &sv yhigh - [max %yhigh% [value y%i%] ]
&end
MAPEXTENT %xlow% %ylow% %xhigh% %yhigh%
/*
/* Redraw screen
/*
CLEAR
ARCLINES STREETS 14
POINTMARKERS SITES 2
POLYGONLINES ZONING 3
/*
/* Draw the boundary of the selected polygon, if you want to.
/* Use the line command with the point buffer.
/* Use MAPE_COUNTLIMIT variable again to push the vertices into the
/* buffer.  Finally, push in %keylast%, %xlast%, and %ylast% for the
/* last vertex.  End the line command by pushing key of 9, x of 0,
/* and y of 0 into the buffer.  The 9 key is important in this
/* last push for the line command, but the x and y coordinates of 0
/* will be ignored.
/*
&flushpoints
&do i = 1 &to %mape_countlimit%
  &pushpoint [value key%i%]  [value x%i%]  [value y%i%]
&end
&pushpoint %keylast%  %xlast%  %ylast%
&pushpoint 9 0 0
LINESYMBOL 6
LINE *
&end  /* &if %count% gt 3
```

```
&else &do
  &type Not enough points to form a polygon!!
  &type Exiting with no action performed!!
&end
/*
/* CLEANUP - Clear the buffer
/*
&flushpoints
&return
```

Exercise 6.3.1

```
MAPEXTENT 0 0 8 8
LINESET COLOR
UNITS MAP
&type   Choose the central location
&getpoint &map &cursor
/*
&sv center_x = %pnt$x%
&sv center_y = %pnt$y%
MARKERSYMBOL 1
MARKER %center_x% %center_y%
/*
&type Choose point at edge of primary circle
&getpoint &map &cursor
/*
&sv primary_distance = [invdistance %center_x% %center_y% %pnt$x% ~
  %pnt$y%]
/*
LINESYMBOL 2
CIRCLE %center_x% %center_y% %primary_distance%
/*
&sv quarter_dist = %primary_distance% / 4.0
&sv half_dist    = %primary_distance% / 2.0
&sv twice_dist   = %primary_distance% * 2.0
/*
LINESYMBOL 3
CIRCLE %center_x% %center_y% %quarter_dist%
LINESYMBOL 5
CIRCLE %center_x% %center_y% %half_dist%
LINESYMBOL 7
CIRCLE %center_x% %center_y% %twice_dist%
/*
&type Primary distance (RED)    is %primary_distance%
&type Quarter distance (GREEN)  is %quarter_dist%
&type Half distance    (BLUE)   is %half_dist%
&type Twice distance   (YELLOW) is %twice_dist%
&return
```

Exercise 6.3.2

```
/* Displacement solution using trig
/*
  UNITS MAP
  &getpoint &map &cursor
  &sv distance = 1000.0
  /*
  /* 45 degrees up from the x axis (east) is northeast
  /* convert degrees to radians for use with [cos] and [sin]
  /*
  &sv radians = [angrad 45.0]
  /*
  /* Calc x and y displacement
  /*1
  &sv add_x = %distance% * [cos %radians%]
  &sv add_y = %distance% * [sin %radians%]
  /*
  /* Add displacements to original point to get new point that is
  /* 1000 meters at 45.0 degrees (northeast) from original.
  /*
  &sv new_x = %pnt$x% + %add_x%
  &sv new_y = %pnt$y% + %add_y%
  &type New location is %new_x% , %new_y%
  MARKERSYMBOL 1
  MARKER %new_x% %new_y%
  &return
```

Exercise 6.3.3

```
* Displacement solution using trig
MAPEX 0 0 10000 10000
  &sv distance = [response 'Please enter distance in meters']
  &sv angle    = [response 'Please enter angle in degrees']

  &if [null %distance%] or [null %angle%] &then &do
    &type Non-specified distance or angle -- Exiting
    &return
  &end

  &if not [okdistance %distance%]  or not [okangle %angle%] &then &do
    &type Illegal distance or angle -- Exiting
    &return
  &end

  UNITS MAP
  &getpoint &map &cursor
```

```
/* 45 degrees up from the x axis (east) is northeast
/* convert degrees to radians for use with [cos] and [sin]
&sv radians = [angrad %angle%]

/* Calc x and y displacement
&sv add_x = %distance% * [cos %radians%]
&sv add_y = %distance% * [sin %radians%]

/* Add displacements to original point to get new point that is
/* 1000 meters at 45.0 degrees (northeast) from original.
&sv new_x = %pnt$x% + %add_x%
&sv new_y = %pnt$y% + %add_y%

&type New location is %new_x% , %new_y%
MARKERSYMBOL 1
MARKER %new_x% %new_y%

&return
```

Exercise 7.1.1

```
&sv string1 = [response 'Enter the first string']
&sv string2 = [response 'Enter the second string']
&type The concatenated string is: %string1%%string2%
&return
```

Exercise 7.1.2

```
&sv string1 = [response 'Enter the first string']
&sv string2 = [response 'Enter the second string']
&sv string3 = %string1%%string2%
&lv string3
&return
```

The second program shows quotes embedded in the final concatenated string. These surround whichever of the two original strings contained spaces. The [RESPONSE] function automatically quotes strings that contain blanks. (Lesson 7.2 discusses quoting and unquoting strings in detail.)

Exercise 7.1.3

```
&sv cov = [getcover * -all 'Select a coverage']
&do &until [type %scale%] = -1
  &sv scale [response 'Enter a map scale']
  &if [type %scale%] ne -1 &then &type Enter an INTEGER value for a scale.
&end
&if not [null %cov%] &then &do
  MAPEXTENT %cov%
  MAPSCALE %scalc%
&end
&else &type No coverage chosen, run the program from a different workspace.
&return
```

Exercise 7.2.1

```
&sv st1 = [response 'Enter the first string']
&sv st2 = [response 'Enter the second string']
&sv conc = [quote [unquote %st1%] [unquote %st2%]]
&type %conc%
&return
```

Exercise 7.2.2

```
&s num_amls = [unquote [listfile *.aml -file]]
&s num_menus = [unquote [listfile *.menu -file]]
&if [null %num_amls%] &then
  &do
    &ty There are no amls in this directory
  &end
&else
  &do
    &ty There are [token %num_amls% -count] AML(s) in this directory
  &end
&if [null %num_menus%] &then
  &do
    &ty There are no menus in this directory
  &end
&else
  &do
    &ty There are [token %num_menus% -count] MENU(s) in this directory
  &end
&return
```

Exercise 7.2.3

```
&s string = 0117mb0616tb757311463r
/*
&sv section = [substr %string% 7 6]
&sv block = [substr %string% 13 4]
&sv lot = [substr %string% 17 6]
/*
&type section = %section%
&type block = %block%
&type lot = %lot%
&return
```

Exercise 7.2.4

```
&sv cov [getcover]
&format 3
&describe %cov%
&type The bounding coordinates from the coverage [entryname %cov%]~
 are:
&type [format 'Xmin = %1,-15%   Ymin = %2,-15%' %dsc$xmin% %dsc$ymin%]
&type [format 'Xmax = %1,-15%   Ymax = %2,-15%' %dsc$xmax% %dsc$ymax%]
&return
```

Exercise 7.2.5

```
/* Mapextent to the maximum extent of all features across all covers
/*
mapextent arcs streets
&sv streets_extent [show mapextent]
&s xlow  [extract 1 %streets_extent%]
&s ylow  [extract 2 %streets_extent%]
&s xhigh [extract 3 %streets_extent%]
&s yhigh [extract 4 %streets_extent%]
/*
mapextent points sites
&sv sites_extent   [show mapextent]
&s xlow  [min %xlow%  [extract 1 %sites_extent%] ]
&s ylow  [min %ylow%  [extract 2 %sites_extent%] ]
&s xhigh [max %xhigh% [extract 3 %sites_extent%] ]
&s yhigh [max %yhigh% [extract 4 %sites_extent%] ]
/*
mapextent polys  zoning
&sv zoning_extent   [show mapextent]
&s xlow  [min %xlow%  [extract 1 %zoning_extent%] ]
```

```
&s ylow  [min %ylow%  [extract 2 %zoning_extent%] ]
&s xhigh [max %xhigh% [extract 3 %zoning_extent%] ]
&s yhigh [max %yhigh% [extract 4 %zoning_extent%] ]
/*
mapextent %xlow% %ylow% %xhigh% %yhigh%
show mapextent
&return
```

Exercise 7.2.6

```
&sv x1  1.234
&sv y1  10.567
&sv x2  88.7654
&sv y2  100.09876
&sv x3  9876.65
&sv y3  10012.111
&sv x4  1234567.8901
&sv y4  2345678.9012

&format 2
&type [format 'x = %1,-10%   y = %2,-10%' %x1%  %y1%]
&type [format 'x = %1,-10%   y = %2,-10%' %x2%  %y2%]
&type [format 'x = %1,-10%   y = %2,-10%' %x3%  %y3%]
&type [format 'x = %1,-10%   y = %2,-10%' %x4%  %y4%]
/* or:
&format 2
&do i = 1 &to 4
  &type [format 'x = %1,-10%   y = %2,-10%' [value x%i%] [value y%i%]]
&end
```

Exercise 7.3.1

```
&type [before [extract 1 [show version]] .]
```

Exercise 7.3.2

```
&s name = [unquote [response  'Enter a name']]
&type [upcase [quote [substr [extract 1 %name%] 1 1]. [extract 2 ~

%name%]]]
&return
```

Exercise 7.3.3

```
&sv rel = [getchoice [show relates] -prompt 'Select a relate']
&sv type [extract 6 [show relate %rel%]]
&select [upcase %type%]
  &when RO; &type Relate [upcase %rel%] has READ ONLY ACCESS
  &when RW; &type Relate [upcase %rel%] has READ WRITE ACCESS
  &when AUTO;  &type Relate [upcase %rel%] has AUTO ACCESS
&end
```

Exercise 7.3.4

```
&sv resp = [response 'enter a coverage, and a scale~
 (separated by spaces)']
&if [token [unquote %resp%] -count] ne 2 &then
  &return Wrong number of arguments entered.
&sv cover = [extract 1 [unquote %resp%]]
&sv scale = [extract 2 [unquote %resp%]]
&sv mappo = [after %resp% %scale%]
&if not [exists %cover% -cover] &then
  &return %cover% is not a coverage.
&if [type %scale%] ne -1 &then
  &return %scale% is not an integer scale.
MAPEXTENT %cover%
MAPSCALE %scale%
&return
```

Exercise 7.3.5

```
&sv table = [getfile * -info 'Select an INFO file']
&sv item = [getitem %table% -info 'Select and item']
&sv type = [extract 3 [iteminfo %table% -info %item%]]
&type Item type is %type%
&return
```

Exercise 8.1.1

```
/* Start.aml
&run hardware
&run editcov
&run buildcov
&run combine
&return
```

```
/* Hardware.aml
&terminal 9999
&fullscreen &popup
display 9999
DIGITIZER 9100 ttya
COORDINATE DIGITIZER
&return

/* Editcov.aml
ARCEDIT
&sv cov = [getcover * -all 'Choose a cover to edit']
EDIT %cov%
MAPEXTENT %cov%
DRAWENVIRONMENT [unquote [response 'Specify features to ~
 display']]
DRAW
/* Use Editcov.menu to perform editing tasks
&menu editcov &pos &ur &screen &ur
SAVE %cov% %cov%ed
&type /& Your edited coverage is [entryname %cov%ed]
QUIT
&return


/* Buildcov.aml
&sv cov = [getcover * -all 'Choose the cover you wish to build']
COPY %cov% %cov%bd
&sv topology = [getchoice POLY LINE POINT -prompt ~
 'What type of topology to build?']
BUILD %cov%bd %topology%
&type /&Your built coverage is [entryname %cov%bd]
&return

/* Combine.aml
&sv incov = [getcover * -all 'Choose an in-cover']
&sv cov_type = [getchoice  poly line point -prompt 'What type of cover~
 is your in-cover']
&sv identity_cover = [getcover * -poly 'pick the cover to combine']
&describe %identity_cover%
&if %dsc$qedit% &then
  &do
    &popup error.msg
    &sv identity_cover = [getcover * -poly 'Pick the cover to combine']
  &end
IDENTITY %incov% %identity_cover% compositecov %cov_type%
&return
```

Exercise 8.1.2

```
&run 911
```

Exercise 8.2.1

```
/* drawcov.aml-- symbolizing features using attribute values
/* Set a variable to a coverage
&sv .draw$cov = [getcover]
&run checkcov
/* Check the type of toplogy
&select %.draw$topology%
  &when polys
    ----
  &when lines
    ----
  &when points
    ----
  &otherwise
    ----
&end
&return
```

```
/* checkcov.aml
/* Describe the coverage
&describe %.draw$cov%
/* Set a variable to the status of topology and features
/* Polygons and topology present
&if %dsc$qtopology% and %dsc$polygons% > 1 &then
&sv .draw$topology = polys
/* Arcs and topology present
&else &if %dsc$aat_bytes% > 0 and %dsc$arcs% > 0 &then
  &sv .draw$topology = lines
/* Points and topology present
&else &if %dsc$xat_bytes% > 0 and %dsc$points% > 0 &then
  &sv .draw$topology = points
&else
  ----
&return
```

Exercise 8.2.2

```
/* drawcov2.aml
&sv cov = [getcover]
&run checkcov %cov%
```

```
/* checkcov2.aml
&args cover
&describe %cover%
```

Exercise 8.2.3

Establish a naming convention for the global variables that uses a common prefix (e.g., .edit$cov, .edit$extent, etc.).

(a)
```
/* edit.aml
&sv .edit$cov = [getcover]
EDIT %.edit$cov%
&sv .edit$extent = [show mapextent]
&sv .edit$limits = [show maplimits]
&sv .edit$position = [show mapposition]
save %.edit$cov% %.edit$cov%ed
```

```
/* plotmap.aml
MAPEXTENT %.edit$extent%
MAPLIMITS %.edit$limits%
MAPPOSITION %.edit$position%
ARCS %.edit$cov%ed
```

(b) `&dv edit$*`

Exercise 8.2.4

(a)
```
start2.aml
&sv cov = [getcover]
```

(b)
```
main.menu
QUERY      &run identcov %cov%
BUFFER     &run buffer %cov%
OVERLAY    &run overlay %cov%
```

(c) `identcov.aml`
   `&args cov`

   `buffer.aml`
   `&args cov`

   `overlay.aml`
   `&args cov`

Exercise 8.2.5

```
/* Overlay.aml
&args in_cover
&sv union_cover = [getcover]
&sv .new_cover = [substr [entryname %in_cover%] 1 5][substr~
  [entryname %union_cover%] 1 5]
UNION %in_cover% %union_cover% %.new_cover%
&return
```

Note: ARC/INFO coverage names can't be longer than thirteen characters. This solution uses [SUBSTR] and [ENTRYNAME] to concatenate a maximun of five characters from both the <in_cover> and <union_cover> as the value for .NEW_COVER. For example, if the <out_cover> results from the union of SOILS and LANDUSE, the value of %.NEW_COVER% is SOILSLANDU.

Exercise 8.3.1

```
/* zoompan.aml
&args changeview
&select %changeview%
  &when zoom_in
    &call zoom_in
  &when zoom_out
    &call zoom_out
  &when pan
    &call pan
&end
&return
/* Routines
&routine zoom_in
----
----
&return
/*
```

```
&routine zoom_out
----
----
&return
/*
&routine pan
----
----
&return
```

Exercise 8.3.2

```
/* edit.aml
&args coverage
/* If no coverage argument, get a coverage
&if [null %coverage%] &then
  &do
    &call getcover
  &end
ARCEDIT
EDIT %coverage%
DRAWENVIRONMENT [unquote [response 'Specify features to draw']]
EDITFEATURE [unquote [response 'Specify feature class to edit']]
----
----
&return
/* Get a coverage
&routine getcover
&sv coverage = [getcover]
&return
```

Exercise 9.1.1

```
/* This program will prompt the user for the arguments of the buffer
/* command, and then execute the command.
/*
/*
/* Original coding: Nick Frunzi-----may 29, 1996
/*
/*
&sv incov [getcover *  -all 'Please choose a cover to buffer']
&sv outcov [response 'Please enter the name of the coverage to create']
&sv fuzz [response 'Please enter a fuzzy tolerance' #]
&sv featclass [getchoice LINE POLY POINT NODE -prompt 'Please choose a
feature to buffer']
&if [query 'would you like to use a {buffer_item} and ~
```

```
{buffer_table}' .false.] &then
  &do
    &sv item [getitem %incov% -%featclass% ~
     'Please select a buffer item']
    &sv table [getfile * -info 'Please select a buffer table']
    BUFFER %incov% %outcov% %item% %table% # %fuzz% %featclass%
  &end
&else
  &do
    &sv dist [response 'Please enter a buffer distance']
    BUFFER %incov% %outcov% # # %dist% %fuzz% %featclass%
  &end
&return
```

Exercise 9.1.2

```
/* This program will calculate the MIN, MEAN, and MAX depth for the
/* wells in a WELL coverage.
/*
/* The statistics functionality is to be performed by the user though
/* an &TTY
/*
/* Original coding:  Nick Frunzi------May 30, 1997
/*
ARCPLOT
&term 9999
display 9999 1 position ul
MAPEXTENT WELL
POINTS WELL
&type Please use STATISTICS to calculate the MIN, MEAN, and MAX
&type for the item DEPTH on the WELL coverage. /&/&
&type Please enter the commands in the order above
&type Remember to use &RETURN to return to the program /&/&
&tty
&sv min [show statistic 1 1]
&sv mean [show statistic 2 1]
&sv max [show statistic 3 1]
&lv
&pause Hit return to exit ARCPLOT
QUIT
&return
```

Exercise 9.1.3

```
/* This program will calculate the MIN, MEAN, and MAX depth for the
/* wells in a WELL coverage.
/*
/* The program runs silently (no display, no popup, no messages)
/*
/* Original coding:  Nick Frunzi------July 30, 1992
/*
&sv olddis [show display]
&sv oldmess [show &messages]
&messages &off
DISPLAY 0
ARCPLOT
STATISTICS WELL POINT
MIN DEPTH
MEAN DEPTH
MAX DEPTH
END
&sv min [show statistic 1 1]
&sv mean [show statistic 2 1]
&sv max [show statistic 3 1]
QUIT
&type The MINIMUM well depth is: %min%
&type The MEAN well depth is: %mean%
&type The MAXIMUM well depth is: %max%
DISPLAY %olddis%
&messages %oldmess%
&return
```

Exercise 9.2.1

```
&sv cov = [getcover * -ALL 'Select a coverage to draw,~
 or "none" to quit' -none]
&do &while not [NULL %cov%]
&describe %cov%
MAPEXTENT %cov%
/*
/* Create the list of valid feature types
/*in the variable: LIST
/*
&sv list =
&if  %dsc$aat_bytes% gt 0  &then &sv list = %list% LINE
&if  %dsc$pat_bytes% gt 0  &then &sv list = %list% POLY
&if ( %dsc$xat_bytes% gt 0 and not %dsc$qtopology% ) &then
  &sv list = %list% POINT
/*
```

```
/* &do &while not [NULL %cov%]
/*
&if [token [unquote %list%] -count]  eq 1 &then &sv choice = %list%
&else
    &sv choice = [getchoice %list% -prompt 'What kind of display']
/*
&select [locase %choice%]
    &when point; &do
      MARKERSYMBOL 71  /* green star
      POINTS %cov%
    &end /* when
    &when line; &do
      LINECOLOR red
      ARCS %cov%
    &end /* when
    &when poly; &do
      LINECOLOR green
      POLYGONS %cov%
      POLYGONTEXT %cov% [entryname %cov%]-ID
    &end /* when
  &end /* select
/*
/* Get another coverage
/*

&sv cov = [getcover * -all 'Select a coverage to draw, or~
  "none" to quit' -none]
  CLEAR
&end /* do while
&return
```

Exercise 9.2.2

In addition to restoring program environments and environment settings, exit and/or bailout routines can continue executing programs, terminate programs, send messages to the user, run another program, or return.

Exercise 9.2.3

```
&sv cover = [getcover * -poly]
&describe %cover%
&if %dsc$qedit% &then &do
  &type You must reconstruct coverage topology.
  &sv choice = [query 'Do you wish to rebuild polygon topology']
  &if %choice% &then &do
    &severity &error &ignore
    BUILD %cover% POLY
     &if [quote %aml$message%] cn 'Use CLEAN' &then &do
       &if [query 'BUILD failed, do you want to CLEAN the coverage'] ~
       &then &do
         &type Renaming your original coverage to %cover%old
         RENAME %cover% [entryname %cover%OLD]
         CLEAN %cover%OLD %cover%
       &end
     &end
  &end
&end
&messages &on
&return
```

Exercise 9.3.1

```
/* purpose: Draw a coverage that getcover set
/*          get the coverages featureclass from
/*          the user and set that to .feat
/*
&severity &error &routine bailout
&args cover
MAPEXTENT %cover%
&s .feat = [getchoice Point Line Polygon -prompt [quote pick a feature
  class for %cover%]]
CLEAR
&select [locase %.feat%]
 &when point
   &do
     MARKERSYMBOL [getsymbol -marker]
     POINTS %cover%
   &end
 &when line
   &do
     LINESYMBOL [getsymbol -line]
     ARCS %cover%
   &end
 &when polygon
```

```
    &do
      POLYGONSHADES %cover% [getitem %cover% -poly -nonchar]
      LINECOLOR [getsymbol -line]
      POLYGONS %cover%
    &end
&end /* &select

&call exit
&return

&routine exit
&return

&routine bailout
&severity &error &ignore
&call exit
&return &error
```

Exercise 10.1.1

```
&s numobs = [listfile *.lut -info lookup.file]
```

Exercise 10.1.2

```
/* descpoly.aml
/* purpose: Describe all the polygon coverages in a
/*          workspace.
/* Tom Brenneman 3-17-97    Original coding
&severity &error &routine bailout
&do polycov &list [listfile * -cov -polygon]
  DESCRIBE %polycov%
&end

&call exit
&return

&routine exit
&return

&routine bailout
&severity &error &ignore
&call exit
&return &error
```

Exercise 10.1.3

```
/* getlines.aml
&severity &error &routine bailout
/* Get the workspace
&sv workspace [getfile * -workspace 'Please select a workspace']
/* Create the file
&sv numobs [listfile %workspace% ~
  -cover -line %workspace%.line]
/* View the file
&popup %workspace%.line
&return

/*------------------
&routine EXIT
/*------------------
/*
&return

/*------------------
&routinc BAILOUT
/*------------------
&severity &error &ignore
&severity &warning &ignore
&call exit
&return &error An error has occurred in GETLINES.AML
```

Exercise 10.1.4

```
/* readlines.aml
&args file num_lines start
&severity &error &routine bailout
/* Check arguments

&if [NULL %num_lines%] &then
  &call usage

&if not [exists %file%] &then &do
  &type %file% is not a system file.
  &call usage
&end

&if not [access %file% -read] &then &do
  &type you do not have read access to %file%
  &call usage
&end
```

```
&select [type %num_lines%]
   &when 1; &do   /* Character
     &if [upcase %num_lines%] ne ALL &then
        &call usage
   &end
   &when 2, -2   /* Boolean or real
      &return Positive integer value required
   &when -1; &do   /* Integer
     &if %num_lines% lt 1 &then
        &return Positive integer value required
   &end
&end

&if [null %start%] &then
   &sv start = 1

/* Read the whole file in an array to determine if the
/* user hasn't asked for a bad start position or number of lines
&sv unit = [open %file% openstat -read]
&if %openstat% ne 0 &then
   &return File could not be opened.  Error code: %openstat%
&sv i = 1
&sv rec%i% = [read %unit% readstat]
&do &while %readstat% = 0 or %readstat% = 104
   &sv i = %i% + 1
   &sv rec%i% = [read %unit% readstat]
&end
&sv tot = %i% - 1
&call closefile
&if %readstat% ne 102 &then
   &return Error during read of file %file%.  Error code %readstat%
&if %start% gt %tot% &then
   &return Start position: %start%. Only %tot% lines in file.

&if [type %num_lines%] = 1 &then &do
   &if [upcase %num_lines%] = ALL &then
     &sv range = %tot%
&end
&else &do
   &sv range = %start% + %num_lines% - 1
   &if %range% gt %tot% &then &do
     &type Cannot read %num_lines% lines from position %start%.
     &type File has only %tot% lines
     &return
   &end
&end

&do i = %start% &to %range%
```

```
  &type [value rec%i%]
&end

&return
/*------------------
&routine USAGE
/*------------------
&type Usage: readlines <file> <num_lines | ALL> {start_num}
&return &inform
/*------------------
&routine CLOSEFILE
/*------------------
/* Delete any global variables and other threads set by this tool
&if [variable unit] &then &do /* If file has been opened.
  &s closestat = [close %unit%]
  &if %closestat% ne 0 &then
    &type FILE CLOSE ERROR %closestat%.
  &dv unit
&end
&return
/*------------------
&routine BAILOUT
/*------------------
&severity &error &ignore
&severity &warning &ignore
&call closefile
&return &error An error has occurred in READLINES.AML
```

Exercise 10.1.5

```
/* minmaxpoly.aml
&severity &error &routine bailout

&s cov_list [unquote [listfile * -cover -polygon]]
&s num_covers = [token %cov_list% -count]

&if %num_covers% eq 0 &then
  &do
    &return No polygon coverages in this workspace
  &end

&s covcount = 1
&do cover &list %cov_list%
  &describe %cover%
  &if %covcount% = 1 &then
    &do
/* Set values for the first coverage.
```

```
        &s lowx %dsc$xmin%
        &s lowxcov %cover%
        &s lowy %dsc$ymin%
        &s lowycov %cover%
        &s highx %dsc$xmax%
        &s highxcov %cover%
        &s highy %dsc$ymax%
        &s highycov %cover%
        &s covcount 0
      &end
  &else
    &do
/* Compare current coverage values against lowest (so far)
/* Update (if necessary new coordinates and coverages
      &if %dsc$xmin% le %lowx% &then
        &do
          &s lowxcov %cover%
          &s lowx %dsc$xmin%
        &end
      &if %dsc$ymin% le %lowy% &then
        &do
          &s lowycov %cover%
          &s lowy %dsc$ymin%
        &end
      &if %dsc$xmax% ge %highx% &then
        &do
          &s highxcov %cover%
          &s highx %dsc$xmax%
        &end
      &if %dsc$ymax% ge %highy% &then
        &do
          &s highycov %cover%
          &s highy %dsc$ymax%
        &end
    &end /* else
&end /* &list %cov_list%
/* Report results back to the user
&type Coverage %lowxcov% has the lowest X-Coordinate: %lowx%
&type Coverage %lowycov% has the lowest Y-Coordinate: %lowy%
&type Coverage %highxcov% has the highest X-Coordinate: %highx%
&type Coverage %highycov% has the highest Y-Coordinate: %highy%
&call exit
&return
/*-------------------
&routine EXIT
/*-------------------
&if [VARIABLE fileunit] &then /* If file has been opened.
  &do
```

```
    &s closestat = [CLOSE %fileunit%]
    &select %closestat%
      &when 101
        &do
          &type FILE CLOSE ERROR %closestat%: Bad unit number.
          &return
        &end
      &when 102
        &do
          &type FILE CLOSE ERROR %closestat%: Unit not open.
          &return
        &end
      &end /* &select %closestat%
  &end /* &if [VARIABLE fileunit]
&return
/*------------------
&routine BAILOUT
/*------------------
&severity &error &ignore
&severity &warning &ignore
&call exit
&return &error An error has occurred in MINMAXPOLY.AML
```

Exercise 10.2.1

```
/* savapset.aml
&args cover
&severity &error &routine bailout
/* Check arguments
&if [null %cover%] &then
  &call usage

&if not [exists %cover% -cover] &then
  &return [entryname %cover%] is not a coverage.
/* Create the settings file

&sv unit = [open [entryname %cover%]_apset.aml openstat -write]
&if %openstat% ne 0 &then
  &return Unable to open file, error code: %openstat%

&sv writestat = [write %unit% [quote MAPEXTENT %cover%]]
&sv writestat = [write %unit% [quote MAPANGLE [show mapangle]]]
&sv writestat = [write %unit% [quote MAPLIMITS [show maplimits page]]]
&sv writestat = [write %unit% [quote MAPPOSITION [show mapposition]]]
&sv writestat = [write %unit% [quote MAPSCALE [show mapscale]]]
&sv writestat = [write %unit% [quote MAPUNITS [show mapunits]]]
&sv writestat = [write %unit% [quote PAGESIZE [show pagesize]]]
```

```
&sv writestat = [write %unit% [quote PAGEUNITS [show pageunits]]]
&sv writestat = [write %unit% &return]

&call exit
&return
/*------------------
&routine USAGE
/*------------------
&type Usage: savapset <cover>
&return &inform
/*------------------
&routine EXIT
/*------------------
/* Close the new AML
&if [variable unit] &then &do
  &sv closestat = [close %unit%]
  &if %closestat% ne 0 &then
    &return FILE CLOSE ERROR %closestat%
&end
&return
/*------------------
&routine BAILOUT
/*------------------
&severity &error &ignore
&severity &warning &ignore
&call exit
&return &error An error has occurred in SAVAPSET.AML
```

Exercise 10.2.2

```
/* delfiles.aml
&args string
&severity &error &routine bailout
/* Check arguments
&if [null %string%] or [index %string% *] = 0 &then
  &call usage
/* Get the list of file names

&s files = [unquote [listfile %string% -file]]
&s numobs = [token %files% -count]
&if %numobs% = 0 &then &do
  &return No matches for the search string
&end

&do name &list %files%
  &ty %name%
&end
```

```
/* Make sure that the user wants to delete the chosen files.
/* If so, delete them.
&if [query 'Are you sure you want these files deleted' .FALSE.] &then &do
  &do name &list %files%
    &sv delstat = [delete %name%]
    &if %delstat% ne 0 &then
      &type File %fn% not deleted
  &end
&end
&return


/*------------------
&routine USAGE
/*------------------
&type Usage: delfiles <*search_string | search*string | search_string*>
&return &inform
/*------------------
&routine BAILOUT
/*------------------
&severity &error &ignore
&return &error An error has occurred in DELFILES.AML
```

Exercise 10.3.1

```
/* topo_desc.aml
/* purpose: generate a list of coverages that need to have
/*          topology created.  Give the user a pulldown
/*          menu of these coverages and let them describe
/*          them.
/*
/* Tom Brenneman 3-19-97  Original coding
/*
&severity &error &routine bailout
&fullscreen &popup

&call getcov
&call create
&call menu

&call exit
&return

/**************
&routine GETCOV
/**************
&s cov_list
```

```
&do cov &list [listfile * -cov]

&describe %cov%
&if %dsc$qedit% &then
   &s cov_list = %cov_list% %cov%
&end /* do list

&if [token %cov_list% -count] = 0 &then
   &return No coverages need to be built or cleaned!

&return /* getcov

/**************
&routine CREATE
/**************
&s menu_file = [scratchname]
&s menu_file_unit = [open %menu_file% openstat -write]
&s writestat = [write %menu_file_unit% 1]
&s writestat = [write %menu_file_unit% 'Describe']

&do dsc &list %cov_list%
   &s writestat = [write %menu_file_unit% ~
      [format ' %1,13% describe %2' %dsc% %dsc%]]
&end

&s writestat = [write %menu_file_unit% 'Log']

&do log &list %cov_list%
   &s writestat = [write %menu_file_unit% ~
      [format ' %1,13% log %2' %log% %log%]]
&end

&s writestat = [write %menu_file_unit% 'Exit     &return']

&if [variable menu_file_unit] &then
   &do
      &s closestat = [close %menu_file_unit%]
      &dv menu_file_unit
   &end
&return /* getcov

/**************
&routine MENU
/**************
&m %menu_file%
&return /* getcov
```

```
/***************
&routine EXIT
/***************
&if [variable menu_file_unit] &then
  &do
    &s closestat = [close %menu_file_unit%]
    &dv menu_file_unit
  &end
&if [exists %menu_file%] &then
  &s delstat = [delete %menu_file%]

&fullscreen &off

&return

/***************
&routine BAILOUT
/***************
&severity &error &ignore
&call exit
&return &error
```

Exercise 11.1.1

```
/* res.aml
&severity &error &routine bailout
&if [show program] ne ARCPLOT &then
  &return Program must be run from ARCPLOT
&sv file = [getfile * -info 'Choose a file to RESELECT from' -none -other]
&if [null %file%] &then
  &return No file chosen, exiting from RES.AML
&sv item = [getitem %file% -info 'Choose an item to RESELECT on']
&sv value = [getunique %file% -info %item% ~
  'Choose a value to RESELECT for']
&if [type %value%] lt 0 &then
  &sv formval = %value%
&else &sv formval [quote %value%]
RESELECT [entryname %file% -info] info %item% = %formval%
  list [entryname %file% -info] info
&return
/*-------------------
&routine EXIT
/*-------------------
/* Delete any global variables
&dv .template$*
&return
/*-------------------
```

```
&routine BAILOUT
/*------------------
&severity &error &ignore
&severity &warning &ignore
&call exit
&return &error An error occurred in RES.AML
```

Exercise 11.1.2

```
/* select.aml
&severity &error &routine bailout
&if [show program] ne ARCPLOT &then
  &return Program must be run from ARCPLOT
&sv file = [getfile * -info 'Choose a file to SELECT from' -none -other]
&if [null %file%] &then
  &return No file chosen, exiting from SELECT.AML
&do &until %task% = QUIT
  &sv task = [getchoice RESELECT 'Clear selected set' LIST ITEMS ~
    'Get new file' QUIT -prompt 'Choose a command or task']
  &select %task%
    &when RESELECT,            ASELECT, UNSELECT; &call selection
    &when 'Clear selected set'; ASELECT [entryname %file% -info] info
    &when LIST;                LIST [entryname %file% -info] info
    &when ITEMS;               ITEMS [entryname %file% -info] info
    &when 'Get new file';      &sv file = [getfile * -info ~
      'Choose a file to SELECT from' -other]
    &when QUIT;                &return
  &end /* &select
&end /* &do &until
&return
/*------------------
&routine SELECTION
/*------------------
&sv item = [getitem %file% -info [quote Choose an item to %task% on]]
&if [extract 3 [iteminfo %file% -info %item%]] ne C &then
  &sv op = [getchoice GT GE EQ LE LT -prompt 'choose an operator']
&else &sv op = EQ
&sv value = [getunique %file% -info %item% [quote ~
  Choose a value to %task% for]]
&if [type %value%] lt 0 &then &sv formval = %value%
&else &sv formval [quote %value%]
%task% [entryname %file% -info] info %item% %op% %formval%
&return
/*------------------
&routine EXIT
/*------------------
/* Delete any global variables and other threads set by this tool
```

```
&dv .template$*
&return
/*-------------------
&routine BAILOUT
/*-------------------
&severity &error &ignore
&severity &warning &ignore
&call exit
&return &error An error occurred in routine: %routine% ~
  (SELECT.AML)
```

Exercise 11.1.3

```
/* join.aml
&args out_file
&severity &error &routine bailout
/* Check arguments
&if [NULL %out_file%] &then
  &call usage
&if [exists %out_file% -info] &then &do
  &type File already exists.
  &call usage
&end
&sv in_file  = [getfile * -info 'Select a file']
&sv i_file   = [entryname %in_file% -info]
&sv jn_file  = [getfile * -info 'Select a file to join to %i_file%']
&sv rel_item = [getitem %in_file% -info 'Select a relate item']
&sv st_item  = [getitem %in_file% -info 'Select a start item']
&sv type     = [getchoice LINEAR ORDERED LINK -prompt ~
  'Select a relate type']
JOINITEM %i_file% [entryname %jn_file% -info] ~
%out_file% %rel_item% %st_item% %type%
&call exit
&return
/*-------------------
&routine USAGE
/*-------------------
&type Usage: join <out_file>
&return &inform
/*-------------------
&routine EXIT
/*-------------------
/* Delete any global variables and other threads set by this tool
&dv .template$*
&return
/*-------------------
&routine BAILOUT
```

```
/*------------------
&severity &error &ignore
&severity &warning &ignore
&call exit
&return &error An error occurred in (TEMPLATE.AML)
```

Exercise 11.1.4

```
/* joinnew.aml
&args out_file
&severity &error &routine bailout
/* Check arguments
&if [NULL %out_file%] &then
  &call usage
/* Check for existence of out file.
&if [exists %out_file% -info] &then &do
  &type File already exists.
  &call usage
&end
/* Make sure the program is running in ARC.
&if [upcase [show program]] ne ARC &then &do
  &type This program must be run from ARC
  &call usage
&end

&s list

/* establish in and join files
&sv in_file = [getfile * -info 'Select a file']
&sv i_file  = [entryname %in_file% -info]
&sv jn_file = [getfile * -info 'Select a file to join to %i_file%']

/* loop through all of the items in the in file.
/* If any of these items have the same name
/* as the join-file items.  If so then
/* check to see if they have the same
/* definitions.  If they do then attatch them to
/* the list of possible joinitems.
/* as the items in the joinfile then
&do item &list [listitem %in_file% -info]
  &if [keyword %item% [listitem %jn_file% -info]] > 0 &then &do

    /* If the definitions are the same then add the item to the
    /* list.
    &s in_item_def = [iteminfo %in_file% -info %item% -definition]
    &s jn_item_def = [iteminfo %jn_file% -info %item% -definition]
```

```
      /* Token 4 is deleted because number of decimal places does not
      /* have to be the same in both files.
      /* Quote is used to because token changes commas to spaces.
      &if [quote [token %in_item_def% -delete 4]] = ~
      [quote [token %jn_item_def% -delete 4]] &then
        &s list = %list% %item%
  &end /* if item
&end

/* if there are no possible join items then exit the AML.
&s string = [unquote %list%]
&if [token %string% -count] = 0 &then &do
  &type I'm sorry these files have no items in common
  &call exit
  &return
&end

/* Pick the relate and list choices out of the list of possibilities
/* and do the joinitem.
&s rel_item = [getchoice %list% -prompt 'Select the item to join by']
&sv st_item = [getitem %in_file% -info 'Select a start item']
&sv type    = [getchoice LINEAR ORDERED LINK -prompt ~
  'Select a relate type']
JOINITEM %i_file% [entryname %jn_file% -info] ~
%out_file% %rel_item% %st_item% %type%
&call exit
&return
/*-------------------
&routine USAGE
/*-------------------
&type Usage: join <out_file>
&return &inform
/*-------------------
&routine EXIT
/*-------------------
/* Delete any global variables
&dv .template$*
&return
/*-------------------
&routine BAILOUT
/*-------------------
&severity &error &ignore
&severity &warning &ignore
&call exit
&return &error An error occurred
```

Exercise 11.2.1

```
ARCPLOT: ARC MAPJOIN SOIL3
/* You will be prompted to enter additional information:
Enter Coverages to be MAPJOINed (Type END or a blank line when done):
=================================================================

Enter the 1st coverage: soil1
Enter the 2nd coverage: soil2
Enter the 3rd coverage:
Done entering coverage names (Y/N)? y
Do you wish to use the above coverages (Y/N)? y

 Appending coverages...
 Sorting...
 Partial process enabled. 100% of the coverage will be processed.
 Intersecting...
 Assembling polygons..
 Creating PAT...
Arcplot:/* You will return to ARCPLOT.

/* Shading polygons for the coverage SOIL using the item TYPE and
/* lookup table SOIL.EXP
Arcedit: APC POLYGONSH SOIL TYPE SOIL.EXP
```

Exercise 11.2.2

```
/* lease2.aml
&severity &error &routine bailout
&sv oldmess [show &messages]
&messages &off &all
&messages &off &info
MAPEXTENT PROPERTY
POLYS PROPERTY
&sv oldlc [show linecolor [show linesymbol] 1]
LINECOLOR RED
RESELECT PROPERTY POLY LANDUSE = 14
POLYGONLINES PROPERTY 3
/* After selecting the city-owned parcels (LANDUSE = 14), get a
/* coordinate for one of the selected parcels from the user. If the
/* coordinate is a valid one, call the INFO accessing routine. List
/* the global variables the routine creates. Keep getting points
/* until the user enters a 9.
&type Enter a '9' to quit.
&flushpoints
&getpoint &map &push
&do &while %pnt$key% ne 9
```

```
   RESELECT PROPERTY POLY ONE *
   &if [extract 1 [show select property poly]] ne 0 &then &do
     &call getdays
     &type Lease Date: %.lease$date%
     &type Days till expiration: %.lease$numdays%
 &end
  &else
     &type No parcel selected.  Select again.
   ASELECT PROPERTY POLY LANDUSE = 14
   &getpoint &map &push
&end
&call exit
&return
/*------------------
&routine GETDAYS
/*------------------
/* Get the parcel number (APN) and enter INFO. In INFO select the
/* PROPERTY.PAT file and relate the APN.CITYOWN file to it. This file
/* contains the lease dates for all city-owned parcels. Create a
/* temporary AML with [SCRATCHNAME] and store that file name in the
/* variable AMLFILE.  Write the lines to set the global variables equal
/* to the LEASE_DATE and the number of days until expiration. This value
/* is stored in the INFO variable $NUM1 which is created using date math
/* on the parcel's lease date and today's date ($TODAY)
&sv apn = [show select property poly 1 item apn]
&sv amlfile = [scratchname]
&sv oldws = [show &workspace]
&wo info
&data info
ARC
COMO -NTTY
SELECT PROPERTY.PAT
RELATE APN.CITYOWN BY APN
RESELECT APN = [quote %apn%]
OUTPUT ../%amlfile% INIT
CALC $NUM1 = $1LEASE_DATE - $TODAY
PRINT '&sv .lease$numdays = ',$NUM1
PRINT '&sv .lease$date = ',$1LEASE_DATE
PRINT '&return'
Q STOP
&end
&wo %oldws%
&type /&/&
&r %amlfile%
&return
/*------------------
&routine EXIT
/*------------------
```

```
/* Delete global variables and temporary file.
/* Reset the linecolor
&messages %oldmess%
&sv delstat = [delete %amlfile%]
&dv .lease$*
LINECOLOR %oldlc%
&return
/*------------------
&routine BAILOUT
/*------------------
&severity &error &ignore
&severity &warning &ignore
&call exit
&return &error An error occurred in (LEASE2.AML)
```

Exercise 11.2.3

```
/* genpoints.aml
&severity &error &routine bailout
/* Go to INFO an define the file. Use the ADD FROM command to load the
/* data contained in the file: WELL.DAT. Output the id and
/* coordinates to a new file: GEN.FILE. Add a line with the word
/* 'END' -- GENERATE requires this.
&sv oldws [show &workspace]
&wo info
&data info
ARC
DEFINE WELL.EXP
WELL-ID,4,5,B
X-COORD,4,12,F,3
Y-COORD,4,12,F,3
WELL_LOC,25,25,C
DEPTH,3,3,I
DIAMETER,7,7,C
WELL_TYP,5,5,C
HP,3,3,I
GPM,4,4,I
DISCHARGE,20,20,C
~
ADD FROM ../WELL.DAT
OUTPUT ../GEN.FILE INIT
PRINT WELL-ID,X-COORD,Y-COORD
PRINT 'END'
Q STOP
&end
&wo %oldws%
/* GENERATE the coverage WELL using the file GEN.FILE as input.
```

```
GENERATE WELL
INPUT GEN.FILE
POINTS
Q
/* Create topology
BUILD WELL POINT
/* Remove the coordinate items from the expansion file: WELL.EXP
DROPITEM WELL.EXP WELL.EXP
X-COORD
Y-COORD
END
&call exit
&return
/*-------------------
&routine EXIT
/*-------------------
/* Delete the temporary file
&sv delstat = [delete gen.file]
&return
/*-------------------
&routine BAILOUT
/*-------------------
&severity &error &ignore
&severity &warning &ignore
&call exit
&return &error An error occurred in: GENPOINTS.AML
```

Exercise 11.2.4

```
/*-----------------------------------------------------------------
/*             Environmental Systems Research Institute
/*-----------------------------------------------------------------
/*    Program: JUSTIFY.AML
/*    Purpose: This program left justifies a character item in an INFO /*
file.
/*             Exercise 11.2.4
/*
/*-----------------------------------------------------------------
/*      Usage: justify <info_file> <character_item> <start_column>
/*
/* Arguments: file - name of INFO file to be processed.
/*            item - name of item to left justify.  It must be
/*                 CHARACTER!
/*            start - Start column of requested item.
/*
/*    Globals:
/*-----------------------------------------------------------------
```

```
/*      Calls:
/*---------------------------------------------------------------
/*      Notes: There is no error checking for the correctness of the /*
start column.  The only provision for this is the presentation /*of the
temporary file to the user for evaluation.  If the user
/*              passed an incorrect start column, it will show! The -
FULLDEF
/*              option to the [ITEMINFO] function will return the start
column
/*              of a given item.  With this, the start argument could be
/*              removed to make the program need less information to be
run
/*              successfully.
/*---------------------------------------------------------------
/*              Input:
/*              Output:
/*---------------------------------------------------------------
/* History: N. Frunzi - 12/15/92 - Original coding
/*===============================================================
/*
&args file item start
&severity &error &routine bailout
/* Change the file and item to uppercase and check arguments
&sv file = [upcase %file%]
&sv item = [upcase %item%]
&sv have_pg = .FALSE.
&if [NULL %start%] &then
  &call usage
&if [type %start%] ne -1 &then
  &return Bad start value.  Must be an integer.
&if not [exists %file% -info] &then
  &return %file% does not exist
&if not [iteminfo %file% -info %item% -exists] &then
  &return Item %item% does not exist on table %file%
/* Get the item definition and insure that it is a character item.
&sv itmdef = [iteminfo %file% -info %item% -definition]
&if [extract 3 %itmdef%] NE C &then
  &return %item% is not a CHARACTER item.
&sv itmwid = [extract 1 %itmdef%]
&sv newwid = %itmwid% - 1
&sv oldmess = [show &messages]
&messages &off
/* Create a temporary file for processing to insure no corruption to
/* the original if the program fails.
&if [exists XX%file% -info] &then
  &sv delstat [delete xx%file% -info]
copyinfo %file% xx%file%
/* Enter INFO and add the redefined items (LC,RC,LEFT,RIGHT) to the
```

```
/* temporary file.
/*      &sv have_pg = .TRUE.
&if [exists xxjustify.pg] &then &sv have_pg = .TRUE.
&sv oldws = [show &workspace]
&wo info
&data info
ARC
COMO -NTTY
SEL XX%file%
REDEFINE
%start%,FC,1,1,C
[calc %start% + %newwid%],LC,1,1,C
%start%,LEFT,%newwid%,%newwid%,C
[calc %start% + 1],RIGHT,%newwid%,%newwid%,C
~
REM
REM Delete the INFO program if it happens to exist.
REM
&if %have_pg% &then &do
DELETE XXJUSTIFY.PG
Y
&end
REM
REM Write the INFO program to process the item in the selected file.
REM
PROGRAM XXJUSTIFY.PG
SELECT XX%file%
PROGRAM SECTION TWO
IF %item% NE ' '
  DO WHILE FC EQ ' '
    MOVE RIGHT TO LEFT
    MOVE ' ' TO LC
  DOEND
ENDIF
PROGRAM END
~
REM
REM Run the program. Delete it when finished. The program compiles
REM and runs. If the program fails here, AML won't know.
REM
RUN XXJUSTIFY.PG
DELETE XXJUSTIFY.PG
Y
Q STOP
&end
&wo %oldws%
/* Present the temporary file to the user. If it's acceptable, delete
/* the original, copy the temporary file to the original, drop the
```

```
/* redefined items, and delete the temporary file.
&messages %oldmess%
LIST xx%file%
&if [query 'Is this file acceptable' .TRUE.] &then &do
  &if [query 'Are you sure' .TRUE.] &then &do
    &sv delstat = [delete %file% -info]
    copyinfo xx%file% %file%
    DROPITEM %file% %file%
    FC
    LC
    LEFT
    RIGHT
    END
    &sv delstat = [delete xx%file% -info]
  &end
&end
&call exit
&return
/*------------------
&routine USAGE
/*------------------
/*
&return Usage: justify <info_file> <character_item> <start_column>
/*------------------
&routine EXIT
/*------------------
/* Make sure that the temporary file is deleted and that &messages
/* is restored to its initial setting.
&if [exists xx%file% -info] &then
  &sv delstat = [delete xx%file% -info]
&messages %oldmess%
&return
/*------------------
&routine BAILOUT
/*------------------
&severity &error &ignore
&severity &warning &ignore
&call exit
&return &error An error occurred in JUSTIFY.AML
```

Exercise 11.3.1

```
/* book.aml
/*
/*
&severity &error &routine bailout
&args routine turnto
```

```
/*
/* Check to see if the program is being run from the menu
/*
&if not [null %routine%] &then &do
  &call %routine%
  &return
&end
/*
/* Display the image and lot polygons.  Shade the buildings
/*
MAPEXTENT LOTS
SHADESET COLOR
IMAGE WOODLAND
LINECOLOR 1
ARCS LOTS
CLEARSELECT
POLYGONSHADES BUILDINGS 7
POLYGONS BUILDINGS
/*
/* Allow the user to select a building
/*
&type Sclect a building
&getpoint &map &push
RESELECT BUILDINGS POLY ONE *
&if [extract 1 [show select buildings poly ]] = 0 &then
    &type No building located.
&else &do
    /*
    /* Make sure that the file ATTRIB exists.  This file contains the
    /* locations of the appropriate images. If it does, select all of
    /* the images associated with the selected parcel.
    /*
    &if ^ [exists attrib -info] &then
        &return Cannot open attrib.
    POLYGONSH BUILDINGS 2
    RESELECT ATTRIB INFO ~
        BUILDINGS-ID = [show select buildings poly 1 item buildings-id]
/*
/* Create imageview windows
/*
IMAGEVIEW CREATE 'The Other' SIZE 300 400 POS LR SCREEN LR
IMAGEVIEW CREATE 'One Page' SIZE 300 400 POS LEFT WINDOW 'The Other'
/*
/* Use threads to create the menu that will allow the user to turn pages
/* in the "book" of images.
/*
&sv origthread [show &thread &self]
&thread &create bookthread ~
```

```
&menu book &stripe ~
    'Turn Page' &pos &ll &window 'One Page' &ul ~
    &pinaction 'imageview destroy ''One Page''; imageview destroy ~
'The Other' ; cursor bookcursor remove; &thread &delete bookthread'
/*
/* Declare and open a cursor named BOOKCUR on the INFO file ATTRIB
/*
CURSOR BOOKCURSOR DECLARE ATTRIB INFO
CURSOR BOOKCURSOR OPEN
/*
/* Call the routine that gets the images and allow the user to
/* interact with the menu.
/*
&s .page 1
&call getimg
&thread &focus &off &self
&return
/*
/*
&routine trnpg
/*
/* Controls two imageview windows, emulating page turning
/* Arguments
/*    turnto - next or previous.  displays images corresponding to next and
/*             prev page.
/*
/* If the user chose to go to the next page, and the current page is
/* less than the number of selected images (in ATTRIB) -1 then get the image.
/* If there are no more pages, infor the user.
/*
&if %turnto% = next &then &do
    &if %.page% < [calc %:bookcursor.aml$nsel% - 1] &then &do
        &s .page = %.page% + 1
        &call getimg
    &end   /* if
    &else &s .bookmes = No more pages.
&end
&else &if %.page% > 1 &then &do
        &s .page = %.page% - 1
        &call getimg
    &end
    &else &s .bookmes No more pages.
&return
/*
/*
&routine getimg
/*
/* Gets an image from an info file and displays it and the next one
```

```
/*
&s messvar [show &messages]
/* &message &off
CURSOR BOOKCURSOR FIRST
/*
/* If the page isn't the first one, move the cursor to the correct page.
/*
&s pageno = %.page% - 1
&if ^ %pageno% = 0 &then
    &do i = 1 &to %pageno%
        CURSOR BOOKCURSOR NEXT
    &end
/*
/* Now that the cursor is at the correct page, use the IMAGEVIEW command
/* with the image stored in the item IMAGE in the file ATTRIB.
/*
IMAGEVIEW %:bookcursor.image%  # # 'One Page'
/*
/* Move the cursor to the next page.
/*
CURSOR BOOKCURSOR NEXT
/*
/* Use IMAGEVIEW to disply the image stored in the item IMAGE.
/*
IMAGEVIEW %:bookcursor.image%  # # 'The Other'
&message %messvar%
&return
/*
&routine bailout
&severity &error &fail
/*
/* If the program fails, determine if BOOKCURSOR exists, if so remove it.
/*
&if not [null [show cursor bookcursor]] &then
   CURSOR BOOKCURSOR REMOVE
&return &error Bailing out of BOOK.AML
```

Exercise 12.1.1

1. Variety in graphical appearance of widgets.
2. Variables passed between widgets make form menus dynamic.
3. Supports data verification.
4. Supports help messages and help text files.
5. Can obtain information for an entire operation in one form.
6. Menu file can be created using graphical editor as well as text editor.

Exercise 12.1.2

In FormEdit, reference widgets are located on the menu <u>palette</u> and dragged to the menu <u>canvas</u> to create a new form menu.

Exercise 12.1.3

Parameters that apply to the entire form menu are established with preferences. Property sheets are used to set parameters for a single widget.

Exercise 12.1.4

Yes, comments can (and should) be added to form menus. From the Edit menu, select Comments to open a window that receives typed comments.

Exercise 12.1.5

The snapping grid is always active, whether or not it's displayed. This can cause your widgets to snap to an unexpected location. Display the grid by selecting Show grid from the View menu.

Exercise 12.2.1

(a) Assign the variable on the form menu preferences sheet in the Initialization string: field.
(b) Run an AML program that assigns the variable before executing the menu.

Exercise 12.2.2

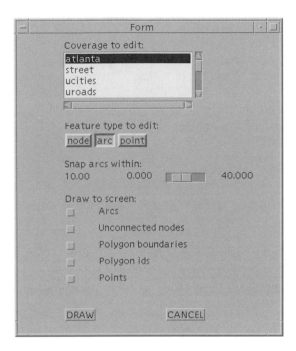

All the headings (e.g., Coverage to edit:) and the feature names next to the check boxes are created with background text.

The scrolling list of coverages is defined with the following property sheet. The selected coverage is assigned to %.setae$cover% and established as the EDITCOVERAGE. Notice that the variables follow a consistent naming convention throughout the application. The variables are global because they're referenced in AML programs called from the Draw and Cancel buttons.

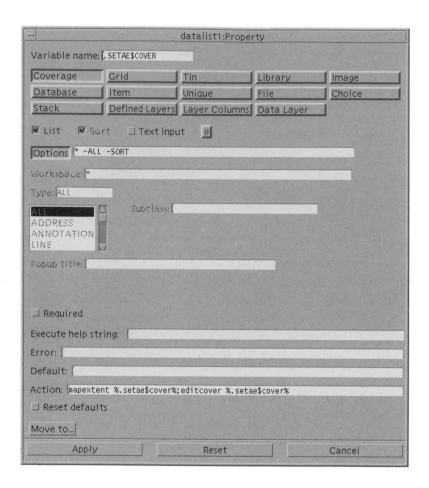

The EDITFEATURE is established with the choice widget defined in this property sheet:

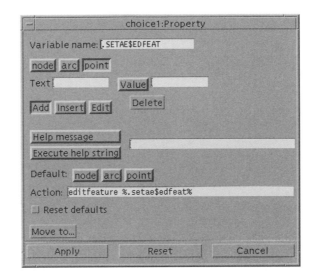

A slider widget gets user input for the arc- and node-snapping tolerance and is defined as follows:

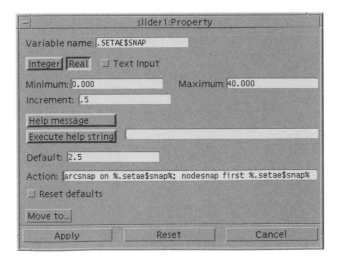

Each check box sets a variable defining a feature type for the DRAWENVIRONMENT. These variables are referenced in the AML program invoked by pressing the Draw button.

The Draw button runs an AML program that determines which features belong in the DRAWENVIRONMENT and then draws the coverage.

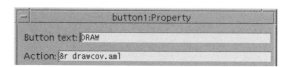

```
/* drawcov.aml
&severity &error &routine bailout
/* Determines which features the user checked on the menu,
/* establishes the drawenvironment, then draws the coverage.
&if %.setae$qdangle% &then drawenv node dangle
    &else drawenv node off
&if %.setae$qarc% or %.setae$qpoly% &then drawenv arc
    &else drawenv arc off
&if %.setae$qpolyid% or %.setae$qpoint% &then drawenv label ids
    &else drawenv label off
DRAW
&return
```

Note: This is a place where your solution might differ. The AML program run by the Draw button could set all the parameters that are currently set on the Action: property of the other widgets (e.g., the EDITCOVERAGE and EDITFEATURE). In this case, the DRAWCOV.AML would incorporate all the code on the Action: properties, and the Action: properties would be cleared.

The Cancel button is for dismissing the menu. The AML program cleans up after the application by deleting the global variables and resetting environments.

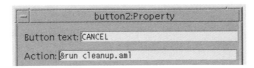

```
/* cleanup.aml
&severity &error &routine bailout
&call exit
&return
/********
&routine exit
&dv .setae*
drawenv all off
removeedit all
&return
/*******
&routine bailout
&severity &error &ignore
&call exit
&return &error Error encountered in Cleanup.aml
```

Exercise 12.3.1

1. Use the Help message option on the property sheet. This provides a single line of help at the bottom of the menu when the user presses either the 2 or 3 button (depending on the windowing system).

2. Use the Execute help string and pop up a customized text file or ARC/INFO help file when the user presses either the 2 or 3 button.

3. Use a button named HELP. The Action: property could be set to invoke more sophisticated help, such as a help menu, a tutorial, or a document page.

4. Use a help message and a help button—you can't use both help message and execute help string.

Exercise 12.3.2

Possible improvements:

1. Instead of having the user type a coverage name, use a scrolling data list. This is easier for the user, reducing typing errors and invalid names. Using the mouse to select a coverage is also an improvement because it keeps the method of input consistent for the next operation (i.e., the method of input doesn't shift from keyboard to mouse).

2. A choice widget is more appropriate than using a series of check boxes for the feature type. Only one feature can be chosen. A check box doesn't enforce a single choice—the user can check more than one. A choice enforces a single choice and is best to use when the choices are mutually exclusive.

3. The two icons look too much alike. If they were displayed any smaller, they'd be difficult to distinguish. You could modify the graphics (e.g., replace arrows with a + or – sign inside the inner rectangle) or use text instead of a picture.

4. The purpose of the CHANGE button is ambiguous. It's not clear whether it changes the line symbol, the coverage, or the extent displayed. Clearly name the button and maybe even move it to a more appropriate location.

5. Assuming the OK button draws the coverage (and either dismisses the menu or not), there's no button to cancel the operation (i.e., dismiss the menu and *not* draw the coverage). It's always important to give the user an escape route. Although the user could dismiss the window through the window manager, a better design places a CANCEL or an EXIT button on every menu.

Exercise 12.3.3

```
%choice1 CHOICE choice1 PAIRS KEEP ~
        HELP 'Select a county'  ~
        RETURN '&sv county_flag = 1' ~
        'San Bernardino' '1' 'Riverside' '2' 'Orange' '3'
```

Exercise 12.3.4

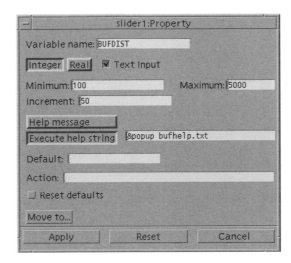

Exercise 13.1.1

```
&thread &create edit_arc &menu edit_arc.menu
```

Exercise 13.1.2

```
&do thread &list edit_arc edit_select drawenv
  &if [show &thread &exists %thread%] &then
    &thread &delete %thread%
&end
```

Exercise 13.1.3

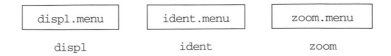

When the calling program deletes its thread, the user can interact with all three threads.

Exercise 13.1.4

```
&thread &delete &others

&thread &delete &all
```

Exercise 13.2.1

When you dismiss the two menus, AML issues the following warning and creates a TTY thread because it can't return control to the AML program:

```
AML WARNING - Can't select a thread, creating new one
```

If you issue &TB, you'll see that the original thread still exists but isn't focused:

```
Arc: &tb

Thread thread0001 (unfocused)
Unit 0       Line 132     Terminal     tty
Unit 1       Line 2       AML File     /balboa1/nick/workbook/edit.aml

* Thread thread0003
Unit 0       Line 1       Terminal        tty
```

Exercise 13.2.2

To ensure that a menu is the only menu focused when it's executed, use the &MODAL option to &THREAD &CREATE:

```
&thread &create modmenu &modal &menu modmenu.menu
```

Exercise 13.2.3

Use the &MODAL option when you need to reset environments. When a modal thread is dismissed, the thread environment is reset to the way it was before the modal thread was invoked.

Exercise 13.2.4

```
&if not [null %.create$newcov%] &then
   &thread &focus &on menu2
```

Exercise 13.3.1

```
&thread &create select &menu select.menu ~
&position &ul &thread disp &ur
```

Exercise 13.3.2

```
/*------------------
&routine INIT
/*------------------
/*    .
/*    .
/*    .
&if [SHOW &thread &exists tool$getattribute] &then
   &thread &delete tool$getattribute
&thread &create tool$getattribute &modal ~
   &menu getattribute.menu ~
   &position [UNQUOTE %position%] ~
   &stripe [QUOTE [UNQUOTE %stripe%]] ~
   &pinaction '&run getattribute cancel'
&return
```

Exercise 13.3.3

The resulting thread environment contains four threads. SOME.AML is on top of one, while the menus sit on top of the other three. Only thread MENU3 is focused.

If the user dismisses MENU3, the thread is deleted and control returns to the AML program. If the program ends with an &RETURN, the user is left with a prompt (tty), and MENU1 and MENU2 are unfocused.

To ensure that this doesn't happen, add the following line to SOME.AML:

```
&thread &delete &self
```

Exercise 14.2.1

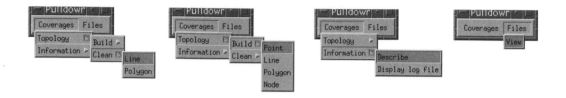

```
8
&BEGIN_MENU
&BEGIN_BLOCK "Coverages"
   &BEGIN_BLOCK "Topology"
      &BEGIN_BLOCK "Build"
         &MENUITEM "Point" build [getcover * -all] point
         &MENUITEM "Line" build [getcover * -all] line
         &MENUITEM "Polygon" build [getcover * -all] poly
         &MENUITEM "Node" build [getcover * -all] node
      &END_BLOCK
      &BEGIN_BLOCK "Clean"
         &MENUITEM "Line" Clean [getcover * -all] # # # line
         &MENUITEM "Polygon" Clean [getcover * -all] # # # poly
      &END_BLOCK
   &END_BLOCK
   &BEGIN_BLOCK "Information"
      &MENUITEM "Describe" describe [getcover * -all]
      &MENUITEM "Display log file" log [getcover * -all]
   &END_BLOCK
&END_BLOCK
&BEGIN_BLOCK "Files"
   &MENUITEM "View" &popup [getfile * -file]
&END_BLOCK
&END_MENU
```

Exercise 14.3.1

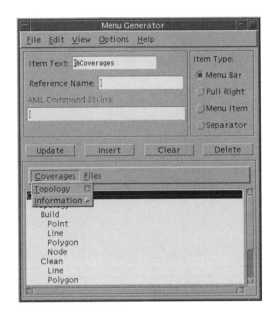

```
8
&BEGIN_MENU
&BEGIN_BLOCK "&Coverages" &REF %coverages
    &BEGIN_BLOCK "&Topology"
        &BEGIN_BLOCK "&Build"
            &MENUITEM "&Point" &REF %point build %cov% point
            &MENUITEM "&Line" build %cov% line
            &MENUITEM "P&olygon" &REF %poly build %cov% poly
            &MENUITEM "&Node" build %cov% node
        &END_BLOCK
        &BEGIN_BLOCK "&Clean"
            &MENUITEM "&Line" Clean %cov% %cov%cl # # line
            &MENUITEM "P&olygon" &REF %poly Clean %cov% %cov%cl # # poly
        &END_BLOCK
    &END_BLOCK
    &BEGIN_BLOCK "&Information"
        &MENUITEM "&Describe" describe %cov%
        &MENUITEM "Disp&lay log file" log %cov%
    &END_BLOCK
&END_BLOCK
&BEGIN_BLOCK "&Files" &REF %files
    &MENUITEM "&View" &popup %file%
&END_BLOCK
&END_MENU
```

Exercise 14.3.2

The following graphics show the FormEdit property sheets for BROWSER.MENU:

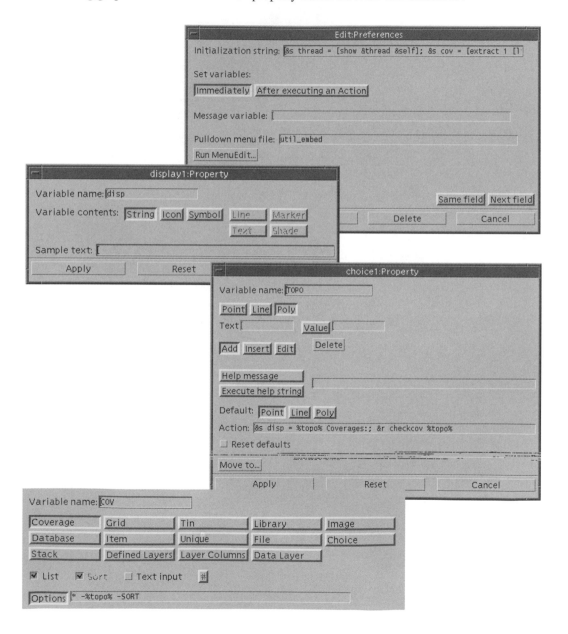

The following graphic shows the general format for the enhanced pulldown menu
UTIL_EMBED.MENU:

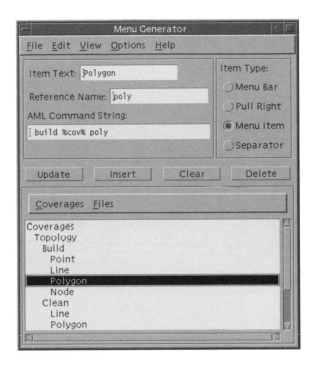

```
7 browser.menu
 Pick the type of
 coverage to display:
 %choice1
 %display1                Files:
 %datalist1               %datalist2




   %button1
 %choice1 CHOICE TOPO SINGLE KEEP ~
   INITIAL 'Point' ~
   RETURN '&s disp = %topo% Coverages:; &r checkcov %topo%' ~
   'Point' 'Line' 'Poly'
 %display1 DISPLAY disp 18 VALUE
```

```
%datalist1 INPUT COV 22 TYPEIN NO SCROLL YES ROWS 6 ~
   KEEP ~
   RETURN '&enable &on %thread% coverages' ~
   COVER ~
   * -%topo% -SORT
%datalist2 INPUT FILE 20 TYPEIN NO SCROLL YES ROWS 6 ~
   KEEP ~
   RETURN '&enable &on %thread% files' ~
   FILE ~
   * -FILE -SORT
%button1 BUTTON KEEP 'Dismiss' &return
%FORMOPT SETVARIABLES IMMEDIATE MENU util_embed
%FORMINIT &s thread = [show &thread &self];~
 &s cov = [extract 1 [listfile * -cov]];   ~
   &s file = [extract 1 [listfile * -file]]; &full &popup; &r checkcov point

/* UTIL_EMBED.MENU
&BEGIN_MENU
&BEGIN_BLOCK "&Coverages" &REF %coverages
   &BEGIN_BLOCK "&Topology"
      &BEGIN_BLOCK "&Build"
         &MENUITEM "&Point" &REF %point build %cov% point
         &MENUITEM "&Line" build %cov% line
         &MENUITEM "P&olygon" &REF %poly build %cov% poly
         &MENUITEM "&Node" build %cov% node
      &END_BLOCK
      &BEGIN_BLOCK "&Clean"
         &MENUITEM "&Line" Clean %cov% %cov%cl # # line
         &MENUITEM "P&olygon" &REF %poly Clean %cov% %cov%cl # # poly
      &END_BLOCK
   &END_BLOCK
   &BEGIN_BLOCK "&Information"
      &MENUITEM "&Describe" describe %cov%
      &MENUITEM "Disp&lay log file" log %cov%
   &END_BLOCK
&END_BLOCK
&BEGIN_BLOCK "&Files" &REF %files
   &MENUITEM "&View" &popup %file%
&END_BLOCK
&END_MENU
```

Exercise 15.1.1

The routine deletes the existing thread.

Exercise 15.1.2

No. In some cases the widget just defines or accesses a variable used by the routines in the program.

Exercise 15.1.3

The menu is visible on the screen but remains busy while the routine executes. &RETURN never executes because the EXIT routine (called by the OK routine) deletes the menu's thread. Other standard control buttons that use &RETURN are Cancel and Dismiss.

Exercise 15.2.1

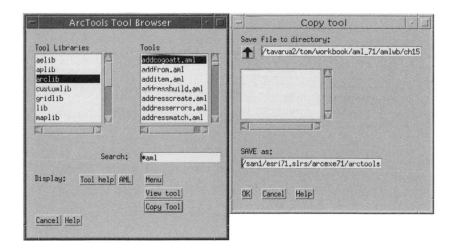

The TOOLBROWSER.MENU should be modified to add the Copy Tool button with the return string:

```
&r toolbrowser copy_tool
```

The following is the COPY_TOOL routine for TOOLBROWSER.AML:

```
/*-----------
&routine COPY_TOOL
/*-----------
/* Copy a tool to a new location.
/*
&if [null %.toolbrowser$tool%] &then
  &return
```

```
&set toolname = [before [entryname %.toolbrowser$tool%] .]
&set toolfile = [joinfile [dir %.toolbrowser$tool%] %toolname% -file]
&set menu_file = [joinfile %toolfile% menu -ext]
&set aml_file = [joinfile %toolfile% aml -ext]

&if [exists %menu_file%] and [exists %aml_file%] &then
  &s fileexists = both
&else &do
  &if [exists %aml_file%] &then &do
    &s .toolbrowser$msg Only AML file available
    &s fileexists = aml
  &end
  &else &do
    &s .toolbrowser$msg Only MENU file available
    &s fileexists = menu
  &end
&end

&select %fileexists%
  &when both
    &do
      &r savefileas init %menu_file% .toolbrowser$filesave # ~
      '&ul &thread tool$toolbrowser &ur' 'Copy tool'
      &if ^ [null %.toolbrowser$filesave%] &then &do
        &s deletestat = [delete %.toolbrowser$filesave% -file]
        &s copystat = [copy %menu_file% %.toolbrowser$filesave% -file]
        &s deletestat = [delete ~
    [joinfile [dir %.toolbrowser$filesave%] [entryname %aml_file%] -file]]
        &s copystat = [copy %aml_file% ~
    [joinfile [dir %.toolbrowser$filesave%] [entryname %aml_file%] -file]]
        &s .toolbrowser$msg TOOL copied to %.toolbrowser$filesave%
      &end
    &end
  &when aml
    &do
      &r savefileas init %aml_file% .toolbrowser$filesave # ~
      '&ul &thread tool$toolbrowser &ur' 'Copy tool (AML only)'
      &if ^ [null %.toolbrowser$filesave%] &then &do
        &s deletestat = [delete %.toolbrowser$filesave% -file]
        &s copystat = [copy %aml_file% %.toolbrowser$filesave% -file]
        &s .toolbrowser$msg AML copied to %.toolbrowser$filesave%
      &end
    &end
  &when menu
    &do
      &r savefileas init %menu_file% .toolbrowser$filesave # ~
      '&ul &thread tool$toolbrowser &ur' 'Copy tool (MENU only)'
      &if ^ [null %.toolbrowser$filesave%] &then &do
```

```
        &s deletestat = [delete %.toolbrowser$filesave% -file]
        &s .toolbrowser$msg MENU copied to %.toolbrowser$filesave%
        &s copystat = [copy %menu_file% %.toolbrowser$filesave% -file]
      &end
    &end
&end /* select
&return
```

Exercise 15.3.1

Copy ATPREFS.AML to the current directory and add the current directory to the &AMLPATH and &MENUPATH:

```
/*****************************************************************
/*   USER SPECIFIED AML AND MENU PATHS   *
/*************************************
/*
/* Sets user specified AML and MENU paths.  These paths will be
/* searched
/* prior to the ArcTool paths.  By setting the following two lines,
/* ArcTools can be customized with system, user, or project specific
/* paths for AMLs and menus.  Multiple paths can be set using spaces
/* between the paths.  Either full paths or local paths can be used.
/* For example, /tmp/projects/tools could be specified, or if
/* arctools is started from /tmp/projects, then just "tools" could
/* be used as the AMLPATH and MENUPATH.  These paths will be appended
/* with the ArcTools paths, which are set by
/* $ATHOME/lib/setpaths.aml.
/* Uncomment these lines to set:
&amlpath [show &workspace]
&menupath [show &workspace]
```

Exercise 16.1.1

Software packages either act as a client or a server as they communicate. A client makes requests of a server, while the server processes the requests. The ARC/INFO and ArcView GIS packages can be both a server for and a client to other packages at the same time.

Exercise 16.1.2

It's important to know that ARC/INFO supports ONC RPC because it can directly communicate with any other software package that supports ONC RPC.

Exercise 16.1.3

Windows NT doesn't come with an RPC package. If you want to use RPC to have ARC/INFO for Windows NT communicate with other programs (such as C or Visual Basic), you must obtain an RPC package that allows such communication.

Exercise 16.2.1

There's no interrupt capability because there can only be one input source at a time to the AML processor. If you were able to interrupt an IAC command to the server, the AML processor would process commands and give unpredictable results.

Exercise 16.2.2

Clients need the connection information (i.e., hostname, program, and version number) to connect to the server. If these parameters are saved in a file, it simplifies the connection process.

Exercise 16.2.3

The returned value from [IACOPEN] of 105 indicates that the connection file can't be created. This may result from not having proper write privileges to the current workspace.

Exercise 16.2.4

Windows NT must be prepared to use the IAC environment. The proper environment must be set up before Windows NT can support RPC connections. Use the PORTMAP command to set this environment.

Exercise 16.2.5

Yes, both ARC/INFO and ArcView GIS can be clients to an ARC/INFO server at the same time. Whichever client sends the request first will get priority. Requests are processed on a first-come, first-served basis.

Exercise 16.2.6

You do n't need to specify a connect file when you're closing the ARC/INFO server. The proper syntax would be:

&type [IACCLOSE}

Exercise 16.3.1

Both the [GETCOVER] function and MARKER * command require user input. The problem is that the user is asked to interact with ARC/INFO from the server, not the client's terminal. The client is sending the request from the client's machine but then is asked to enter data on the server's machine—an impossible task unless the server and client are on the same machine. Be careful with the requests you send. If you need user input, get it before you pass the request to the server.

Exercise 16.3.2

The second line of code is incorrect. When retrieving the results from a server, you must use a <procedure_number> of 2 instead of 1.

Therefore, the request is sent to the server with job2:

```
&sv job2 = [IACREQUEST %server_id% 1 '&sv num = [calc 2 + 2];~
       &iacreturn %num%' reqstat]
```

Also, you must use %job2% in conjunction with a <procedure_number> of 2:

```
&type [IACREQUEST %server_id% 2 %job2% reqstat]
```

Exercise 16.3.3

The code doesn't allow time for the server to finish the ARC RESELECT. The client will try to buffer the "majroad" coverage before it's completely created by the server. An error will result saying something like, "Majroad coverage does not exist."

Exercise 16.3.4

The two programming techniques you can use to avoid problems with ARC/INFO's asynchronous processing are:

1. Build error checking into your AML. Error checking makes sure that the server process is completed before the client tries to process results from the server.

2. Use the AIREQUEST atool. This atool forces the client to wait for the server to respond with the results. Since the client AML is suspended, no problems can result from the client's trying to process data that's not yet available.

Exercise 16.3.5

The five errors in this AML are as follows:

```
/* iacloop.aml

/* Allow client to easily type commands to send to server
[1]    &sv server_id = [IACCONNECT connectfile connectstatus]
[2]    &if %connectionstat% eq 101 &then &type Connect successful...
[3]       &else &return Connection not successful, error number
%connectionstat%
[4]    &sv cmd = [RESPONSE [QUOTE Enter command for A/I server ("<cr>" to
end)] ~
               .TRUE.]
[5]    &do &while [quote %cmd%] nc [quote .FALSE.]
[6]       &sv job = [IACREQUEST %server_id% 2 %cmd% reqstat]
[7]       &if %reqstat% ne 0 &then &return Client request error, number
%reqstat%
[8]       &sv cmd = [RESPONSE [quote Enter command for A/I server ("<cr>" to
end)]~
               .TRUE.]
[9]    &end
[10]   &sv disconnectstat = [IACCLOSE %server_id%]
[11]   &if %disconnectstat% eq 0 &then &type Disconnect was successful...
       &else &return Disconnect not successful, error number %disconnectstat%
[11]   &return
```

Lines [2] and [3] use a variable called `connectionstat`, but the variable set in line [1] is called `connectstatus`. Fix this error by renaming variable in line [1] `connectionstat`.

Line [2] checks the value of `connectionstat`. If the connection is successful, the value should be 0 (zero), not 101. Fix this error by changing the 101 to a 0.

Lines [4] and [8] set the default value for the cmd variable to .TRUE.. This means that the loop will never end because in line [5], the way to stop the loop is to enter a .FALSE.. Fix this error by changing line [5] from .FALSE. to .TRUE..

Line [6] uses [IACREQUEST] with a procedure number of 2. When sending requests to the server, you must use a procedure number of 1. A procedure number of 2 means that you'll be querying the results from the server. Fix this error by changing line [6] from a procedure number of 2 to 1.

Line [10] uses the [IACCLOSE] function, while it really should use [IACDISCONNECT]. Fix this error by changing the [IACCLOSE] function to [IACDISCONNECT].

Alphabetical index

Special characters

Index of AML programs & menus